A User's Guide to Vacuum Technology

A User's Guide to Vacuum Technology

Second Edition

John F. O'Hanlon

Associate Professor of Electrical and Computer Engineering
Director, Center for Microcontamination Control
The University of Arizona

WILEY

A WILEY-INTERSCIENCE PUBLICATION

JOHN WILEY & SONS
New York • Chichester • Brisbane • Toronto • Singapore

This book was set in press type on an Autologic APS-5 phototypesetter driven by an IBM 3080 system. The formatting was done with Termtext/Format, The Yorktown Formatting Language, and the Yorktown Mathematical Formula Processor.

Library of Congress Cataloging in Publication Data:

O'Hanlon, John F., 1937–
   A user's guide to vacuum technology/John F. O'Hanlon.— 2nd ed.
     p.   cm.

   "A Wiley-Interscience publication."
   Includes bibliographical references and index.
   ISBN 0-471-81242-0
   1. Vacuum technology—Handbooks, manuals, etc.   I. Title.

TJ940.037   1989   88-27327
621.5′5—dc19   CIP

Printed in the United States of America

10 9 8 7 6 5 4 3 2 1

For Jean, Carol, Paul, and Amanda

Preface

This book is intended for the vacuum system user—the university student, technician, engineer, manager, or scientist—who wishes a fundamental understanding of modern vacuum technology and a user's perspective of current vacuum practice.

Vacuum technology is largely secondary in that it is a part of other technologies that are central to analysis, research, development, and manufacturing. It is used to provide a process environment. Many advances in vacuum technique have resulted from the demands of other technologies, although scientists and engineers have studied vacuum for its own sake. The average user is process-oriented and becomes immersed in vacuum technique only when problems develop with a process or new equipment purchases become necessary.

A User's Guide to Vacuum Technology focuses on the understanding, operation and selection of equipment for processes used in semiconductor, optics and related technologies. It emphasizes subjects not adequately covered elsewhere while avoiding in-depth treatments of topics interesting only to the designer or curator. Residual gas analysis is an important topic whose treatment here differs from the usual explanation of mass filter theory. Components such as turbomolecular and helium gas refrigerator cryogenic pumps are widely used but not so well understood as diffusion pumps. The discussion of gauges, pumps, and materials is a prelude to the central discussion of the total system. Systems are grouped according to their common vacuum requirements of speed, working pressure, and gas throughput. The suitability of each pump is examined for several classes of systems, and basic operational procedures are given for each high vacuum pumping system.

In this edition, material has been added on components, lubrication, pump fluids and other materials. The remainder of the material has been reviewed and updated. Important formulas have been denoted with a ▶ for emphasis. Questions have been added in each chapter. The easier questions have been emphasized with a †.

Thanks are due to Dr. Webster Howard of IBM Research for permission to print the revised edition on the Yorktown IBM photocomposer and to Dr. John Baker of IBM Research for assistance with the logistics.

J. F. O'Hanlon

Tucson, Arizona
November 1988

Contents

MEASUREMENT

MATERIALS

SYSTEMS

A User's Guide to
Vacuum Technology

Its Basis

An understanding of how vacuum components and systems function begins with an understanding of the behavior of gases at low pressures. Chapter 1 discusses the nature of vacuum technology. Chapter 2 reviews basic gas properties. Chapter 3 describes the flow of gases at reduced pressures, and Chapter 4 discusses how gas is evolved from the surfaces of materials. Together, this material forms the basis of vacuum technology.

CHAPTER 1

Vacuum Technology

Galileo was the first person to create a partial vacuum, and he did so with a piston. His seventeenth-century discovery was followed in 1643 by the invention of the mercury barometer by Torricelli and in 1650 the first pump by von Guericke. Interest in properties of gases at reduced pressures remained at a low level for more than 200 years, when a period of rapid discovery began with the invention of the compression gauge by McLeod. In 1905 Gaede, a prolific inventor, designed a rotary pump sealed with mercury. These developments were followed by the development of the thermal conductivity gauge, the diffusion pump, the ion gauge and pump, helium liquefication and refinement of organic pumping fluids. They formed the basis of a technology that has made possible everything from light bulbs to the simulation of outer space. The significant discoveries of this early period of vacuum science and technology have been summarized recently in a series of papers [1-5].

A vacuum is a space from which air or other gas has been removed. We know we cannot remove all the gas. The amount we need to remove depends on the application, and we do this for many reasons. At atmospheric pressure surfaces are constantly bombarded by molecules. These molecules can either bounce from the surface, attach themselves to the surface, or perhaps chemically react with a surface. The air or other gas surrounding a surface quickly contaminates it as soon as it is cleaned. A clean surface, for example a freshly cleaved crystal, will remain clean in an ultrahigh vacuum chamber for long periods of time because the rate at which it is bombarded by molecules is very small.

Molecules are crowded together very closely at atmospheric pressure and travel in every direction much like people in a crowded plaza. It is impossible for a molecule to travel from one wall of a chamber to another without colliding with many molecules. By reducing the pressure to a

3

suitably low value a molecule from one wall can travel to another without a collision. Many effects become possible if molecules can travel long distances between collisions. Metals can be evaporated from a pure source without reacting in transit. Molecules or atoms can be accelerated to high energies to sputter away or be implanted in the bombarded surface. Electrons or ions can be scattered from surfaces and collected. The energy changes they undergo on scattering or release from a surface can be used to probe or analyze the surface or underlying layers.

The degree to which we must reduce the pressure is a function of the application. For convenience we divide the pressure scale below atmospheric into several ranges to relate phenomena and processes to these ranges. Table 1.1 lists the ranges currently in use. Epitaxial growth of semiconductor films (reduced pressure epitaxy) and laser etching of metals takes place in the low vacuum range. Sputtering, plasma etching and deposition, low pressure chemical vapor deposition, ion plating and gas filling of encapsulated heat transfer modules are examples of processes performed in the medium vacuum range.

Pressures in the high vacuum range are needed for the manufacture of microwave, power, cathode ray and photomultiplier tubes, light bulbs, aluminizing of mirrors, glass coating, decorative metallurgy, gas display panels and ion implantation. The pressure must be reduced to the very high vacuum range for most thin-film preparation, electron microscopy, mass spectroscopy, crystal growth, and x-ray and electron beam lithography. For ease of reading, we call the very high vacuum region *high vacuum* and the pumps, *high vacuum pumps*. Pressures in the ultrahigh

Table 1.1   Vacuum Ranges

Degree of Vacuum	Pressure Range (Pa)		
Low	10^5	$> P >$	3.3×10^3
Medium	3.3×10^3	$\geq P >$	10^{-1}
High	10^{-1}	$\geq P >$	10^{-4}
Very high	10^{-4}	$\geq P >$	10^{-7}
Ultrahigh	10^{-7}	$\geq P >$	10^{-10}
Extreme ultrahigh	10^{-10}	$> P$	

Source: Reprinted with permission from *Dictionary for Vacuum Science and Technology*, M. Kaminsky and J. M. Lafferty, Eds., American Vacuum Society, New York, 1980.

vacuum range are necessary for surface analytical techniques and material studies. It is also necessary in particle accelerators, fusion machines, space simulation systems and systems used for the growth of films by molecular beam epitaxy. Contamination control, or the reduction of impurities and defects, has resulted in the development of systems which can pump to the ultrahigh vacuum (UHV) range but operate in the low to high vacuum range.

A vacuum system is a combination of pumps, valves and pipes which creates a region of low pressure. It can be anything from a simple mechanical pump or aspirator for exhausting a vacuum desiccator to a complex system such as an underground accelerator with miles of piping that is exhausted to ultrahigh vacuum pressures.

Removal of air at atmospheric pressure is usually done with a displacement pump. A displacement pump is one which removes the air from the chamber and expels it to the atmosphere. Rotary vane and piston pumps are examples of pumps used to exhaust gases at atmospheric pressure. Liquid nitrogen capture pumps or sorption pumps have also been designed for exhausting gases at atmospheric pressure. They are used only on small chambers because of the limited amount of gas they can sorb.

Rotary vane, piston and sorption pumps have low pressure limits in the range 10^{-1} to 10^{-3} Pa. To pump a system below this range, we require a pump which will function in this rarefied atmosphere. There are several displacement and capture pumps which will remove air at these low pressures. The diffusion pump was the first high vacuum pump. It is a displacement pump also, but it cannot be exhausted to atmospheric pressure. The turbomolecular pump, a system of high speed rotating turbine blades, can also exhaust gas at low pressures. The exhaust pressures of these two pumps need to be kept in the range 0.5 to 50 Pa, so they have to exhaust into a rotary vane or piston "backing" pump, often called a "backing" or "fore" pump. If the diffusion or turbomolecular pump is big enough, a lobe blower will be placed between the exhaust of the high vacuum pump and the inlet of the rotary pump to help push the gas along at a faster rate in the intermediate pressure range.

Capture pumps can effectively remove gas from a chamber at low pressure. They do so by freezing molecules on a wall (cryogenic pump), chemically reacting with the molecules (getter pump), or accelerating the molecules to a high velocity and burying them in a metal wall (ion pump). Capture pumps are more useful as high vacuum pumps than as atmospheric exhaust pumps because the number of molecules to be captured at high vacuum is less than the number removed during initial evacuation from atmosphere. The gas in a high vacuum chamber can be reduced to pressures ranging from 10^{-2} to 10^{-12} Pa.

Air is the most important gas which we pump, because it is in every system. It contains at least a dozen constituents, whose concentrations

Table 1.2   Components of Dry Atmospheric Air

Constituent	Content (vol %)	ppm	Pressure (Pa)
N_2	78.084±0.004		79,117
O_2	20.946±0.002		21,223
CO_2	0.033±0.001		33.437
Ar	0.934±0.001		946.357
Ne		18.18±0.04	1.842
He		5.24±0.004	0.51
Kr		1.14±0.01	0.116
Xe		0.087±0.001	0.009
H_2		0.5	0.051
CH_4		2.	0.203
N_2O		0.5±0.1	0.051

Source: Reprinted with permission from *The Handbook of Chemistry and Physics*, 59th ed., R. C. Weast, Ed., Copyright 1978, The Chemical Rubber Publishing Co., CRC Press, Inc., West Palm Beach, FL 33409.

are given in Table 1.2. The differing ways in which pumps remove air and gauges measure its pressure can be understood in terms of the partial pressures of its components. The concentrations listed in Table 1.2 are those of dry atmospheric air at sea level (total pressure 101,323.2 Pa or 760 Torr). The partial pressure of water vapor is not given in this table because it is constantly changing. At 20°C a relative humidity of 50% is equivalent to a partial pressure of 1165 Pa (8.75 Torr) which makes it the third largest constituent of air. The total pressure changes rapidly with altitude, as shown in Fig. 1.1; its proportions, slowly but significantly. In outer space the atmosphere is thought to be mainly hydrogen with some helium [6]. In the pressure region below 10 Pa gases, which evolve from the surface of the materials, contribute more molecules per second to the pump than do the gases in the volume of the chamber. We need to choose the correct pump, but we will not achieve low pressures unless we construct the chamber from suitable materials and use proper techniques for joining components and cleaning surfaces. In the remaining chapters we discuss the pumps, gauges, and materials to construct systems and describe their operation in terms of kinetic theory. We focus on the understanding and operation of systems for many applications.

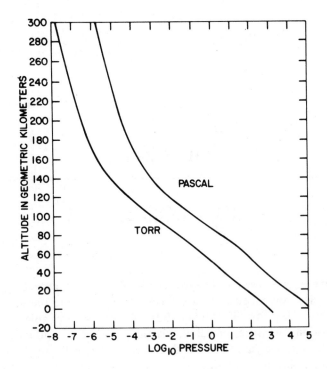

Fig. 1.1 Relation between the atmospheric pressure and the geometric altitude. Reprinted with permission from *The Handbook of Chemistry and Physics*, 59th ed., R. C. Weast, Ed. Copyright 1978, The Chemical Rubber Publishing Co., CRC Press, Inc., West Palm Beach, FL 33409.

We will use SI units in all formulas unless noted differently. Pumping speeds are often given in L/s (high vacuum pumps and conductances) and in m^3/h (mechanical pumps) instead of m^3/s.

REFERENCES

1. T. E. Madey, *J. Vac. Sci. Technol. A*, **2**, 110 (1984).
2. M. H. Hablanian, *J. Vac. Sci. Technol. A*, **2**, 118 (1984).
3. J. H. Singleton, *J. Vac. Sci. Technol. A*, **2**, 126 (1984).
4. P. A. Redhead, *J. Vac. Sci. Technol. A*, **2**, 132 (1984).
5. T. E. Madey and W. C. Brown, Eds., *History of Vacuum Science and Technology*, American Institute of Physics, New York, 1984.
6. D. J. Santeler, et al., *Vacuum Technology and Space Simulation*, NASA SP-105, National Aeronautics and Space Administration, Washington, D.C., 1966, p.34.

CHAPTER 2

Gas Properties

In this chapter we discuss the properties of gases at reduced pressures. The properties developed here are based on the kinetic picture of a gas. Kinetic theory has its limitations, but with it we are able to describe particle motion, pressure, effusion, viscosity, thermal conductivity, diffusion and thermal transpiration of ideal gases. We will use these ideas as the starting point for our discussions of gas flow, gauges, pumps and systems.

2.1 KINETIC PICTURE OF A GAS

The kinetic picture of a gas is based on several assumptions. (i) The volume of gas under consideration contains a large number of molecules. A cubic meter of gas at a pressure of 10^5 Pa and a temperature of 22°C contains 2.48×10^{25} molecules, while at a pressure of 10^{-7} Pa, a very high vacuum, it contains 2.5×10^{13} molecules. Indeed, any volume and pressure normally used in the laboratory will contain a large number of molecules. (ii) Adjacent molecules are separated by distances that are large compared with their individual diameters. If we could stop all molecules instantaneously and place them on the coordinates of a grid, the average spacing between them would be about 3.4×10^{-9} m at atmospheric pressure (10^5 Pa). The diameter of most molecules is in the 2×10^{-10} to 6×10^{-10} m range, and they are separated by distances of about 6 to 15 times their diameter at atmospheric pressures. For extremely low pressures, say 10^{-7} Pa, the separation distance is about 3×10^{-5} m. (iii) Molecules are in a constant state of motion. All directions of motion are equally likely and all velocities are possible, although not equally probable. (iv) Molecules exert no force on one another except

8

when they collide. If this is true, then the molecules will be uniformly distributed throughout the volume and travel in straight lines until they collide with a wall or with one another.

Many interesting properties of ideal gases have been derived by using these assumptions. Some elementary properties are reviewed here.

2.1.1   Velocity Distribution

As the individual molecules move about, they collide with one another. These collisions are elastic; that is, they conserve energy, while the particles change velocity with each collision. We stated that all velocities are possible, but not with equal probability. The distribution of particle velocities calculated by Maxwell and Boltzmann is

$$\frac{dn}{dv} = \frac{2N}{\pi^{1/2}}\left(\frac{m}{2kT}\right)^{3/2} v^2 e^{-mv^2/(2kT)} \tag{2.1}$$

where m is the mass of each particle whose temperature T is in kelvin. The relation between the absolute, or kelvin, scale and the celsius scale is $T(K) = 273.16 + T(°C)$. In (2.1) N is the total number of particles, and k is Boltzmann's constant. Figure 2.1 illustrates (2.1) for nitrogen molecules (air) at three temperatures. It is a plot of the relative number of molecules between velocity v and $v+dv$. We see that there are no molecules with zero or infinite velocity and that the peak or most probable velocity v_p, is a function of the average gas temperature. The particle velocity also depends on the molecular mass; the peak velocity can be

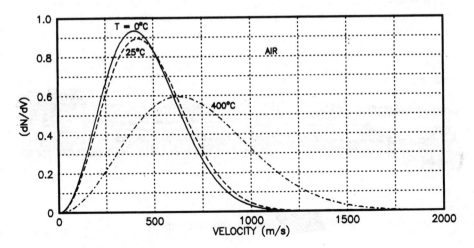

Fig. 2.1   Relative velocity distribution of air at 0, 25 and 400°C.

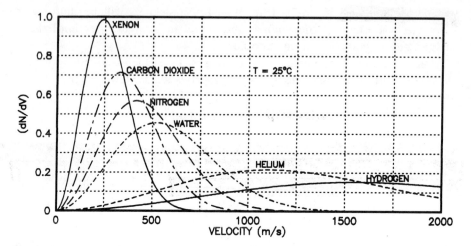

Fig. 2.2 Relative velocity distribution of several gases at 25°C.

expressed as $v_p = (2kT/m)^{1/2}$. The arithmetic mean or average velocity v, is useful when describing particle flow:

$$v = \left(\frac{8kT}{\pi m} \right)^{1/2}$$

▶ (2.2)

The average velocities of several gas and vapor molecules are given in Appendix B.2. The root of the mean square velocity, or rms velocity, is $v_{rms} = (3kT/m)$. The rms velocity is the square root of the average, or mean, of each velocity squared times the number of particles with that velocity. For Maxwell-Boltzmann statistics the average velocity is always 1.128 times as large as v_p, while $v_{rms} = 1.225 v_p$. In Fig. 2.1 we illustrated the temperature dependence of the velocity distribution. As the temperature is increased, the peak is broadened and shifted to a higher velocity. We may also plot (2.1) for different gases having the same temperature. Figure 2.2 illustrates the velocity distribution for H_2, He, H_2O, N_2, CO_2 and Xe. There are two concepts illustrated in Figs. 2.1 and 2.2. First the average velocity of a particle is proportional to $(T/m)^{1/2}$. An increase in temperature or decrease in mass causes an increase in a particle's velocity and frequency with which it collides with other particles or nearby walls. Second, not all the particles in a distribution have the same velocity. The Maxwell-Boltzmann distribution is quite broad—over 5% of the molecules travel at velocities greater than two times the average velocity.

2.1.2 Energy Distribution

Maxwell and Boltzmann also derived an energy distribution which is

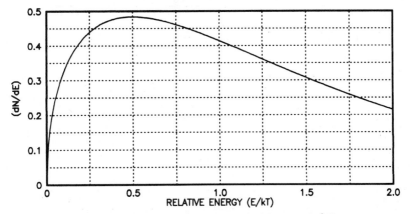

Fig. 2.3   Relative energy distribution of a gas at $25°C$.

based on the same assumptions as the velocity distribution. It is

$$\frac{dn}{dt} = \frac{2N}{\pi^{1/2}} \frac{E^{1/2}}{(kT)^{3/2}} e^{-E/(kT)} \tag{2.3}$$

From this the average energy can be calculated to be $E_{ave} = 3kT/2$, and the most probable energy to be $E_p = kT/2$. Notice that neither the energy distribution nor the average energy are a function of the molecular mass. Each is only a function of temperature, as illustrated in Fig. 2.3. For example, all the gases illustrated in Fig. 2.2 have the same energy distribution because they all have the same average temperature. See Fig. 2.3.

2.1.3   Mean Free Path

The fact that molecules are randomly distributed and move with different velocities implies that each travels a different straight-line distance (a free path) before colliding with another. As illustrated in Fig. 2.4, not all free paths are the same length. The average, or mean, of the free paths λ, according to kinetic theory, is

$$\lambda = \frac{1}{2^{1/2} \pi d_o^2 n} \qquad \blacktriangleright (2.4)$$

where d_o is the molecular diameter in meters and n is the gas density in molecules per cubic meter. The mean free path is clearly gas density dependent. If the temperature is constant, it is also pressure dependent. See (2.13). For air at room temperature the mean free path is easily remembered by

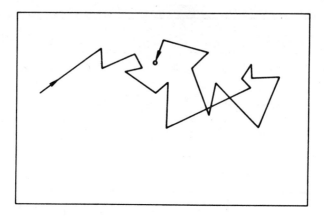

Fig. 2.4 Random motion of a molecule.

$$\lambda(\text{mm}) = \frac{6.6}{P}$$
▶(2.5)

where λ has units of millimeters, and P is the pressure in pascals. Kinetic theory also describes the distributions of free paths:

$$N = N'e^{-x/\lambda}$$
(2.6)

N' is the number of molecules in the volume and N is the number of molecules that traverse a distance x before suffering a collision. Equation (2.6) states that 63% of the collisions occur in a distance $0 \leq x \leq \lambda$, while about 37% of the collisions occur in the range $\lambda \leq x \leq 5\lambda$. Only about 0.6% of the particles travel distances greater than 5λ without suffering a collision.

For the case of two gases, a and b, the mean free path of a in b is

$$\lambda_a = \frac{1}{\left[2^{1/2}\pi n_a d_a^2 + \left(1 + \frac{v_b^2}{v_a^2}\right)^{1/2} n_b \frac{\pi}{4}(d_a + d_b)^2 \right]}$$
(2.7)

2.1.4 Particle Flux

The concept of particle flux is helpful in understanding gas flow, pumping, and evaporation. According to kinetic theory the flux Γ of an ideal gas striking a unit surface or crossing an imaginary plane of unit area from one side is

$$\Gamma(\text{particles}/(\text{m}^2 - s)) = nv/4$$
▶(2.8)

where n is the particle density and v is the average velocity. On substituting (2.2) we see that

$$\Gamma = n\left(\frac{kT}{2\pi m}\right)^{1/2} \tag{2.9}$$

The particle flux is directly proportional to the particle density and the square root of T/m.

2.1.5 Monolayer Formation Time

The time to saturate a surface with one layer of molecules is a function of the molecular arrival rate Γ and the size of a molecule. Assuming each molecule sticks and occupies surface area d_o^2, the time to form a monolayer is

$$t_{\mathrm{ml}} = \frac{1}{\Gamma d_o^2} = \frac{4}{nvd_o^2} \tag{2.10}$$

At ambient temperature, a monolayer of air ($d_o = 0.372$ nm, $v = 467$ m/s), will form in about 2.5 s at a pressure of 10^{-4} Pa. The formation time will be longer if the sticking coefficient is less than unity.

2.1.6 Pressure

The pressure on a surface is defined as the rate at which momentum mv is imparted to a unit surface. A molecule incident on a surface at an angle Θ from the normal will impart a total impulse or pressure of $2mv \cos \Theta$. By integrating over all possible angles in the half-plane we find that the pressure is

$$P = \frac{1}{3}nmv_{\mathrm{rms}}^2 \tag{2.11}$$

The total energy of a molecule, however, is proportional to its temperature

$$E = \frac{mv_{\mathrm{rms}}^2}{2} = \frac{3kT}{2} \tag{2.12}$$

Equations (2.11) and (2.12) may be combined to form the ideal gas law:

$$P = nkT \qquad\blacktriangleright (2.13)$$

If n is expressed in units of m^{-3}, k in joules per kelvin, and T in kelvins, then P will be given in units of pascals (Pa). A pascal is a newton per square meter and the fundamental unit of pressure in System International (SI). Simply divide the number of pascals by 133.32 to convert to units of Torr or by 100 to convert to units of millibars. A table for converting between different systems of units is included in Appendix

Table 2.1   Low Pressure Properties of Air[a]

Pressure (Pa)	n (m^{-3})	d' (m)	λ (m)	Γ (m^{-2}-s^{-1})
1.01×10^5 (760 Torr)	2.48×10^{25}	3.43×10^{-9}	6.5×10^{-8}	2.86×10^{27}
100 (0.75 Torr)	2.45×10^{22}	3.44×10^{-8}	6.6×10^{-5}	2.83×10^{24}
1 (7.5 mTorr)	2.45×10^{20}	1.6×10^{-7}	6.6×10^{-3}	2.83×10^{22}
10^{-3} (7.5×10^{-6} Torr)	2.45×10^{17}	1.6×10^{-6}	6.64	2.83×10^{19}
10^{-5} (7.5×10^{-8} Torr)	2.45×10^{15}	7.41×10^{-6}	664	2.83×10^{17}
10^{-7} (7.5×10^{-10} Torr)	2.45×10^{13}	3.44×10^{-5}	6.6×10^4	2.83×10^{15}

[a] Particle density, n; average molecular spacing, d'; mean free path, λ; and particle flux on a surface, Γ. $T = 22°C$.

A.3. Values of n, d', λ, and Γ for air at $22°C$ are tabulated in Table 2.1 for pressures ranging from atmospheric to ultrahigh vacuum. The pressure dependence of the mean free path is given for several gases in Appendix B.1.

2.2   GAS LAWS

Kinetic theory, as expressed in (2.13), summarizes all the earlier experimentally determined gas laws. However, we review several of the experimentally verified laws here, because they are especially helpful to those with no experience in gas kinetics. When using kinetic theory we need to remember that the primary assumption of a gas at rest in thermal equilibrium with its container is not always valid in practical situations; for example, a pressure gauge close to and facing a high vacuum cryogenic pumping surface will register a lower pressure than when it is close to and facing a warm surface in the same vessel [1]. This and other nonequilibrium situations will be discussed as required.

2.2.1   Boyle's Law

In 1662 Robert Boyle demonstrated that the volume occupied by a given quantity of gas varied inversely as its pressure when the gas temperature remained the same:

$$P_1 V_1 = P_2 V_2 \quad (N, T \text{ constant}) \tag{2.14}$$

This is easily derived from the general law by multiplying both sides by the volume V and noting that $N = nV$.

2.2.2   Amontons' Law

Amontons discovered the pressure in a confined chamber increased as the temperature increased. Amontons' law can be expressed as

$$\frac{P_1}{T_1} = \frac{P_2}{T_2} \quad (N, V \text{ constant}) \tag{2.15}$$

In 1703 he constructed an air thermometer based on this relationship. This later came to be known as the law of Gay-Lussac.

2.2.3   Charles' Law

The French chemist Charles found in 1787 that gases expanded and contracted to the same extent under the same changes of temperature provided that no change in pressure occurred. Again by the same substitution in (2.13) we obtain

$$\frac{V_1}{T_1} = \frac{V_2}{T_2} \quad (N, P \text{ constant}) \tag{2.16}$$

2.2.4   Dalton's Law

Dalton discovered in 1801 that the total pressure of a mixture of gases was equal to the sum of the forces per unit area of each gas taken individually. By the same methods for a mixture of gases, we can develop the relation

$$P_t = n_t kT = n_1 kT + n_2 kT + n_3 kT + \dots + n_i kT \tag{2.17}$$

which reduces to

$$P_t = P_1 + P_2 + P_3 + \dots + P_i \qquad \blacktriangleright (2.18)$$

where P_t, N_t are the total pressure and density and P_i, n_i are the partial pressures and densities. Equation 2.18 is called Dalton's law of partial pressures and is valid for pressures below atmospheric [2].

2.2.5   Avogadro's Law

In 1811 Avogadro observed that the pressure and number of molecules were proportional for a given temperature and volume:

$$\frac{P_1}{N_1} = \frac{P_2}{N_2} \quad (T, V \text{ constant}) \tag{2.19}$$

At standard temperature and pressure (STP), 0°C and 1 atmosphere, Avagadro's number of any kind of gas molecule is $N_o = 6.02252 \times 10^{26}$, and it occupies the molar volume of $V_o = 22.4136$ m^3. Avogadro's number of molecules is also known as a mole. One kg-mole of a substance is equal to the molecular weight in kilograms. For example one kg-mole of oxygen contains 6.02252×10^{26} molecules and weighs 32 kg. Its density at STP is therefore 32 kg/22.4136 m^3, or 1.45 kg/m^3.

2.2.6   Graham's Law

In the nineteenth century Graham studied the rate of effusion of gases through very small holes in porous membranes. He observed the rate of effusion to be inversely proportional to the square root of the density of the gas provided that the pressure and temperature were held constant. Since the density of a gas is proportional to its molecular weight, Graham's law can be stated as

$$\frac{\text{effusion rate}_a}{\text{effusion rate}_b} = \left(\frac{M_b}{M_a} \right)^{1/2} \tag{2.20}$$

Grahams' law describes how a helium-filled balloon looses its gas more quickly than an air-filled balloon.

2.3   ELEMENTARY GAS TRANSPORT PHENOMENA

In this section approximate views of viscosity, thermal conductivity, diffusion, and thermal transpiration are discussed. We state results from kinetic theory without derivation.

2.3.1   Viscosity

A viscous force is present in a gas when it is undergoing shear. Figure 2.5 illustrates two plane surfaces, one fixed and the other traveling in the x-direction with a uniform velocity. The coefficient of absolute viscosity η is defined by the equation

$$\frac{F_x}{A_{xz}} = \eta \frac{du}{dy} \tag{2.21}$$

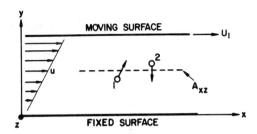

Fig. 2.5   Origin of the viscous force in a gas.

where F_x is the force in the x-direction, A_{xz} is the surface area in the x-z plane, and du/dy is the rate of change of the gas velocity at this position between the two surfaces.   Because the gas stream velocity increases as the moving plate is approached, those molecules crossing the plane A_{xz} from below (1 in Fig. 2.5) will transport less momentum across the plane than will those crossing the same plane from above (2 in Fig. 2.5).   The result is that molecules crossing from below the plane will, on the average, reduce the momentum of the molecules from above the plane, in the same manner molecules crossing from above the plane will increase the momentum of those molecules below the plane.   To an observer this viscous force appears to be frictional; actually it is not.   It is merely the result of momentum transfer between the plates by successive molecular collisions.   Again, from kinetic theory the coefficient of viscosity is

$$\eta = \frac{1}{3}nmv\lambda \tag{2.22}$$

When the gas density is measured in units of m^{-3}, the molecular mass in kg, the velocity in m/s and the mean free path in m, η will have units of (N-s)/m^2, or Pa-s.   One Pa-s is equal to 10 P.   A more rigorous treatment of viscosity [3] yields a result with a slightly different numerical coefficient:

$$\eta = 0.499nmv\lambda \qquad\qquad \blacktriangleright (2.23)$$

Substituting (2.2) and (2.4) into this result yields

$$\eta = \frac{0.499(4mkT)^{1/2}}{\pi^{3/2}d_o^2} \tag{2.24}$$

From (2.24) we see that kinetic theory predicts that viscosity should increase as $(mT)^{1/2}$ and decrease as the square of the molecular diameter.   An interesting result of this simple theory is that viscosity is independent of gas density or pressure.   This theory, however, is valid only in a limited pressure range.   If there were a perfect vacuum between the two

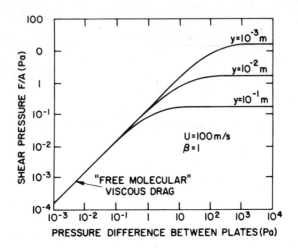

Fig. 2.6 Viscous shear between two plates in air at 22°C.

plates, there would be no viscous force because there would be no mechanism for transferring momentum from one plate to another. This leads to the conclusion that (2.24) is valid as long as the distance between the plates is of the order of the mean free path or greater.

For a rarefied gas in which the ratio of the mean free path to plate separation $\lambda/y \gg 1$, the viscous force can be expressed as

$$\frac{F}{A_{xz}} = \left(\frac{Pmv}{4kT} \right) \frac{U_1}{\beta} \qquad (2.25)$$

where the term in parentheses is referred to as the *free-molecular viscosity*. The viscous force is directly proportional to the pressure or number of molecules available to transfer momentum between the plates and is valid in the region $\lambda \gg y$. The constant β in (2.25) is related to the slip of molecules on the surface of the plates. For most surfaces and gases involved in vacuum work $\beta \sim 1$.

Figure 2.6 illustrates the magnitude of the viscous force caused by air at 22°C between two plates moving with a relative velocity of 100 m/s for three plate separations. Equation (2.24) was used to calculate the asymptotic value of the viscous drag at high pressures and (2.25) was used to calculate the free-molecular limit. A more complete treatment of the intermediate or viscous slip region is given elsewhere [4]. The viscous shear force is independent of the plate separation as long as the mean free path is larger than the largest dimension in question. This concept was used by Langmuir [5] to construct a viscosity gauge in which the damping was proportional to the pressure.

2.3.2   Thermal Conductivity

Heat conductivity is explained by kinetic theory in a manner analogous to that used to explain viscosity. The diagram in Fig. 2.6 could be relabeled to make the top plate stationary at temperature T_2, while the lower plate becomes a stationary plate at a temperature T_1 where $T_1 < T_2$. The phenomenon of heat conduction can be modeled by noting that the molecules moving across the plane toward the hotter surface carry less *energy* than those moving across the plane toward the cooler surface. The heat flow can be expressed as

$$H = AK\frac{dT}{dy} \tag{2.26}$$

where H is the heat flow and K the heat conductivity. The simple theory predicts that the heat conductivity K is expressed by $K = \eta c_v$, where η is the viscosity and c_v, the specific heat at constant volume. This simple theory is correct only to an order of magnitude. A more detailed analysis, which accounts for the rotational and vibrational energy of the molecules, yields

$$K = \frac{1}{4}(9\gamma - 5)\eta c_v \qquad \blacktriangleright (2.27)$$

where γ is the ratio of specific heats at constant pressure and constant volume c_n/c_v. When η has the units of Pa-s and c_v has units (J/kg)/K, then K will have units of (W/m)/K. At room temperature the heat conductivity increases approximately as $(mT)^{1/2}/d^2$, as does the viscosity. Heat conductivity does not depend on pressure as long as the mean free path is smaller than the dimension of the chamber. This means that an object in a vacuum system can be thermally equilibrated with its surroundings by adding just enough gas to reach the knee of the curve in Fig. 2.6; any further increase in pressure will not hasten equilibration by collisional energy transfer. At very low pressures, at which the mean free path is much greater than the dimension of the system, molecules will bounce back and forth between surfaces, carrying heat from the hot surface to the cold surface without making any intermolecular collisions. The heat transfer in this case [6] has been calculated as

$$E_o = \alpha \Lambda P(T_2 - T_1)$$

where

$$\alpha = \frac{\alpha_1 \alpha_2}{\alpha_1 + \alpha_2 - \alpha_1 \alpha_2} \qquad \Lambda = \frac{1}{8}\frac{(\gamma + 1)v_1}{(\gamma - 1)T_1} \tag{2.28}$$

This equation has the same general form as (2.25) for free-molecular viscosity. Λ is the free-molecular heat conductivity, and α_1, α_2, α are the accommodation coefficients of the cold surface, hot surface, and system, respectively. If the molecule is effective in thermally equilibrating with the surface, say by making many small collisions on a rough surface, α will have a value that approaches unity. If, however, the same surface is smooth and the molecule recoils without gaining or losing energy, α will approach zero. The kinetic picture of heat conductivity is rather like viscosity except that the heat conductivity is determined by *energy* transfer, and the viscosity is determined by *momentum* transfer. Even so, Fig. 2.5 can be sketched for thermal conductivity where the vertical axis has dimensions of heat flow. Both thermocouple and Pirani gauges operate in a region in which the heat conduction to the wall is linearly dependent on pressure. From an equivalent of Fig. 2.5 for heat conduction the relation between the diameter of the pressure-gauge tube and the range of the gauge can be observed. The tube size must be reduced to increase the high pressure limit of operation.

In SI Λ has units of $W\text{-}m^{-2}\text{-}K^{-1}\text{-}Pa^{-1}$, while E_o has units of W/m^2. Tables of the accommodation coefficient are given elsewhere [3,6]. The accommodation coefficient of a gas is not only dependent on the material but on its cleanliness, surface roughness, and gas adsorption as well.

2.3.3   Diffusion

The general phenomenon of diffusion is complex. This discussion has been simplified by restricting it to the situation in a vessel that contains two gases whose compositions vary slightly throughout the vessel but whose total concentration is everywhere the same. The coefficient of diffusion D of the two gases is defined in terms of the particle fluxes $\Gamma_{1,2}$:

$$\Gamma_1 = -D\frac{dn_1}{dx} \qquad \Gamma_2 = -D\frac{dn_2}{dx} \qquad (2.29)$$

These fluxes are due to the partial pressure gradients of the two gases. The result from kinetic theory, when corrected for the Maxwellian distribution of velocities and for velocity persistence, is [7]

$$D_{12} = \frac{8\left(\dfrac{2kT}{\pi}\right)^{1/2}\left(\dfrac{1}{m_1}+\dfrac{1}{m_2}\right)^{1/2}}{3\pi(n_1+n_2)(d_{01}+d_{02})^2} \qquad \blacktriangleright (2.30)$$

where D_{12} is the constant of interdiffusion of the two gases and, in SI, has units of m^2/s. For the special case of similar molecules the coefficient of self-diffusion is

$$D_{11} = \frac{4}{3\pi n d_o^2}\left(\frac{kT}{\pi m}\right)^{1/2} \qquad \blacktriangleright (2.31)$$

If the density n is replaced by P/kT, it becomes apparent that the diffusion constant is approximately proportional to $T^{3/2}$ and P^{-1}.

The diffusion coefficient tells us a lot about what is happening during the diffusion process.   Let us look at the problem of out-diffusion of a system of $2N$ molecules distributed uniformly throughout the x-y plane at the location $z = 0$.   At the time $t = 0$ the molecules start to diffuse through the gas, half in the $-z$-direction, half in the $+z$-direction.   At any time t the number of molecules dn located in the region between z and $z + dz$ is given by

$$dn = \frac{Ndz}{(\pi Dt)^{1/2}} e^{-z^2/(4Dt)} \tag{2.32}$$

From this, the fractional number of molecules f, located in the region between z_o and ∞, is

$$f = \text{erfc} \frac{z_o}{2(Dt)^{1/2}} \tag{2.33}$$

A plot of (2.33) is given in Fig. 2.7.   From this solution we see that the fractional number of particles between some point z_o and $+\infty$ increases with time.   Furthermore, if the value of z_o is increased, more time will elapse before the arrival of particles from the source.   We can also view this solution by calculating the values of z_o, t such that, say, 10% of the particles will have diffused beyond z_o.   That condition is satisfied when the argument of the complementary error function (erfc) is 1.16:

$$\frac{z_o}{2(Dt)^{1/2}} = 1.16 \tag{2.34}$$

or

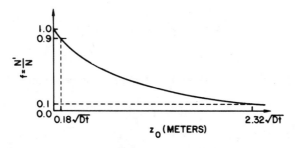

Fig. 2.7   The fractional number of particles located between z_o and $+\infty$ that have diffused from a plane source.

$$z_o = 2.32(Dt)^{1/2}$$

Equation (2.31) is basic to the solution of many diffusion problems in that it contains a critical distance proportional to $(Dt)^{1/2}$. The diffusion "front" moves through the gas as $(Dt)^{1/2}$. For $(Dt)^{1/2} \ll z_o$ the diffusion front will not have arrived at z_o, while for $(Dt)^{1/2} \gg z_o$ the diffusing species will have equilibrated with the carrier at z_o. Values of viscosity, thermal conductivity and diffusion in air of several common gases at $0°C$ and atmospheric pressure are given in Appendix B.2.

Examination of (2.31) shows the diffusion coefficient will become infinitely large as the density of molecules goes to zero. This does not happen. When the pressure becomes low enough so that the mean free path is much larger than the dimensions of the container, say the diameter of a pipe, gas diffusion is limited by molecules bouncing off the walls rather than off each other. At low pressures, the diffusion coefficient is given by

$$D = \frac{2}{3}rv \qquad\qquad \blacktriangleright (2.35)$$

where r is the radius of the pipe and v is the thermal velocity. This is called the Knudsen diffusion coefficient for a long capillary [8].

2.3.4 Thermal Transpiration

When two chambers of different temperatures are connected by a tube or orifice, their relative pressures are a function of the ratio of mean free path to diameter (λ/d) in the connecting tubing. For $\lambda \ll d$ the pressure is everywhere the same, and the densities are

$$\frac{n_2}{n_1} = \frac{T_1}{T_2} \qquad\qquad \blacktriangleright (2.36)$$

When the orifice diameter is such that $\lambda \gg d$, the flux of gas through the orifice is given by (2.8):

$$\Gamma = \frac{n}{4}\left(\frac{8kT}{\pi m}\right)^{1/2} = \frac{P}{(2kT\pi m)^{1/2}} \qquad\qquad (2.37)$$

In equilibrium the flux from each chamber must be equal, with the result

$$\frac{P_1}{P_2} = \left(\frac{T_1}{T_2}\right)^{1/2} \qquad\qquad \blacktriangleright (2.38)$$

Equations (2.36) and (2.38) are used to calculate the actual pressures within furnaces or cryogenic enclosures when the pressure gauge is located outside the enclosure at a different temperature. Equation (2.36) is used at high pressure ($\lambda < d/10$), and (2.38) is used at low pressure (λ

> 10d).

Thermal transpiration was discovered by Neumann [9] and studied by Maxwell [10], who predicted the square-root dependence given in (2.38). Deviations are introduced by the geometry and reflectivity of the walls. Siu [11] has studied these effects theoretically and has shown that the ratio predicted in (2.38) is obtained in short tubes only for specular reflection and in long tubes only for diffuse reflection.

REFERENCES

1. R. W. Moore, Jr., *Proc. 8th Nat. Vac. Symp. 1961*, **1**, Pergamon, New York, 1962, p. 426.

2. E. H. Kennard, *Kinetic Theory of Gases*, McGraw-Hill, New York, 1938, p. 9.

3. Ref. 1, pp. 135-205 and 291-337.

4. S. Dushman, *Scientific Foundations of Vacuum Technique*, 2nd ed., J. M. Lafferty, Ed., Wiley, New York, 1962, pp. 35.

5. I. Langmuir, *Phys. Rev.*, **1**, 337 (1913).

6. Ref. 4, p. 6.

7. Ref. 4, p. 67.

8. For example, see L. M. Lund and A. S. Berman, *J. Appl. Phys.*, **37**, 2489 (1966).

9. C. Neumann, *Math Phys. K.*, **24**, 49 (1872).

10. J. C. Maxwell, *Philos. Trans. R. Soc. London*, **170**, 231 (1879).

11. M. C. I. Siu, *J. Vac. Sci. Technol.*, **10**, 368 (1973).

PROBLEMS

2.1 †State the assumptions that form the basis of kinetic theory.

2.2 Compute (a) the mean, (b) the most probable and (c) the rms speeds of the following groups of molecules: Four molecules whose velocity is 1 m/s, 6 at 2 m/s, 10 at 3 m/s, 3 at 4 m/s and 1 at 5 m/s.

2.3 Calculate the average velocities of the following gases at 25°C: (a) N_2, (b) O_2, (c) Ar, and (d) He.

2.4 †Room temperature nitrogen molecules are being directed toward a surface 100 cm from their entrance. To what pressure must the chamber be evacuated for a molecule to reach the target, on average, without first colliding with another nitrogen molecule?

2.5 †At room temperature and pressure one cm³ is expanded to a pressure of 10⁻⁴ Pa. How many 2-liter soda bottles will it fill?

2.6 What is the mass (in kg) of 1 m³ of air at 0°C? What is the mass of the air in the room in which you are?

2.7 Explain why the viscosity of a gas should increase with increasing particle mass and temperature.

2.8 †An object is cooled by the air surrounding it. If the object is heated by a constant-power source, what would happen to its temperature if the surrounding gas were replaced (a) by helium and (b) by argon?

2.9 What is the heat flow by thermal conduction between two 0.1-m² sheets of copper with a temperature difference of 100°C and which are separated by 0.1 cm of CO_2 at pressures of (a) 10^5 Pa, (b) 10^3 Pa, and (c) 100 Pa.

2.10 Two chambers are interconnected by a 0.5-cm-diameter tube. The left hand chamber is heated to a temperature of 250°C, while the right hand chamber remains at room temperature. Over what pressure range will the pressure in the two chambers be the same? Over what pressure range will the transpiration equation apply? Sketch a plot of P_{hot}/P_{cold} versus pressure of the cold chamber.

CHAPTER 3

Gas Flow

In this chapter we discuss the flow of gas at reduced pressures as it is encountered in a vacuum system. Gas flow is complex and the nature of the solution depends on the flow rate, gas properties, geometry, and surface properties of the duct. We begin by defining the flow regimes and introducing the concepts of throughput, mass flow and conductance. We describe the gas throughput and conductance for several kinds of flow. We show how approximation techniques and probability methods are used to solve complex problems such as flow in ducts containing entrance and exit orifices, aperture plates, or other irregular shapes.

3.1 FLOW REGIMES

Gas flow regimes are characterized by the *nature* of the gas and by the *relative quantity* of gas flowing in a pipe. The nature of the gas is determined by examining Knudsen's number, while the relative flow is described by Reynolds' number. In the viscous gas region (high pressures) the flow is called continuum flow and it can be turbulent or viscous. Turbulent flow is chaotic, like the flow behind a moving vehicle or the rising smoke some distance from a cigarette. Laminar viscous flow commences when the velocity and surface irregularities are small enough for the gas to flow gently around obstructions in laminar streamlines. In the molecular gas region, the mean free path is so long in comparison to the pipe size that the flow is entirely determined by gas-wall collisions. The flow in this region is called molecular flow. Between the continuum flow region and the molecular flow region is the transition region. In this region the gas molecules collide with each other and with the walls.

A viscous gas is characterized by a Knudsen number Kn < 0.01. Knudsen's number Kn, which is dimensionless, is the ratio of the mean free path to a characteristic dimension of the system, say, the diameter of a pipe:

$$Kn = \frac{\lambda}{d} \qquad \blacktriangleright (3.1)$$

In continuum flow the diameter of the pipe is much greater than the mean free path and the character of the gas flow is determined by gas-gas collisions. The flow has a maximum velocity in the center of the channel and zero velocity at the wall. Continuum flow can be either turbulent or laminar viscous. The boundary between turbulent and viscous flow can be expressed in terms of Reynolds' dimensionless number **R** for round pipes:

$$\mathbf{R} = \frac{U\rho d}{\eta} \qquad (3.2)$$

where ρ is the mass density (kg/m^3) of the gas of viscosity η flowing with stream velocity U in a pipe of diameter d. Reynolds' number is used to characterize the relative quantity of gas flow. It is a ratio of the shear stress due to turbulence to the shear stress due to viscosity. Alternatively, it tells something about the forces necessary to drive a gas system in relation to the forces of dissipation due to viscosity. Reynolds [1] found two flow situations dynamically similar when this dimensionless number was the same. When **R** > 2200, the flow was always turbulent and when **R** < 1200 the flow was always viscous [2]. In the region 1200 < **R** < 2200 the flow was viscous or turbulent, depending on the geometry of the inlet and outlet and on the nature of the piping irregularities.

Laminar viscous flow, the ordered flow of a gas in streamlines, occurs in the region bounded by a Reynolds number lower than 1200 and a Knudsen number lower than 0.01.

When the mean free path is equal to or greater than the pipe diameter, say Kn > 1, and when **R** < 1200, the gas is said to be a molecular gas, and the flow is called molecular flow. To be precise, Reynolds' number does not have any meaning for a molecular gas because classical viscosity cannot be defined. The nature of molecular flow is very different from laminar viscous flow. Gas-wall collisions predominate and the concept of viscosity is meaningless. For most surfaces diffuse reflection at the wall is a good approximation; that is, each particle arrives, sticks or rattles around in a surface imperfection, and is re-emitted in a direction independent of its incident velocity. Thus there is a chance that a particle entering a pipe in which $\lambda \gg d$ will not be transmitted, but will be returned to the entrance. In molecular flow gas molecules do not collide with one another, and gases can flow in opposite directions without interaction.

In the region $1 > Kn > 0.01$ the gas is neither viscous nor molecular. Flow in the transition region is not well understood. In this range, called the transition or slip flow range, where the pipe is several mean free paths wide, the velocity at the wall is not zero as in viscous flow and the reflection is not diffuse as in free-molecular flow. Now let us define throughput, mass flow and conductance and develop some practical gas flow formulas.

3.2   THROUGHPUT, MASS FLOW AND CONDUCTANCE

Throughput is the quantity of gas (the volume of gas at a known pressure) that passes a plane in a known time; $d/dt(PV) = Q$. In SI throughput has units of Pa-m^3/s. Because 1 Pa=1 N/m^2, and 1 J=1 N-m, the units may be more simply expressed as J/s or watts; 1 Pa-m^3/s = 1 W. Thus throughput is the *energy* per unit time crossing a plane. The energy in question is not the kinetic and potential energy contained in the gas molecules, but rather the energy required to *transport* the molecules across a plane. Expressing gas flow in units of watts seems awkward, but it does help explain the concept that throughput is energy flow. Throughput does not describe the concept of mass flow unless the temperature is specified. It is in many ways unfortunate that vacuum technologists have chosen to use a volumetric unit which conveys little information. Volumetric flow does not conserve mass.

Mass (molar or molecular) flow is the quantity of substance, respectively, in units of kg, kg-moles, or molecules that passes a plane in a known time. Molar flow and throughput are related:

$$N'(\text{kg} - \text{moles/s}) = \frac{Q}{N_o kT} = \frac{Q}{RT} \qquad (3.3)$$

Also, we can relate mass flow and throughput; $N'(\text{kg/s}) = MQ/N_o kT$. Throughput can be related to the mole flow, or mass flow, only if the temperature is constant and known. A spatial change in the temperature can alter the throughput without altering the mass flow. We discuss applications of mass flow in Chapter 6 (flow meters) and in Chapter 15 where we examine cryogenic pumping.

The flow of gas in a channel is dependent on the pressure drop across the tube as well as the geometry of the channel. Division of the throughput by the pressure drop across a channel held at constant temperature yields a property known as the intrinsic conductance of the channel:

$$C = \frac{Q}{P_2 - P_1} \qquad \blacktriangleright (3.4)$$

In SI the units of flow are Pa-m³/s and the units of conductance or pumping speed are m³/s; however, flow units of Pa-L/s and conductance units of L/s are also used. Unless explicitly stated, all formulas in this chapter use the cubic meter as the volumetric unit.

The pressures P_1 and P_2 in (3.4) refer to the pressures measured in large volumes connected to each end of the channel or component. According to (3.4) conductance is the property of the object between the points at which the two pressures are measured. For those whose first introduction to flow was with electricity, (3.4) is analogous to an electrical current divided by a potential drop. As in electrical charge flow, there are situations (viscous flow) in which the gas conductance is nonlinear, that is, a function of the pressure in the tube. Unlike electrical charge flow, there are cases in which the molecular conductance depends not only on the object, but on the nature of adjacent objects as well. We will explore this last issue in detail when we describe methods for combining conductances in the molecular flow regime.

3.3   CONTINUUM FLOW

A gas is called a viscous gas when Kn < 0.01. The flow in a viscous gas can be either turbulent, **R** > 2200, or viscous, **R** < 1200. Equation (3.2) can be put in a more useful form by replacing the stream velocity with

$$U = \frac{Q}{AP} \tag{3.5}$$

and if we replace the mass density, using the ideal gas law, (3.2) becomes

$$\mathbf{R} = \frac{4m}{\pi kT\eta} \frac{Q}{d} \qquad \blacktriangleright (3.6)$$

For air at 22°C this reduces to

$$\mathbf{R} = 8.41 \times 10^{-4} \frac{Q(\text{Pa} - \text{L/s})}{d(\text{m})} \tag{3.7}$$

In ordinary vacuum practice turbulent flow occurs infrequently. Reynolds' number can reach high values in the piping of a large roughing pump during the initial pumping phase. For a pipe 250 mm in diameter connected to a 47-L/s pump, **R** at atmospheric pressure is 1.6×10^4. Turbulent flow will exist whenever the pressure is greater than 1.5×10^4 Pa (100 Torr). In practice roughing lines are often throttled during the initial portion of the roughing cycle to prevent the sudden out-rush of gas from scattering any process debris that may reside on the work chamber floor. The flow in the throttling orifice is turbulent at high pressures.

In the high flow limit of the turbulent flow region the velocity of the gas may reach the velocity of sound in the gas. Further reduction of the downstream pressure cannot be sensed at the high pressure side so that the flow is choked or limited to a maximum or critical value of flow. The value of critical flow depends on the geometry of the element, for example, orifice, short tube, or long tube, and the shape of the entrance. A detailed discussion of critical flow has been given by Shapiro [3].

Rather than divide the discussion of continuum flow into viscous, turbulent and critical, it is easier to discuss the flow in terms of the geometry of the pipe. We divide this discussion into orifice flow, long tube flow and short tube flow, and give equations for each region.

3.3.1 Orifices

For tubes of zero length (e.g., a small, thin orifice) the flow versus pressure is a rather complicated function of the pressure. Consider a fixed high pressure, say atmospheric pressure, on one side of the orifice with a variable pressure on the downstream side. As the downstream pressure is reduced, the gas flowing through the orifice will increase until it reaches a maximum. At this ratio of inlet to outlet pressure (the critical pressure ratio), the gas is flowing at the speed of sound in the gas. The gas flow through the orifice is given by

$$Q = AP_1 C' \left(\frac{2\gamma}{\gamma-1} \frac{kT}{m} \right)^{1/2} \left(\frac{P_2}{P_1} \right)^{1/\gamma} \left[1 - \left(\frac{P_2}{P_1} \right)^{(\gamma-1)/\gamma} \right]^{1/2} \blacktriangleright (3.8)$$

$$\text{for } 1 > P_2/P_1 \geq (2/(\gamma+1))^{\gamma/(\gamma-1)}$$

C' is a factor which reduces the crosssectional area because the high speed gas continues to decrease in diameter after it passes through the orifice. This phenomenon is called the *vena contracta*. For thin orifices C' is about 0.85. If the downstream pressure is further reduced, the gas flow will not increase because the gas in the orifice is traveling at the speed of sound and cannot communicate further with the high pressure side of the orifice to tell it that the pressure has changed. In this region the value of P_2 is of no meaning as long as $P_2/P_1 < (2/(\gamma+1))^{\gamma/(\gamma-1)}$. The ratio of specific heats is γ, whose values are given in Appendix B.4. The flow is then given by

$$Q = AP_1 C' \left(\frac{kT}{m} \frac{2\gamma}{\gamma+1} \right)^{1/2} \left(\frac{2}{\gamma+1} \right)^{1/(\gamma-1)} \blacktriangleright (3.9)$$

$$\text{for } P_2/P_1 \leq (2/(\gamma+1))^{\gamma/(\gamma-1)}$$

This value is called the critical, or choked, flow. See Fig. 3.1. For air, $\gamma = 1.4$ and $P_2/P_1 = 0.525$. This limit is important in describing flow

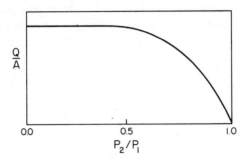

Fig. 3.1 Throughput vs. pressure ratio in a circular orifice.

restrictors which control gas flow and the rate of pumping or venting in a
vacuum system, choked flow in air-to-air load locks, and small leaks from
atmosphere.   In (3.8) and (3.9) the conductance can be found from
$Q/(P_1 - P_2)$. The conductance for air at 22°C is given by

$$C(\text{L/s}) = \frac{7.66 \times 10^5 C' A(\text{m}^2)}{1 - P_2/P_1} \left(\frac{P_2}{P_1}\right)^{0.714} \left[1 - \left(\frac{P_2}{P_1}\right)^{0.286}\right]^{1/2} \quad (3.10)$$

for $1 > P_2/P_1 \geq 0.52$

and

$$C(\text{L/s}) \sim 2 \times 10^5 \frac{A(\text{m}^2)C'}{1 - P_2/P_1} \qquad \blacktriangleright (3.11)$$

for $P_2/P_1 \leq 0.52$

3.3.2 Long Round Tubes

A general mathematical treatment of viscous flow results in the Navier-
Stokes equations, which are most complex to solve.   The simplest and
most familiar solution for long straight tubes is the equation due independently to Poiseuille and Hagen and called the Hagen-Poiseuille equation

$$Q = \frac{\pi d^4}{128 \eta l} \frac{P_1 + P_2}{2} (P_1 - P_2) \qquad \blacktriangleright (3.12)$$

The conductance for air at 0°C is

$$C(\text{L/s}) = 1.38 \times 10^6 \frac{d^4}{l} \frac{P_1 + P_2}{2} \qquad (3.13)$$

This specific solution is valid when four assumptions are met: (1) fully
developed flow—the velocity profile is not position dependent, (2)

laminar flow, (3) zero wall velocity, and (4) incompressible gas. Assumption 1 holds for long tubes in which the flow lines are fully developed. The criterion for fully developed flow was determined by Langhaar [4], who showed that a distance of $l_e = 0.0568d\mathbf{R}$ was required before the flow streamlines developed into their parallel, steady-state profile. For air at 22°C this reduces to l_e(meters) $= 0.0503Q$ when Q is given in units of Pa-m^3/s. Assumptions 2 and 3 are satisfied if $\mathbf{R} < 1200$ and if Kn < 0.01. The assumption of incompressibility holds true, provided that the Mach number U, the ratio of gas to sound velocity, is < 0.3.

$$U = \frac{U}{U_{sound}} = \frac{4Q}{\pi d^2 P U_{sound}} < \frac{1}{3} \tag{3.14}$$

For air at 22°C

$$Q(\text{Pa} - \text{L/s}) \; < \; 9.0 \times 10^5 d^2 P \qquad \blacktriangleright (3.15)$$

This is a value of flow that may be exceeded in many cases and would render the results of the Poiseuille equation incorrect.

Relationships for viscous flow between long, coaxial cylinders and long tubes of elliptical, triangular and rectangular crosssection have been tabulated by Holland, Steckelmacher and Yarwood [5]. Williams et al. [6] give the relation for flow in a long rectangular duct for air at 20°C:

$$Q(\text{Pa} - \text{L/s}) = 4.6Y\frac{b^2a^2}{l}\frac{(P_1 + P_2)}{2}(P_1 - P_2) \tag{3.16}$$

where the duct crosssectional dimensions a, b and the length l are given in cm. The function $Y(b/a)$ is obtained from the following table:

b/a	Y	b/a	Y	b/a	Y
1.0	0.4217	0.4	0.30	0.05	0.0484
0.8	0.41	0.2	0.175	0.02	0.0197
0.6	0.31	0.1	0.0937	0.01	0.0099

In the limit $b \ll a$ the air flow reduces to the one-dimensional solution of Sasaki and Yasunaga [7]:

$$Q(\text{Pa} - \text{L/s}) = 4.6\frac{b^3a}{l}\frac{P_1 + P_2}{2}(P_1 - P_2) \tag{3.17}$$

Again b, a and l are given in cm. The flow in (3.16) and (3.17), like (3.13), is inversely proportional to viscosity and may be accordingly scaled for other gases. These relations for long tubes are of limited use.

They are of use in components such as mass flow meter tubes, controlled leaks, and piping connecting chambers with remotely located pumps and gas tanks. In most practical cases we connect chambers with as short a duct as possible to reduce unwanted pressure drops, and need to know relationships which are valid for these cases.

3.3.3   Short Round Tubes

As we noted above, the flow in short tubes does not obey the Poiseuille equation. The flow may switch from viscous to critical flow without there being any pressure region in which the Poiseuille equation is valid. This problem has been treated in several ways. Dushman [8] gives a non-linear relation for flow in short, round tubes. It is valid only for un-choked flow. Santeler [9] has devised a technique in which he considers the short tube to be an aperture or orifice in series with a short tube of length l'. The problem is formatted by assuming an unknown pressure P_k between the "tube" and the "aperture." This is the pressure that would be measured by a gauge just inside the end of the tube which was pointing upstream. The flow through the tube [(3.13)] is equated to the flow through the aperture when it is either choked [(3.9)] or un-choked [(3.10)]. P_x is assumed to be the inlet pressure to the aperture and the outlet pressure of the tube. The problem can be solved with increased accuracy by fitting measurements to the calculations using an adjustable length l' and an adjustable orifice coefficient C'.

3.4   MOLECULAR FLOW

A gas is called molecular when Kn>1.0 (or Pd<6.6 Pa-mm for air at $22\,^{\circ}$C). In this region the flow is called molecular flow. For completeness we could say that **R** < 1200; however, we really cannot define a Reynolds number in the region where viscosity is not defined. The molecular flow region is theoretically the best understood of any flow type. This discussion focuses on orifices, infinite tubes, finite tubes, and other geometries and including combinations of components in molecular flow.

3.4.1   Orifices

If two large vessels are connected by an orifice of area A and the diameter of the orifice is such that Kn > 1, then the gas flow from one vessel (P_1, n_1) to the second vessel (P_2, n_2) is given by

$$Q = \frac{kT}{4}vA(n_2-n_1) = \frac{v}{4}A(P_2-P_1) \qquad (3.18)$$

and the conductance of the orifice is

$$C = \frac{v}{4}A \qquad \blacktriangleright (3.19)$$

which for air at 22°C has the value

$$C(m^3/s) = 116A(m^2) \qquad (3.20)$$

or

$$C(L/s) = 11.6A(cm^2) \qquad \blacktriangleright (3.21)$$

From (3.18) we see an interesting property of the molecular flow regime. Gas can flow from vessel 2 to vessel 1; at the same time gas is flowing from vessel 1 to vessel 2 without either of the gases colliding with the other.

3.4.2 Long Round Tubes

The diffusion method of Smoluchowski [10] and the momentum transfer method of Knudsen [11] and Loeb [12] were the first used to describe gas flow through very long tubes in the free-molecular-flow region. For circular tubes both derivations yield conductances of

$$C_{tube} = \frac{\pi}{12} v \frac{d^3}{l} \qquad (3.22)$$

For air at 22°C this becomes

$$C_{tube} \ (m^3/s) = 121\frac{d^3}{l} \qquad (3.23)$$

Steckelmacher [13] has derived formulas for the conductance of other long tubes with uniform but non-circular crosssections.

3.4.3 Short Round Tubes

The flow equation for long tubes [(3.22)] indicates the conductance becomes infinite as the length tends toward zero, while in Section 3.4.1 we showed the conductance actually becomes $vA/4$. Dushman [14] developed a solution to the problem of short tubes by considering the total conductance to be the sum of the reciprocal conductances of an aperture and a section of tube of length l:

$$\frac{1}{C_{total}} = \frac{1}{C_{tube}} + \frac{1}{C_{aperture}} \qquad (3.24)$$

As $l/d \rightarrow 0$, this equation reduces to (3.19), and as $l/d \rightarrow \infty$ it reduces to (3.22). Although this equation gives the correct solution for the extreme cases, it is not correct for the intermediate. It can be in error by as much as 12 to 15% for intermediate cases.

The difficulty in performing calculations for short tubes lies in the nature of the gas-wall interaction. Lorentz [15] assumed the walls of a pipe are molecularly rough, that is, molecules are scattered according to the cosine law (diffuse reflection). Molecules hit a wall, oscillate in potential wells, and recoil in a direction that is independent of their arrival angle. In diffuse reflection scattered molecules have the greatest probability of recoiling at an angle of 90° from the surface. Particles not scattering at 90° have as much likelihood of going forward through the tube as going backward toward the source. See Fig. 3.2. Clausing [16] solved this problem by calculating the probability that a molecule entering the pipe at one end will escape at the other end after making diffuse collisions with the walls. Clausing's solution is in the form of an integral that is difficult to evaluate. For simple cases such as round pipes it has been solved approximately by Clausing and others, and the solutions tabulated are found in many standard texts. The solution is usually given in the form of a transmission probability a that a molecule entering the pipe will leave the pipe at the other end. The conductance of a pipe is then found from (3.25), where A is the crosssectional area of the pipe and v is the thermal velocity of the gas:

$$C = \frac{a'v}{4}A \qquad \qquad \blacktriangleright (3.25)$$

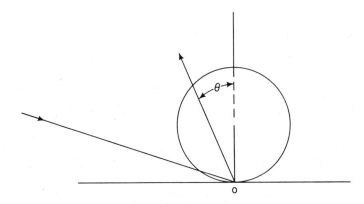

Fig. 3.2 A molecule making diffuse collisions with a wall is scattered in a direction independent of its original path. The molecule loses its sense of direction and is emitted with a probability proportional to the cosine of the angle from the normal to the surface. The probability of a particle leaving the surface at angle ϑ is proportional to the distance from the point o to the circle described by $\cos \vartheta$. The most probable angle is 0° from the normal, and the particle has an equal probability of going forward as it does in returning to the direction from which it came.

For air at 22°C

$$C(\text{L/s}) = 1.16 \times 10^5 aA$$

$$C(\text{L/s}) = 11.6aA(\text{cm}^2) \tag{3.26}$$

Equation (3.26) tells us the molecular conductance per unit area of any structure in molecular flow has a maximum value [11.6 L/(s-cm²) for air at 22°C] and that any structure (thicker than a thin aperture) will have a conductance less than this value.

DeMarcus [17] used a variational principle to solve the Clausing equation with improved accuracy. Berman [18] made a polynomial fit to DeMarcus' solution and extended it to larger l/d values. Values of the transmission probability for round pipes obtained from his equations are given in Table 3.1 for a range of l/d values. The DeMarcus-Berman results agree with very precise calculations done by Cole [19] to between 4 and 5 decimal places. These values of a can be used in (3.25) and (3.26). "Exit effects," which we shall discuss shortly, are included in these tabulations of probability values.

3.4.4 Other Short Structure Solutions

The calculation of molecular conductance in an arbitrarily complex short tube structure is not possible in closed form. The molecular flow conductance has been solved analytically for non-circular crosssections in only a few cases. Where exact solutions are not possible, probabilistic methods have been used.

Analytical Solutions

One interesting geometry that is not circular in crosssection is the thin, rectangular, slit-like pipe. This geometry is often encountered in differentially pumped feedthroughs and joints between differentially pumped chambers. It consists of a thin gap of thickness b, length l, and width a with the condition $a \gg b$, where a, b, and l are defined in Fig. 3.7. Berman [18] developed a polynomial fit to solutions for the transmission coefficient, and values calculated with the use of his formula are given in Table 3.2. His results agree with those of Neudachin et al. [20]. The conductance of a slit can be calculated from (3.25) using the transmission probability from Table 3.2 and an inlet area of ab. "Exit conductance" drops are also included in these transmission probabilities.

In addition to the short round pipe and the slit-like tube discussed above, solutions exist for the annular cylindrical pipe [21], the rectangular pipe [20], the elliptical tube and the triangular tube [5]. Other short tube crosssections and complex structures have been treated with statistical techniques such as the Monte Carlo technique described here.

Table 3.1   Transmission Probability *a* for Round Pipes

l/d	a	l/d	a
0.00	1.00000	1.6	0.40548
0.05	0.95240	1.7	0.39195
0.1	0.90922	1.8	0.37935
0.15	0.86993	1.9	0.36759
0.2	0.83408	2.0	0.35658
0.25	0.80127	2.5	0.31054
0.3	0.77115	3.0	0.27546
0.35	0.74341	3.5	0.24776
0.4	0.71779	4.0	0.22530
0.45	0.69404	4.5	0.20669
0.5	0.67198	5.0	0.19099
0.55	0.65143	6.0	0.16596
0.6	0.63223	7.0	0.14684
0.65	0.61425	8.0	0.13175
0.7	0.59737	9.0	0.11951
0.75	0.58148	10.0	0.10938
0.8	0.56655	15.0	0.07699
0.85	0.55236	20.0	0.05949
0.9	0.53898	25.0	0.04851
0.95	0.52625	30.0	0.04097
1.0	0.51423	35.0	0.03546
1.1	0.49185	40.0	0.03127
1.2	0.47149	45.0	0.02796
1.3	0.45289	50.0	0.02529
1.4	0.43581	500.0	0.26479×10^{-2}
1.5	0.42006	5000.0	0.26643×10^{-3}

Monte Carlo Technique

The Monte Carlo statistical methods developed for the calculation of molecular flow conductance by Davis [22] and Levenson, Milleron, and Davis [23] were a major breakthrough in the calculation of complex, but practical, vacuum system elements such as elbows, traps, and baffles. The Monte Carlo technique uses a computer to simulate the individual trajectories of a large number of randomly chosen molecules in calculating transmission probabilities. Figure 3.3 is a computer graphical model of the trajectories of 15 random molecules entering an elbow. It yielded a transmission probability of 0.222. When a large number of particles

Table 3.2   Transmission Probability a
for Thin, Rectangular, Slit-like Tubes

l/b	a	l/b	a
0.0	1.00000	15	0.18664
0.1	0.95245	20	0.15425
0.2	0.90958	30	0.11648
0.3	0.87097	40	0.09471
0.4	0.83617	50	0.08035
0.5	0.80473	60	0.07008
0.6	0.77620	70	0.06234
0.7	0.75021	80	0.05627
0.8	0.72643	90	0.05136
0.9	0.70457	100	0.04731
1.0	0.68438	200	0.02722
2.0	0.54206	500	0.01276
3.0	0.45716	1000	0.70829×10^{-2}
4.0	0.39919	2000	0.38914×10^{-2}
5.0	0.35648	5000	0.17409×10^{-2}
6.0	0.32339	10000	0.94000×10^{-3}
7.0	0.29684	20000	0.50472×10^{-3}
8.0	0.27496	50000	0.22023×10^{-3}
9.0	0.25655	100000	0.11705×10^{-3}
10	0.240805	200000	0.61994×10^{-4}

was used, the transmission probability of 0.31 was calculated. This points out one difficulty in the Monte Carlo technique; its accuracy depends on the number of molecular trajectories used in the calculation. A great deal of computational time is required for accurate solutions to complex problems. Figures 3.4 through 3.11 contain examples of the Monte Carlo technique for some structures of interest [22-24]. The molecular conductance is the product of the probability and conductance of an aperture identical in shape to the entrance of the structure under consideration. End effects are included in these formulas as well. The great computational time required to perform a transmission probability calculation by the Monte Carlo technique has driven others to approximate complex systems by combining cylindrical tubes, orifices, and baffle plates. See, for example, Füstöss and Tóth [25], Harries [26], Stekelmacher [27], Oatley [28], Haefer [29] and Ballance [30]. However, molecular conductances must be combined in series with great care.

3.4.5   Combining Molecular Conductances

Implicit in the definition of conductance (3.4) is the understanding that molecules will arrive at the entrance to the component distributed in a Maxwell-Boltzmann fashion and depart into a void without colliding with another surface.   This is possible only if there are no other walls in the vicinity of the entrance and exit of the component.   It can be accomplished by connecting the component between two large reservoirs so that the pressures in the vessels will be unaffected by the flow through the component.   In practice this condition is rarely met.   Typically traps, baffles, or elbows, whose lengths are of the order of the pipe diameter, are interconnected by the shortest possible lengths of pipe that will connect the vacuum chamber and pump.

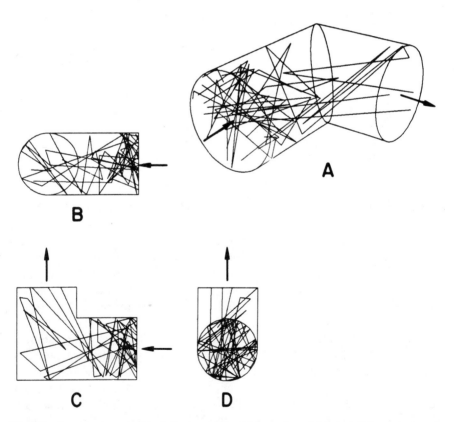

Fig. 3.3   A computer graphical display of the trajectories of 15 molecules entering an elbow in free molecular flow.   Courtesy of A. Appel, IBM T. J. Watson Research Center.

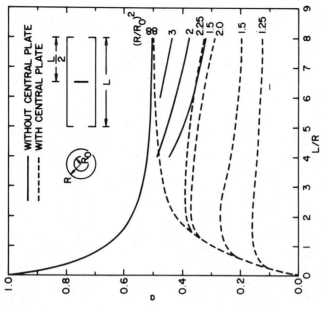

Fig. 3.5 Molecular transmission probability of a round pipe with entrance and exit apertures. Reprinted with permission from *J. Appl. Phys.*, **31**, p.1169, D. H. Davis. Copyright 1960, The American Institute of Physics.

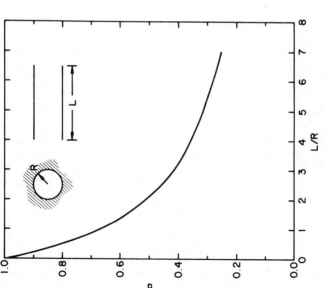

Fig. 3.4 Molecular transmission probability of a round pipe. Reprinted with permission from *Le Vide*, No.103, p. 42, L. L. Levenson et al. Copyright 1963, Societe Francaise des Ingeneirs et Techniciens du Vide.

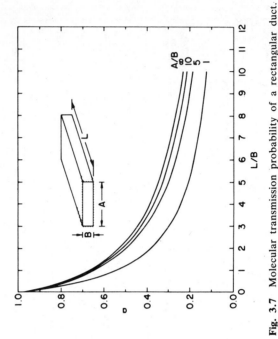

Fig. 3.6 Molecular transmission probability of an annular cylindrical pipe. Reprinted with permission from *Le Vide*, No. 103, p. 42, L. L. Levenson et al. Copyright 1963, Societe Francaise des Ingeneirus et Techniciens du Vide.

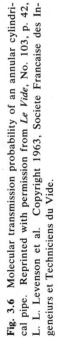

Fig. 3.7 Molecular transmission probability of a rectangular duct. Reprinted with permission from *Le Vide*, No. 103, p. 42, L. L. Levenson et al. Copyright 1963, Societe Francaise des Ingeneirus et Techniciens du Vide.

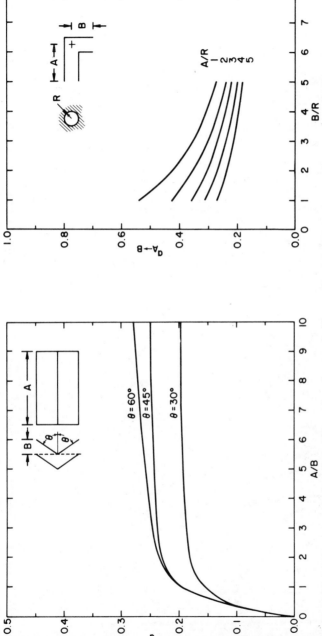

Fig. 3.8 Molecular transmission probability of a chevron baffle. Reprinted with permission from *Le Vide*, No. 103, p. 42, L. L. Levenson et al. Copyright 1963, Societe Francaise des Ingeneiurs et Techniciens du Vide.

Fig. 3.9 Molecular transmission probability of an elbow. Reprinted with permission from *J. Appl. Phys.*, **31**, p. 1169, D. H. Davis. Copyright 1960, The American Institute of Physics.

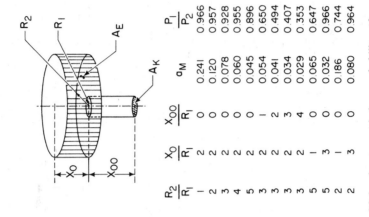

R_2/R_1	X_0/R_1	X_{00}/R_1	a_M	P_1/P_2
1	2	0	0.241	0.966
2	2	0	0.120	0.957
3	2	0	0.078	0.928
4	2	0	0.060	0.955
5	2	—	0.045	0.896
3	2	2	0.054	0.650
3	2	3	0.041	0.494
3	2	4	0.034	0.407
3	2	—	0.029	0.353
5	1	0	0.065	0.647
5	3	0	0.032	0.966
2	1	0	0.186	0.744
2	3	0	0.080	0.964

Fig. 3.11 Molecular transmission probability of a parallel plate model. Reprinted with permission from *Trans. 9th Nat. Vac. Symp.*, J. D. Pinson and A. W. Peck. Copyright 1962, Macmillan, New York, p. 407.

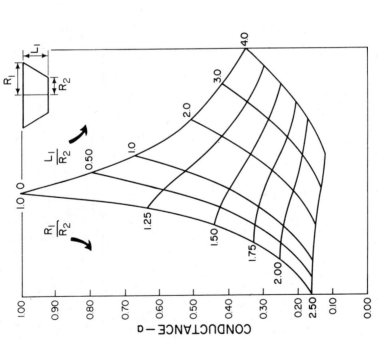

Fig. 3.10 Molecular transmission probability of a frustrum of a cone. Reprinted with permission from *Trans. 9th Nat. Vac. Symp.*, J. D. Pinson and A. W. Peck. Copyright 1962, Macmillan, New York, p. 407.

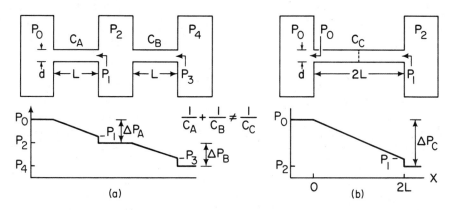

Fig. 3.12 The series conductance of two elements: (*a*) the pipes are isolated by a large volume and (*b*) the pipes are connected directly together. The pressure readings are those measured by a gauge in the gas stream pointing upstream parallel to the flow.

Parallel Conductances

The conductance of tubes connected in parallel can be obtained from the simple sum and is independent of any end effects:

$$C_T = C_1 + C_2 + C_3 + \ldots \qquad \blacktriangleright (3.27)$$

Series Conductances

Series conductances of truly independent elements in molecular flow will yield a total conductance of

$$\frac{1}{C_T} = \frac{1}{C_1} + \frac{1}{C_2} + \frac{1}{C_3} + \ldots \qquad \blacktriangleright (3.28)$$

Equation (3.28) gives the value of conductance we would measure if the elements were isolated from each other by large volumes. See Fig. 3.12*a*. The large volume provides a place for the distribution of molecules exiting the prior conductance to completely randomize or assume the distribution of a rarefied Maxwell-Boltzmann gas.

Exit and Entrance Effects

The simple reciprocal rule does not work where we combine two conductances directly. See Fig. 3.12*b*. Let us consider two tubes each of length-to-diameter ratio $l/d = 1$. From Table 3.1 we obtain the transmission probability of each tube as $a=0.51423$. If we combine them according to the reciprocal rule in (3.28), we will obtain a net transmission probability of $a=0.25712$. We know that the actual transmission of this structure (a tube with $l/d=2$) can be found from the data in Table 3.2 as $a=0.35685$. The error in using the simple reciprocal rule is 27.9%. Why do we have this large error? The reason is the pressure distribution in the two tubes is not the same. For the case of the isolated conductances, the pressure in the (imaginary) large chamber between the two pipes can be defined. It is single valued and could be measured with a gauge in the chamber. If we were to point a directional gauge in any direction in this chamber, we would measure the same pressure. In an analogous fashion, we could measure a voltage at the junction of two series resistors which are carrying a current.

When we combine two pipes in series without the large volume in between, the situation changes drastically. The pressure's at the exit of the first tube and the entrance to the second tube are now the same but not easy to define. If we were to place a directional gauge at the junction of the two pipes and point it upstream, it would read higher than downstream. Also, were it to face sideways, it would also depend on whether it were in the center of the tube or off-axis. Clearly, the pressure is anisotropic. This is to be contrasted to the case of the two tubes connected by a chamber which has a known pressure. In that case the only place where the pressure will differ is at the exit of the first tube as it enters the chamber as measured with a gauge looking upstream. That pressure will be higher than in the chamber.

Recall the definition of conductance given in (3.4). If we examine Fig. 3.12a, we see that the pressure difference in (3.4) is measured in the large chambers at each end of the tube. By measuring the pressure in this way we include the pressure drop schematically shown at the end of the tube. When we connect the two tubes directly, we have eliminated the pressure drop at the exit of the tube that would be measured by an upstream-directed gauge. For this reason we say the conductances tabulated here and in other sources include what Santeler [31] calls the *exit loss*.

There is another effect which is present to a small extent—gas beaming. Note that the gas entering the second tube in Fig. 3.12a is randomly distributed, while the gas entering the second half of the tube in Fig. 3.12b is beamed. See, for example, the molecular exit angles depicted in Fig. 3.13. In normal cases this is a small correction - a few percent - which we will describe shortly. Beaming effects (entrance effects) in most real molecular flow situations do not introduce major errors because real systems are made up of short tubes connected by elbows, traps and so on. We observe in Fig. 3.13 that short tubes have near-cosine exit flux. Any component containing an elbow, interior baffle or chevron, will also scatter molecules and shift the distribution toward cosine.

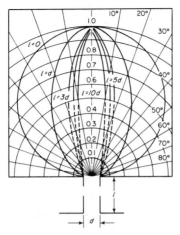

Fig. 3.13 Angular distribution of particles exiting tubes of various ratios of length to diameter. Reproduced with permission from Atom and Ion Sources, 1977, p. 86, L. Vályi. Copyright 1977, Akadémiai Kiadó, Budapest.

Series Calculations

Harries [26], Stekelmacher [27], Oatley [28] and Haefer [29] have each used the concept of a probability factor a to calculate the conductance of a series combination of vacuum elements in free molecular flow.   We examine here the method developed by Oatley.   Figure 3.14 illustrates the concept with a single component; Γ molecules per second enter at the left hand side, $a\Gamma$ molecules per second exit at the right-hand side, and $(1 - a)\Gamma$ molecules per second are returned to the source vessel. The conductance is expressed by

$$C = v\frac{A}{4}a \qquad (3.29)$$

For two tubes in series Oatley developed a technique for calculating a combined probability, the results of which are illustrated in Fig. 3.15. Among the Γ molecules per second that enter the first tube $a_1\Gamma$ enter the second; $\Gamma(1 - a_2)a_1$ of these are returned to the first tube and $\Gamma a_1 a_2$ enter the second.   From the group $\Gamma (1 - a_2)a_1$ molecules returned the first

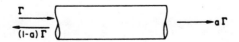

Fig. 3.14 Model for calculating the transmission probability of a single element.

tube $\Gamma a_1(1 - a_2)(1 - a_1)$ are returned to the second, and so on, until an infinite series expression was developed that simplified to

$$\frac{1}{a} = \frac{1}{a_1} + \frac{1}{a_2} - 1 \qquad (3.30)$$

The last term represents the exit pressure drop which is subtracted in this formula. This expression, when generalized to several elements in series, is

$$\frac{1-a}{a} = \frac{1-a_1}{a_1} + \frac{1-a_2}{a_2} + \dots \qquad \blacktriangleright (3.31)$$

Now let us use (3.30) or (3.31) and calculate the series conductance of the two pipes of $l/d=1$ described earlier. We obtain a value $a=0.3460$. Note that this is closer to the Clausing value of $a=0.35685$ obtained from Table 3.1. It is in error by 2.93%, and this is due to beaming effects for which it cannot account.

Oatley's formula as given in (3.31) applies directly to elements of the same diameters, but it can be extended to elements of differing diameters. If the series components are of different diameters or more complex construction, an addition theorem developed by Haefer will simplify the calculation.

Haefer [29] has developed a useful addition theorem for elements in the molecular flow regime. It relates the total transmission probability of n elements a_{1n}, to the transmission probability a_i and the inlet area A_i of each component. Extra terms are included in the equation whenever a crosssectional area *decreases* upon entering the next element but not when

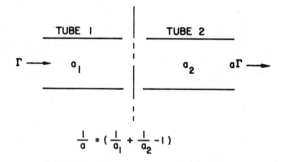

Fig. 3.15 Model for calculating the transmission probability of two elements in series. Reprinted with permission from *Brit. J. Appl. Phys.*, **8**, p. 15, C. W. Oatley. Copyright 1957, The Institute of Physics.

the area increases. It is given here without proof:

$$\frac{1}{A_1}\left(\frac{1-a_{1-n}}{a_{1-n}}\right) = \sum_1^n \frac{1}{A_i}\left(\frac{1-a_i}{a_i}\right) + \sum_1^{n-1}\left(\frac{1}{A_{i+1}} - \frac{1}{A_i}\right)\delta_{i,i+1} \quad \blacktriangleright (3.32)$$

where $\delta_{i,i+1} = 1$ for $A_{i+1} < A_i$, and $\delta_{i,i+1} = 0$ for $A_{i+1} \geq A_i$

The use of this formula is demonstrated with the example given in Fig. 3.16. Shown is a combination of three pipe sections of inlet areas A_1, A_2, and A_3. The pipes have corresponding transmission probabilities a_1, a_2, and a_3. Equation (3.32) then becomes

$$\frac{1}{A_1}\left(\frac{1-a_{1-n}}{a_{1-n}}\right) = \frac{1}{A_1}\left(\frac{1-a_1}{a_1}\right) + \frac{1}{A_2}\left(\frac{1-a_2}{a_2}\right)$$

$$+ \left(\frac{1}{A_3} - \frac{1}{A_2}\right) + \frac{1}{A_3}\left(\frac{1-a_3}{a_3}\right) \quad (3.33)$$

after some simplification (3.33) becomes

$$\frac{1}{a_{1-n}} = \frac{1}{a_1} + \frac{A_1}{A_2 a_2} + \frac{A_1}{A_3}\left(\frac{1}{a_3} - 2\right) \quad (3.34)$$

From (3.34) it can be seen that this answer reduces to (3.31) when the pipe areas are all the same. Haefer's method must be applied with consistency because the total transmission coefficient one calculates is a function of the end of the structure chosen as the origin. If the transmission probability were calculated from right to left, a_{n-1}, we would have to use an inlet area of A_n in calculating the conductance. However, this does not affect the total conductance of a structure as

$$C = \frac{nv}{4} a_{1-n} A_1 = \frac{nv}{4} a_{n-1} A_n \quad (3.35)$$

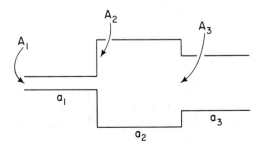

Fig. 3.16 Sample conductance to be evaluated with Haefer's addition theorem.

We can also see that the answer would have been different if the order of the second and third pipes were interchanged. Interchanging the order of components in a series configuration will affect the total transmission probability. As a rule, the conductance of the *actual* configuration should be calculated, because errors can creep in when components are mathematically rearranged to simplify calculation [32]. It is generally true that a complex structure made up of several series elements has the maximum conductance when the elements are arranged in increasing or decreasing order because the exit losses are the smallest and the beaming is the greatest. Arranging conductances in alternating larger and smaller diameters introduces wall scattering which makes the input to the following sections more like a cosine distribution.

Problems arise in conductance calculations because (3.28), which is valid for independently defined C's, is indiscriminately applied to series elements *not isolated* from one another, for example, in which the exit effect is not subtracted and the inlet gas does not obey a cosine distribution. The most serious error is the introduction of the exit impedance of the tube multiple times. In all tabulated transmission coefficients, the exit term has been included. Therefore it is necessary to remove it when combining conductances, and for this reason we use formulas like (3.31) or (3.32). The choice between an exact or approximate formula for the conductance of an individual pipe segment is usually less important than the correction for the exit effect. The Oatley and Haefer formulas remove the biggest error in calculating the conductance of combinations —the exit conductance drop at the end of each junction of equal diameter, but neither formula corrects for entrance effects, that is, non-cosine, or beamed, entrance flux.

Pinson and Peck [24] discuss beaming errors for pipe sections with and without baffles and show the difference between a calculation and Monte Carlo technique is $\leq 10\%$. The greatest differences are seen with no baffle. Beaming corrections for tube combinations have been developed by Santeler [31]. Components like chevron baffles and elbows tend to scatter the gas. For example, the conductance of a non-degenerate elbow can be calculated by using the conductances of the individual arms, obtained from Fig. 3.4, and summing these conductances with (3.30). Saksaganski [33] discusses efficient methods for analyzing complex systems and shows the methods of angular coefficients and integral kinetic method as alternatives to Monte Carlo.

The formulas developed by Haefer and Oatley may be used to calculate the pumping speed at the inlet of pipes connected to a pump. In this case the pump is characterized by its Ho coefficient (see Chapter 7) and inlet area. The pump is simply considered to be a conductance of entrance area A and transmission probability equal to its Ho coefficient.

3.5   THE TRANSITION REGION

The theory of gas flow in the transition region is not well developed. The state of the theory has been reviewed by Thomson and Owens [34] and Loyalka et al. [35]. Some additional work on the transition between molecular and isentropic flow in orifices has been done by DeMuth and Watson [36]. The simplest treatment of this region, due to Knudsen and discussed in many texts, states that

$$Q = Q_{\text{viscous}} + Z' Q_{\text{molecular}} \qquad \blacktriangleright (3.36)$$

where for long circular tubes Z' is given by

$$Z' = \frac{1 + 2.507\left(\dfrac{d}{2\lambda}\right)}{1 + 3.095\left(\dfrac{d}{2\lambda}\right)} \qquad (3.37)$$

3.6   MODELS SPANNING SEVERAL PRESSURE REGIONS

Flow relations which span pressure regions are difficult to construct. Two examples have already been given in this chapter. In addition to Dushman's model for the transition region discussed above, we have discussed Santeler's model for flow in a short tube in the viscous and choke regions. Santeler's [9] model for a short tube separates the tube component from the exit effect. Therefore, it can be applied to a series of tubes by modeling the system as a series of tubes with one exit loss after the last tube section. Other relations have been developed for specialized geometries. One of these is a relation developed by Kieser and Grundner [37] for the thin, slit-like tube. The thin, rectangular slit-like tube with one side in a rarefied gas and the other at atmospheric pressure is encountered in atmosphere-to-vacuum continuous feed systems and reel-to-reel coating systems known as web, or roll, coaters. This relation, which is valid in the molecular, transition and viscous flow region, combines the ideas of Dushman and Knudsen. It is valid for any inlet pressure but only for low exhaust pressures ($P_0 < 0.52 P_i$). Kieser and Grundner begin with Dushman's relationship, which assumes a duct to be composed of a pipe and an entrance aperture. The conductance of the series combination is then

$$\frac{1}{C_{\text{total}}} = \frac{1}{C_{\text{pipe}}} + \frac{1}{C_0} \qquad (3.38)$$

or, for air at 20°C

$$\frac{1}{C_{\text{total}}} = \frac{1}{(0.1106eP_i + Z')C_M} + \frac{1}{C_0} \tag{3.39}$$

In (3.38) C_0 is given by

$$C_0 \text{ (L/s)} = 11.6ew\left[\frac{10 + 0.5(e/\lambda)^{1.5}}{10 + 0.3412(e/\lambda)^{1.5}}\right] \tag{3.40}$$

and

$$C_M \text{ (L/s)} = 11.6ew\left(\frac{a}{1-a}\right) \tag{3.41}$$

The form of (3.40) allows the (air) conductance of the aperture to vary from 11.6 L/s in molecular flow to 17 L/s in the choked limit. Equation (3.40) is an experimental fit to the transition region for air [37].

The pipe conductance given in (3.38) is due to Knudsen and is a superposition of continuum and molecular flow. Since the aperture is included in the C_0 term, we have removed it from (3.41) [38]. The pipe conductance C_M calculated by Berman already contains an end effect C_0. Also the premise on which (3.38) is based, that is, representing the total conductance by a series combination of a "pipe conductance" and an "aperture conductance," is known to be in error by 10 to 15% in the region $l/e = 1$ to 5. For longer pipes there is another small error. Equation (3.39) implicitly lets the average pressure in the "pipe" portion of the conductance be $P_i/2$. This forces the pressure across the viscous conductance to an incorrect value. However, the pipe conductance is in series with a choked orifice, so any error in the "pipe" conductance is greatly attenuated by the series combination. (It is like putting a large resistor in parallel with a very small one—the parallel combination isn't greatly affected by the value of the larger resistor.)

There are few papers on flow relations which are valid over several regions. For example, Schumacher [39] summarizes the flow through small round tubes in graphical form, while Levina [40] develops nomographs for the same problem.

3.7   SUMMARY OF FLOW REGIMES

The values of gas flow, pressure and pipe size discussed in the previous sections each extend over a wide range. We summarize this discussion by sketching a plot of flow divided by pipe size (Q/d) versus pressure times distance (Pd). Figure 3.17 depicts the various regions discussed in this

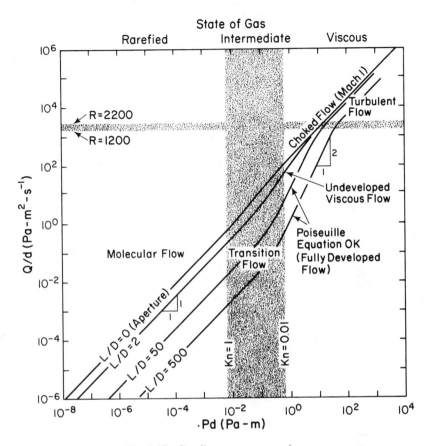

Fig. 3.17 Gas flow—pressure regimes

section.   Molecular flow occurs in the region **R** < 1200, and Kn > 1.
The flow is proportional to the first power of the pressure (slope = 1).
When Kn < 0.01 the gas is viscous, and the flow is either turbulent, fully
developed (Poiseuille), undeveloped or choked.   Observe that fully
developed flow is proportional to the square of the pressure (slope = 2).
The boundary between turbulent and laminar viscous is determined by
the Reynolds number.   The boundary between fully developed and
undeveloped flow is determined by Langhaar's number and the Mach
number.   We illustrate how a short tube can go from molecular to choked
without ever having the Poiseuille equation apply.   The region between
molecular and viscous flow is the transition region.   We see the transition
from completely free molecular flow to completely viscous flow can take
place over a two-decade pressure range.

In this chapter we have considered the equations of flow in each
region.   The equations developed here relate flow to pressure drop in

several pressure regions and for several pipe geometries. When we combine these with the dynamical equations of gas flow from a chamber, we can calculate the time required to pump a chamber to a particular pressure.

REFERENCES

1. O. Reynolds, *Philos. Trans. R. Soc., London*, **174**, (1883).

2. A. Guthrie and R. K. Wakerling, *Vacuum Equipment and Techniques*, McGraw-Hill, New York, 1949, p. 25.

3. A. H. Shapiro, *Dynamics and Thermodynamics of Compressible Fluid Flow*, Ronald, New York, 1953.

4. H. L. Langhaar, *J. Appl. Mech.*, **9**, A-55 (1942).

5. L. Holland, W. Steckelmacher and J. Yarwood, *Vacuum Manual*, E. & F. Spoon, London, 1974. p. 26.

6. B. J. Williams, B. Fletcher and J. A. A. Emery, *Proc. 4th Int. Vac. Congr., 1968*, Inst. of Phys. and the Phys. Soc., London, (1969). p. 753.

7. S. Sasaki and S. Yasunaga, *J. Vac. Soc. Japan*, **25**, 157 (1982).

8. S. Dushman, *Scientific Foundations of Vacuum Technique*, 2nd ed., J. M. Lafferty, Ed., Wiley, New York, 1962, p. 35.

9. D. J. Santeler, *J. Vac. Sci. Technol. A*, **4**, 348 (1986).

10. M. von Smolochowski, *Ann. Phys.*, **33**, 1559 (1910).

11. M. Knudsen, *Ann. Physik*, **28**, 75 (1909); **35**, 389 (1911).

12. L. B. Loeb, *The Kinetic Theory of Gases*, McGraw-Hill, New York, 2nd ed., 1934, Chapter 7.

13. W. Steckelmacher, *J. Phys. D: Appl. Phys.*, **11**, 473 (1978); *Vacuum*, **28**, 269 (1978).

14. Ref. 5, p. 91.

15. H. A. Lorentz, *Lectures on Theoretical Physics*, Vol. 1, Macmillan, London, 1927. Chapter 3.

16. P. Clausing, *Ann. Phys.*, **12**, 961 (1932), English Translation in *J. Vac. Sci. Technol.*, **8**, 636 (1971).

17. W. C. DeMarcus, Union Carbide Corp. Report K-1302, Part 3, 1957.

18. A. S. Berman, *J. Appl. Phys.*, **36**, 3356 (1965), and erratum, *ibid*, **37**, 4598 (1966).

19. R. J. Cole, *Rarefied Gas Dynamics*, **51**, Part 1, of *Progress in Astronautics and Aeronautics*, ed. J. L. Potter, (10th Int'l Symp. Rarefied Gas Dynamics), Am. Inst. of Aeronautics and Astronautics, 1976. p. 261.

20. I. G. Neudachin, B. T. Porodnov and P. E. Suetin, *Soviet Physics, Technical Physics*, **17**, 1036 (1972).

21. A. S. Berman, *J. Appl. Phys.*, **40**, 4991 (1969).

22. D. H. Davis, *J. Appl. Phys.*, **31**, 1169 (1960).

23. L. L. Levenson, N. Milleron, and D. H. Davis, *Le Vide*, **103**, 42 (1963).

24. J. D. Pinson and A. W. Peck, *Trans 9th Nat. Vac. Symp.*, Macmillan, New York, 1962. p. 407.

25. L. Füstöss and G. Tóth, *J. Vac. Sci. Technol.*, **9**, 1214 (1972).

26. W. Harries, *Z. Angew. Phys.*, **3**, 296 (1951).

27. W. Steckelmacher, *Proc. 6th Int. Vacuum Cong.*, Kyoto, *Japan. J. Appl. Phys.*, Sup. 2, Pt. 1, 117 (1974).

28. C. W. Oatley, *Brit. J. Appl. Phys.*, **8**, 15 (1957).

29. R. Haefer, *Vacuum*, **30**, 217 (1980).

30. J. O. Ballance, *Trans 3rd. Int. Vac. Congr.*, Vol. 2, Pergamon, Oxford, 1967, p. 85.

31. D. J. Santeler, *J. Vac. Sci. Technol. A*, **4**, 338 (1986).

32. D. J. Santeler, et al., *Vacuum Technology and Space Simulation*, NASA SP-105, National Aeronautics and Space Administration, Washington, D.C., 1966, p.115.

33. G. L. Saksaganski, *Molecular Flow in Complex Vacuum Systems*, Gordon and Breach, New York, 1988.

34. S. L. Thomson and W. R. Owens, *Vacuum*, **25**, 151 (1975).

35. S. K. Loyalka, T. S. Storvick and H. S. Park, *J. Vac. Sci. Technol.*, **13**, 1188 (1976).

36. S. F. DeMuth and J. S. Watson, *J. Vac. Sci. Technol. A*, **4**, 344 (1986).

37. J. Kieser and M. Grundner, *Proc. VIII Intl. Vac. Congr., Suppl. Rev. Le Vide*, **201**, 376 (1978).

38. J. F. O'Hanlon, *J. Vac. Sci. Technol. A*, **5**, 98 (1987).

39. B. W. Schumacher, *Proc. 8th Nat'l Vac. Symp., 1961*, Vol. 2, Pergamon, New York, 1962. p. 1192.

40. L. E. Levina, *Sov. J. Nondesrtuct. Test.*, **16**, 67 (1980).

PROBLEMS

3.1 †Describe the molecular, transition, laminar viscous and turbulent gas flow regimes.

3.2 †Calculate Reynolds and Knudsen numbers for the following cases: (a) 80 Pa-m^3/s of air through a 2-mm-diameter orifice at a pressure of 101,000 Pa, (b) an air flow of 5 × 10^6 Pa-L/s (1 atm at 50 L/s) in a 5-cm-diameter mechanical pumping line at the time the pump is started, and (c) a flow of 0.1 Pa-L/s of water vapor in a 20-cm-diameter high vacuum pumping line. Assume room temperature. Characterize the flow in each case.

3.3 (a) Calculate the maximum gas flow in a 1-mm-diameter tube which is 15 mm long with atmospheric pressure on one end and 1 Pa on the other end using the Poiseuille equation. (b) Calculate the maximum choked flow through a 1-mm-diameter orifice. Explain the two answers.

3.4 †What is the lowest average pressure necessary to keep the room-temperature air flow predominantly viscous laminar in a 5-cm-diameter line?

3.5 †A 15-cm-diameter pipe, 15 cm long connects a high vacuum chamber and a pumping system. Which of the following modifications will give the greatest increase in conductance in the molecular flow region? (a) Reducing the length of the tube to 7.5 cm, (b) increasing the length of the tube to 17.5 cm, (c) reducing the diameter from 15 to 12 cm, (d) increasing the diameter to 17.5 from 15 cm.

3.6 Two very large vessels are interconnected by a 1-cm-diameter tube, 10 cm long. The vessel A contains nitrogen at 10^{-2} Pa, and vessel B contains nitrogen at 10^{-4} Pa. (a) What is the flow rate of the nitrogen molecules originally in vessel A from vessel A to vessel B? (b) Is there any flow from vessel B to A? If so, how much? (c) What is the net nitrogen flow from vessel A to vessel B?

3.7 The two vessels in problem 3.6 are now connected with one of the thin apertures shown in Fig. 3.18. Calculate the nitrogen conductance of each aperture.

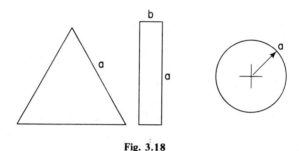

Fig. 3.18

3.8 †Kurtz [*Proc. 4th Int. Vac. Cong. (1968)*, Instute of Physics and the Physical Society, London, 1969, p. 817] published a clever, simplified Monte Carlo technique. His method helps us to visualize the nature of molecular flow. It is easily applied to any two-dimensional structure. First, we divide the entrance to a structure in 6 equal areas and number them serially. We cast a (six-sided) die and randomly determine an entrance position. Second, we cast

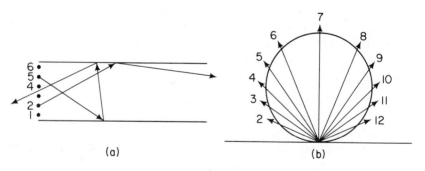

(a)   (b)

Fig. 3.19

a pair of (six-sided) dice and determine the angle or direction of particle motion (explanation below). We draw a ray from the starting point along that angle to a point where the particle "collides" with a wall. See Fig. 3.19a. We cast the dice again, determine a new angle, and repeat the procedure until the particle either leaves the structure at the exit or returns to the entrance. After a number of attempts, the transmission probability can be calculated. It is the ratio of particles successfully navigating the structure to the total number of particles entering the structure.

We can determine the angle of escape from Fig. 3.19b. Seven is the most probable sum obtained when casting two dies; it occurs 6/36 times. Six and 8 each occur 5/36 times, or 5/6 of the probability of obtaining a 7. A sum of 7 is equated with an escape angle of 0° from the normal. A sum of 6 or 8 corresponds to $\cos \phi = 5/6$ or $\phi = \pm 34°$, and so on as sketched in Fig. 3.19b.

†(a) Determine the probabilities for each of the other possible combinations (2 through 12) and make a template of escape angle vs. sum of the dies as sketched in Fig. 3.19b.

†(b) Draw a two-dimensional "pipe" 6 cm diameter, 15 cm long, and determine the transmission probability by Kurtz' technique.

†(c) Why does the Monte Carlo technique require many attempts to calculate the transmission probability accurately?

3.9   †Under what conditions does the gas flux have a cosine distribution at the entrance to a conductance in the molecular flow regime?

3.10 Using Oatley's method, compute the transmission probability of an elbow of diameter 1 and arm length 2. Compare your result with the Monte Carlo analysis given in the text. Why are the answers the same?

CHAPTER 4

Gas Release From Solids

All the gas could be pumped from a vacuum chamber in a very short time if it were located in the volume of the chamber. Consider a 100-L chamber previously roughed to 10 Pa and just connected to a 1000-L/s high vacuum pump. The equations we develop in Chapter 19 show the pressure will drop from 10 Pa, when the high vacuum valve is opened, to 4.5×10^{-4} Pa in 1 s. In practice this will never happen because gases and vapors residing on and in the interior walls desorb slowly and add to the quantity of gas that is to be removed from the chamber. For this example 15 to 60 m will be required to reach the mid-10^{-4}-Pa-range in an unbaked but clean system.

This chapter discusses the mechanisms of gas evolution from solid surfaces and explains how they affect the pumping rate and ultimate pressure in vacuum chambers. Gas is dissolved in and adsorbed on solids. This gas release, collectively referred to as outgassing, is actually a result of several processes. Figure 4.1 shows all the possible sources of gas in addition to the gas located in the volume of the chamber. Gases and vapors released from the surface are a result of vaporization, thermal desorption, diffusion, permeation, and stimulated desorption.

4.1 VAPORIZATION

A vapor is a gas near its condensation temperature and vaporization is the thermally stimulated entry of molecules into the vapor phase. In dynamic equilibrium the rate at which molecules leave the surface of a solid or liquid equals the rate at which they arrive at the surface. The pressure of the vapor over the surface in dynamic equilibrium is the vapor pressure of the solid or liquid, provided the solid or liquid and the vapor

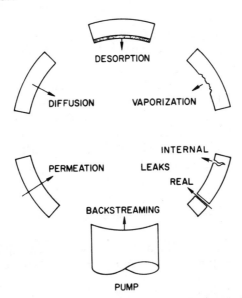

Fig. 4.1 Potential sources of gases and vapors in a vacuum system.

are at the same temperature. In Chapter 2 we stated the molecular flux of vapor crossing a plane was $nv/4$. In equilibrium this is therefore the rate of molecular release from the surface. For the case of free evaporation of a solid from a heated source, (2.9) given here in a different form may be used to calculate the maximum rate of evaporation of a solid from its temperature, vapor pressure, surface area, and molecular weight:

$$\Gamma(\text{molecules/s}) = 2.63 \times 10^{24} \frac{PA}{(MT)^{1/2}} \qquad \blacktriangleright (4.1)$$

The vapor-pressure-temperature curves for many gases are given in Appendixes B.5 and B.6. Appendix C.7 provides the vapor pressures of the solid and liquid elements.

4.2 THERMAL DESORPTION

Thermal desorption is the heat-stimulated release of gases or vapors previously adsorbed on the interior walls of the system. They may have been adsorbed on the chamber surface while it was exposed to the atmospheric environment and then slowly released as the pump removed gas from the chamber. Desorption is also the final step in the processes of diffusion and permeation. The rate of desorption is a function of the molecular binding energy, the temperature of the surface, and the number of monolayers of surface coverage. Gas is sorbed onto surfaces by physisorption and chemisorption. Physisorbed molecules are bonded to the surface by weak van der Waal's forces of energy less than 40 MJ/(kg-mole). Adsorption at energies greater than this value is known as chemisorption. Physisorbed particles are removed quickly from solid surfaces at ambient temperature and do not hinder pumping. Chemisorbed particles desorb slowly unless the surface is heated or bombarded by photons or particles. Chemical desorption is responsible for most of the outgassing encountered in vacuum systems. Let's examine the kinetics of two types of 'idealized' desorption, then comment on how real surfaces desorb.

4.2.1 First Order Desorption

Atoms or molecules which do not dissociate on adsorption desorb at a rate proportional to their surface concentration. This can be expressed mathematically as

$$\frac{dC(t)}{dt} = -K_1 C(t) = -\frac{e^{-E_d/(N_o kT)}}{\tau_o} C(t) \qquad (4.2)$$

Desorption with a rate $dC(t)/dt$ which is proportional to the concentration of atoms or molecules on the surface C is called first order desorption. This is a description of how monoenergetic atoms of, for example, helium or argon desorb from a metal or glass or how water vapor desorbs from itself or a glass. The rate constant K_1 is strongly dependent on the desorption energy E_d and the temperature. It can be described by the equation

$$\frac{1}{K_1} = \tau_r = \tau_o e^{E_d/N_o kT} \qquad \blacktriangleright (4.3)$$

where τ_o is the vibrational frequency of a molecule or atom in an adsorption site and is typically 10^{-12} s: K_1 is also the reciprocal of the average residence time τ_r that a molecule or atom spends on the surface. The desorption rate will also decrease with time as the surface layer becomes depleted. By integrating (4.5) we can show

$$\frac{dC(t)}{dt} = C_o K_1 e^{-K_1 t} = C_o K_1 e^{-t/\tau_r} \qquad (4.4)$$

Equation (4.4) shows the manner in which the desorption rate decreases with time. This equation predicts rapid (exponential) decay of the desorption rate in a few τ_r.

Table 4.1 shows the strong dependence of the residence time on the temperature and desorption energy for three representative gas-metal systems: water weakly bonded to itself, hydrogen strongly bonded on

Table 4.1   Average Residence Time of Chemisorbed Molecules

| System | Desorption Energy (MJ/(kg-mole)) | Residence time at | | |
		77 K (s)	22°C (s)	450°C (s)
H_2O/H_2O	40.6	10^{15}	10^{-5}	10^{-9}
$H_2O/metal$	96	-	10^5	10^{-5}
H_2/Mo	160	-	10^{17}	1

molybdenum, and the intermediate case of water on a metal. These times were calculated using desorption energies given by Ehrlich [1]. The room temperature residence time for water adsorbed on itself is less than a millisecond. Baking is not necessary to remove the bulk of physisorbed water. Hydrogen is strongly chemisorbed at energies of 40,000 cal/(g-mole) [160 MJ/(kg-mole)] and is impossible to remove without a high temperature bakeor the use of stimulated desorption. Water is chemisorbed on a metal with energies in the range 22,000 to 24,000 cal/(g-mole) [92 to 100 MJ/(kg-mole)]. It is a problem because it is bound just strongly enough to stick at room temperature. Although is is hard to remove at room temperature, it is easily desorbed by baking at temperatures of 250 to 300°C. Cooling the surface has a dramatic effect on the residence time of all molecules—physisorbed and chemisorbed. The residence times become very long and cooled surfaces become traps. Water is the most troublesome vapor to desorb from an unbaked system because it evolves in large quantities for long periods of time. Hydrogen is a problem in baked metal systems.

4.2.2   Second Order Desorption

The desorption just described does not apply to cases where gases dissociate on adsorption and must recombine before desorption. Diatomic gases on metals, for example hydrogen on a steel or molybdenum, dissociate to atoms on adsorption. These atoms must recombine on the surface before desorbing (H + H = H_2). A reaction which is proportional to the concentration of each of two species is called a second order reaction. The rate equation obeyed in these cases is

$$\frac{dC(t)}{dt} = -K_2 C(t)^2 \tag{4.5}$$

This equation can be solved to yield

$$\frac{dC(t)}{dt} = \frac{-K_2 C_o^2}{\left(1 + C_o K_2 t\right)^2} \qquad \blacktriangleright (4.6)$$

We see from (4.8) that the time to clean up a surface is longer than for first order because the rate decays as $1/t^2$. Remember that K_2 also contains the energy and temperature dependent term $e^{-E_d/N_o kT}$ so that second order desorption, like first, can go much faster at higher temperatures and weak binding energies.

Figure 4.2 shows a sketch of desorption rate versus time for a first and a second order process. Knize and Cecchi [2] performed an experiment which nicely illustrates the two processes. Their vacuum system contained Zr-Al getters whose surfaces were contaminated with a small amount of deuterium. Since two D atoms need to recombine and form D_2 before desorption can occur, surfaces with a low concentration of deuterium desorbed as a second order process according to (4.6). By flooding the chamber with a high, constant pressure of hydrogen, they created a surface in which the small number of deuterium atoms was flooded with H atoms. Since each D atom was frequently being hit by an H atom, H-D was formed and desorbed at a rate proportional to the concentration of D atoms. Their data showed the deuterium desorption rate to decrease as $1/t^2$ without the addition of hydrogen gas. With the hydrogen flux, the HD desorption rate decreased exponentially, and they effectively converted a second order process into a first order process described by (4.4).

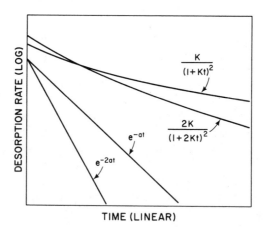

TIME (LINEAR)

Fig. 4.2 Desorption rate (log) vs. time (linear) for first order desorption K_1, and second order desorption K_2. Two activation energies are shown for each type of desoption. In both cases the rate constants are energy dependent, but after long times second order desorption is always slower than first order.

4.2.3 Real Surfaces

Real surfaces are usually more complex than the ideal surfaces containing one monolayer of one ideal gas. For example, water is adsorbed to a depth of many layers, and the binding energy between adjacent molecules is less than between the molecule and a surface. The desorption energy between an atom and a surface is also dependent on the fractional surface coverage. Several desorption energy-surface coverage models have been analyzed by Elsey [3].

Room-temperature outgassing data for most gases sorbed on metals, including water vapor, show the outgassing rate to vary inversely with time, at least for the first 10 h of pumping [4,5]. This can be expressed as

$$q_n = \frac{q_1}{t^\alpha} \qquad \blacktriangleright (4.7)$$

where the subscript n denotes the time in hours for which the data apply. The exponent α will range from 0.7 to 2 with 1 the most common value. Dayton [4] modeled this $1/t$ behavior as the sum of the successive out-diffusion of molecules of a range of energies. He proposed that a uniform distribution of activation energies like that found in a surface oxide containing pores of varying diameters will result in a sum of individual outgassing curves and will appear as one curve with a slope of about -1 on a log-log plot of outgassing rate versus pumping time, as sketched in Fig. 4.3. Each one of the curves in Fig. 4.3 is an out-diffusion curve whose shape we discuss in the next section.

4.3 DIFFUSION

Diffusion is the transport of one material through another. Gas diffusion to the interior wall of a vacuum system followed by desorption into the chamber contributes to the system outgassing. The gas pressure in the solid establishes a concentration gradient that drives molecules or atoms to the surface, where they desorb. Because diffusion is a much slower process than desorption, the rate of transport through the bulk to the surface governs the rate of release into the vacuum. A concentration gradient moves the atoms or molecules to the surface with a flux given by (2.29). The outgassing rate from a solid wall containing gas at an initial concentration C_o is obtained from the diffusion equation [6]. One solution for a uniform initial concentration of dissolved gas is

$$q = C_o\left(\frac{D}{t}\right)^{1/2}\left[1 + 2\sum_0^\infty (-1)^n \exp\left(\frac{-n^2 d^2}{Dt}\right)\right] \qquad (4.8)$$

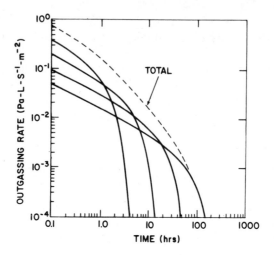

Fig. 4.3 Total outgassing rate as a sum of four rates, each resulting from a single outgass-ing time constant whose value depends on the shape of the surface oxide pores and the activation energy for desorption. Reprinted with permission from *Trans. 8th Vac. Symp. (1961),* p. 42, B. B. Dayton. Copyright 1962, Pergamon Press, Ltd.

where D is the diffusion constant in m^2/s and d is the thickness of the material. If C_o is given in units of Pa, then q will have units of W/m^2. We do not have to derive or solve this equation in order to understand some basic ideas about diffusion. We only need to know its value for short and long times. Let's examine the solutions for these two limits. For short times (4.10) has an asymptotic or limiting value. In the limit as we approach $t=0$, the terms in the sum become zero and the rate of gas release from the surface is

$$q = C_o\left(\frac{D}{t}\right)^{1/2} \qquad \blacktriangleright (4.9)$$

Equation (4.9) mathematically describes the slow decrease in outgassing that we observe experimentally. The initial out-diffusion from a solid containing a uniform gas concentration varies as $t^{-1/2}$. This desorption rate is much slower than first order desorption. It is slower because outdiffusion from the bulk is the rate limiting step.

For long times (4.8) is an infinite series which is not simple to evalu-ate. By the maneuver of placing the mathematical origin at the center of the solid instead of at one surface, the diffusion equation can be solved again to yield

$$q = \frac{2DC_o}{d}\sum_0^\infty \exp\left(-\frac{(2n+1)^2\pi^2 Dt}{2d^2}\right) \qquad (4.10)$$

Equations (4.8) and (4.10) are equivalent but (4.10) is easier to evaluate for large values of t. For long times only the first term is significant, and the solution reduces to

$$q = \frac{2DC_o}{d} \exp\left(-\frac{\pi^2 Dt}{2d^2}\right) \qquad \blacktriangleright (4.11)$$

Equation (4.11) states that the out-diffusion rate decreases as e^{-aDt} at long times. This rapid exponential (first order) decrease in outdiffusion rate is experimentally observed and corresponds to the near depletion of the dissolved gas in the solid. If we were to plot the rate at which gas diffuses from a solid surface into vacuum using either (4.8) or (4.10), we would observe the rate to decrease slowly at first as $t^{-1/2}$, and then after some time begin to decrease very rapidly as e^{-aDt}, when the solid is nearly exhausted of its gas supply. The transition from the early, slow decay to the later, rapid decay occurs at a time of approximately $t = d^2/6D$. The two light curves sketched in Fig. 4.4 for two values of the diffusion constant D each follow this form. The curve with the large diffusion constant has a high initial outgassing rate. As a result the gas supply in the solid is exhausted in a shorter time than shown for the curve with the small diffusion constant.

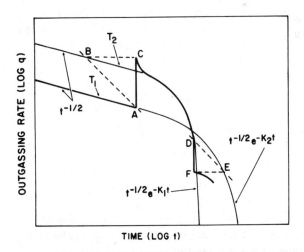

Fig. 4.4 Change in out-diffusion rate for an increase in temperature from T_1 to T_2 for a diffusion process. Reprinted with permission from *Vacuum Technology and Space Simulation,* D. J. Santeler et al., SP-105, 1966, National Aeronautics and Space Administration, Washington, D.C.

4.3.1 Vacuum Baking

The diffusion constant is a function of the thermal activation energy of the diffusing gas in the solid and is given by

$$D = D_o \exp -E_D/kT \tag{4.12}$$

Because of this exponential dependence on temperature, a rather modest increase in temperature will sharply increase the initial out-diffusion rate and reduce the time necessary to unload the total quantity of gas dissolved in the solid.

Figure 4.4 illustrates how vacuum baking reduces the final out-diffusion rate to a level far below that which is possible in the same time without baking. A solid exposed to vacuum at ambient temperature T_1 outdiffuses along the initial portion of the lower curve with a slope of $t^{-1/2}$, as given by (4.9). At a time corresponding to point A, the temperature is increased to temperature T_2. Because the gas concentration in the solid cannot change instantaneously, the out-diffusion rate increases to the value given by point B on the high temperature diffusion curve but shifted in time to point C. The dotted line connecting points A and B is a line of constant concentration. It has a slope of -1 and the value of q_B is given by $q_A t_A = q_B t_B$. Out-diffusion continues along the high temperature curve but is displaced in time. The plot looks curved near point C because of the logarithmic axis. At a time corresponding to point D the baking operation is terminated and the temperature is reduced to T_1. The out-diffusion rate is reduced to a value corresponding to point E on the low temperature curve but at an earlier time given by point F. Out-diffusion continues at a rate given by the low temperature curve but shifted to an earlier point in time. The new pressure at point F is given by [7]

$$P_f = (D_1/D_2)P_d \tag{4.13}$$

where $D_{1,2}$ are the diffusion coefficients at temperatures $T_{1,2}$, respectively. The net effects of baking are the reduction in the outgassing rate and the reduction of the time required to remove the initial concentration of gas dissolved in the solid. Recalling that surface desorption was modeled by Dayton [4] as the sum of diffusive release of gas from the surface oxide pores, we find that baking will reduce the desorption time in the same manner as it reduces the out-diffusion time. The rate of gas release from the surface of a solid containing a dissolved concentration of gas is slow because the rate is determined by the mobility of the gas in the solid. When the solid is nearly exhausted, the rate becomes much more rapid (exponential).

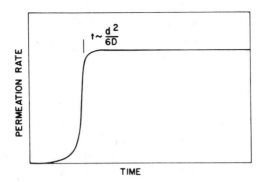

Fig. 4.5 Permeation (linear) of a gas vs time (linear) through a solid wall.

4.4 PERMEATION

Permeation is a three-step process. First gas adsorbs on the outer wall of a vacuum vessel, then diffuses through the bulk, and finally desorbs from the interior wall. The time dependence of the permeation rate of a gas through a solid wall is qualitatively sketched in Fig. 4.5. If the material is initially devoid of the permeating gas, a period of time will pass before the gas is observed to desorb from the vacuum wall. After equilibrium has been established, gas will desorb from the wall at a constant rate. Steady-state permeation therefore behaves like a small leak. The permeability of the wall for a gas is given by

$$K_p = DS' \tag{4.14}$$

where D is the diffusion constant and S' is the solid solubility [3]. The permeability has an exponential dependence similar to the diffusion constant and consequently increases rapidly with temperature.

The steady-state permeation rate of a gas that does not dissociate on adsorption can be expressed as

$$q_k = \frac{K_p P}{d} \qquad \blacktriangleright (4.15)$$

where q_k is the total flux in units of $(Pa\text{-}m^3)/(s\text{-}m^2)$ or W/m^2 and P is the pressure drop across the solid of thickness d. The permeability K_p has units of m^2/s and is the quantity of gas in m^3 at STP flowing through material 1 m^2 in area by 1 m thick when the pressure difference is 1 atm. Units for the measurement of permeation are not well standardized in gas quantity, area, thickness or pressure differential. Some conversion factors are given in Appendix A. Diatomic gases dissociate to atoms when they adsorb, and the steady-state permeation rate of a gas such as hydrogen through a metal is given by

$$q_k = \frac{K_p(P_2^{1/2} - P_1^{1/2})}{d} \qquad \blacktriangleright (4.16)$$

The permeability K_p for gases that dissociate has units of $Pa^{1/2}$-m^2/s.

4.5   STIMULATED DESORPTION

Electrons, atoms, molecules, ions or photons incident on solid surfaces can release adsorbed gases and generate vapors in quantities large enough to limit the ultimate pressure in a vacuum chamber. Many reactions are possible when energetic particles collide with surfaces [8]. Among them are electron-stimulated desorption, ion stimulated desorption, electron- or ion-induced chemical reactions and photodesorption.

4.5.1   Electron–stimulated Desorption

An energetic electron incident on the surface gas layer excites a bonding electron in an adsorbate atom to a non-bonding level. This results in a repulsive effective potential between the surface and atom which allows the atom to desorb in either the neutral or ionized state [9-12]. The electron-stimulated desorbed neutral gas flux can be as high as 10^{-1} atom per electron while the desorbed ion flux is much smaller, of order 10^{-5} ion per electron. This desorption process is specific. That is, it depends on the manner in which a molecule or atom is bonded to the surface. Some molecules are not desorbed by this process. These gas release phenomena have been shown to cause serious errors in pressure measurement with the Bayard-Alpert gauge [13,14] and residual gas analyzer [15] and to increase the background pressure in systems that use high energy electron or ion beams.

4.5.2   Ion–stimulated Desorption

Ion-stimulated desorption has been studied by Winters and Sigmund [16] and Taglauer and Heiland [17]. Winters and Sigmund show that nitrogen chemisorbed on tungsten can be desorbed by noble gas ions in the range up to 500 V. They also showed that the adsorbed atoms were removed by a sputtering process as a result of direct knock-on collisions with impinging and reflecting noble gas ions. Ion-stimulated desorption is responsible for part of the gas release observed in sputtering systems, Bayard-Alpert gauges, and glow discharge cleaning.

Edwards [18] has measured the electron- and ion-stimulated desorption yields of 304 stainless steel and aluminum as a function of cleaning technique. Table 4.2 shows his results for degreased stainless and alumi-

Table 4.2   Desorption Yields for Stainless Steel & Aluminum[a]

Gas	1000-eV Ar^+ ions		500-eV electrons	
	304 SS	Aluminum	304 SS	Aluminum
H_2	2.13	2.38	0.15	0.18
CO	3.22	3.00	0.06	0.05
CO_2	1.55	1.35	0.21	0.16
CH_4	0.075	0.07	0.0025	0.003

Source: Reprinted with permission from *J. Vac. Sci Technol.*, **16**, p. 758, D. Edwards, Jr. Copyright 1979, The American Vacuum Society.
[a] Cleaned by a soap wash, water, acetone and methanol rinses, and air dry followed by 200°C, 60-h bake and 2 days equilibration.

num samples.   The methane desorption is most likely caused by an electron- (or ion-) induced chemical reaction.

4.5.3   Stimulated Chemical Reactions

Ion- and electron-stimulated chemical reactions may also occur at solid surfaces.   For example, hydrogen and oxygen can react with solid carbon to produce methane and carbon monoxide.   An ion-stimulated chemical reaction is responsible for the rapid etch rates observed in reactive-ion etching.   In the reactive-ion etching of silicon high energy ions in a collision cascade process greatly enhance the reaction of neutral F with Si and produce a much greater yield of volatile SiF_4 than is possible without ion bombardment.

4.5.4   Photodesorption

Lichtman et al. [19] observed greater than band gap desorption.   The effect occurs in semiconductors (CrO_2 on SS).   A hole-electron pair is created by an incident photon, after which the hole recombines with the bonding electron and releases the adsorbate.   This is an extremely efficient process.   Grobner et al. [20] and Williams et al. [21] found the desorption of gases from aluminum used in the CERN Large Electron Positron Storage Ring to be affected by high energy photon bombardment.   They measure desorption efficiencies of 0.5 to 0.08 molecules per photon, respectively, for H_2 and CO_2.   CO and CH_4 were the only other gases observed.   Vig [22] has shown that a combination of UV and ozone is more effective in cleaning a surface than UV alone.

4.6 PRESSURE LIMITS

In the introduction to this chapter we alluded to the fact that outgassing, not volume gas removal, determined the final pressure in the high vacuum region. We know this to be true from another viewpoint. One monolayer of gas on the interior of a 1-L sphere, if desorbed instantly, would result in a pressure of 2.5 Pa (18 mT). This is a pressure far above high vacuum. Diffusion, permeation and stimulated desorption are also important. In the previous section we made no attempt to relate numerically these various rates to a real vacuum system. The relative roles of surface desorption, diffusion, and permeation are a function of the materials used for construction (steel, aluminum or glass), the seals (metal or elastomer), and the system history (newly fabricated, unbaked, chemically cleaned, or baked). The mathematical description of this pumping problem can be solved easily with certain approximations. The solution is of the form

$$P = P_o e^{-St/V} + \frac{Q_o}{S} + \frac{Q_D}{S} + \frac{Q_k}{S} \qquad \blacktriangleright (4.17)$$

The first term in the solution represents the time dependence of the pressure that is due to the gas in the chamber volume. The remaining terms represent the contribution of other gas sources. These terms represent outgassing, diffusion and permeation. They are slowly varying functions and become dominant after the initial pumping period has passed. After some time the first term on the right-hand side is zero and the pressure is determined by outgassing, diffusion, desorption and permeation. The first term (outgassing) decays slowly, often as t^{-1}. The second term is the diffusion term, it first decays as $t^{1/2}$, and then as

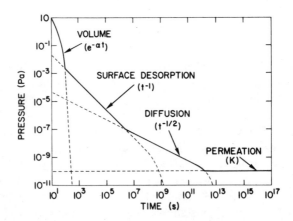

Fig. 4.6 Rate limiting steps during the pumping of a vacuum chamber.

e^{-aDt}. The last term is due to permeation and is a constant. From this equation we can construct a composite pump-down curve that will illustrate the relative roles of these phenomena. Figure 4.6 shows the high vacuum pumping portion of an unbaked, metal-gasketed system. In the initial stages the pressure is reduced exponentially with time as the volume gas is removed. This portion of the pumping curve takes only a short time because a typical system time constant is only seconds. It is expanded here for clarity. In the next phase surface desorption controls the rate of pressure decrease. In a typical unbaked system most of the gas load is water vapor; however, nitrogen, oxygen, carbon oxides and hydrocarbons are also present. The total quantity of gas released is determined by the material and its history. Glass or steel that has been exposed to room ambient for extended periods may contain up to 100 monolayers of water vapor, whereas a carefully vented chamber may contain little water vapor. A slow decrease in pressure is to be expected, if the system has a number of large elastomer O-rings, or a large interior surface area. Unbaked, routinely cycled vacuum systems are never pumped below the outgassing-limited range.

If the system is allowed to continue pumping without baking, the surface gas load will ultimately be removed and the out-diffusion of gases in solution with the solid walls will be observed. The slope of the curve will change from t^{-1} to $t^{1/2}$. For example, hydrogen that diffuses into steel in a short time at high fabrication temperatures diffuses out very slowly at room temperatures. If we were to continue pumping until all the dissolved hydrogen was removed, the pressure would become constant even though the system was leak free. The system would now be at its ultimate pressure given by $P = Q_k/S$. Experimentally, hydrogen permeates the walls of metal systems and helium permeates glass walls. Notice the time required to reach the ultimate pressure in this hypothetical example of an unbaked metal-gasketed system is 10^8 h. This demonstrates clearly the absolute necessity of baking in ultrahigh vacuum technology.

The order of importance of the processes is not always as shown in Fig. 4.6. Elastomer gaskets have a high permeability for atmospheric gases, and if the system contained a significant amount of these materials, the limiting permeation rate would be several orders of magnitude higher than illustrated here. It could be large enough to mask diffusion processes.

The level of outgassing resulting from various electron- and ion-stimulated desorption processes is not shown in Fig. 4.6. These processes play a variable but important role in determining the ultimate pressure in many instances because they desorb atoms that are not baked off.

Electron-stimulated desorption in an ion gauge is easily observable during gauge outgassing. A 10-mA electron flux can desorb 6×10^4

neutrals/s. Because a gas flux of 2×10^{17} atoms/s is equivalent to 1 Pa-L/s, this electron flux can initially provide a desorption flux of 2×10^{-3} Pa-L/s. If the system is pumping on the ion gauge tube at a rate of 10 L/s, this desorption peak can reach a pressure of an order of 10^{-4} Pa in the gauge tube. The scattered electrons from Auger and LEED systems can stimulate desorption in a similar manner, with an increase in background pressure when the beam is operating. Electron-induced desorption is a first order process that should produce a simple exponential decay of the desorbed species. If readsorption is included, the initial rate will decay exponentially to a steady-state level [23]. A 20-mA ion beam can desorb surface gas at an initial rate of 10^{-3} Pa-L/s, assuming a desorption efficiency of two atoms per incident ion.

REFERENCES

1. G. Ehrlich, *Trans. 8th Nat. Vac. Symp. and Proc. 2nd Int'l. Vac. Congr. on Vac. Sci. Technol. (1961),* Vol. 1, Pergamon, New York, 1962, p. 126.

2. R. J. Knize and J. L. Cecchi, *J. Vac. Sci. Technol. A,* 1, 1273 (1983).

3. R. J. Elsey, *Vacuum,* 25 347 (1975).

4. B. B. Dayton, *Trans. 8th Nat. Vac. Symp. and Proc. 2nd Int'l. Congr. on Vac. Sc. Technol. (1961),* Vol. 1, Pergamon, New York, 1962, p. 42.

5. B. B. Dayton, *Trans. 7th Vac. Symp. (1960),* Pergamon, New York, 1961, p. 101.

6. H. S. Carlslaw and J. C. Jaeger, *Conduction of Heat in Solids,* 2nd ed., Oxford, 1959, p. 96.

7. D. G. Bills, *J. Vac. Sci. Technol.,* 6, 166 (1969).

8. See, for example P. A. Redhead, J. P. Hobson, and E. V. Kornelsen, *The Physical Basis of Ultrahigh Vacuum,* Chapman and Hall, London, 1968, Chapter 4.

9. M. J. Drinkwine and D. Lichtman, *Prog. Surf. Sci.,* Vol. 8, Pergamon, New York, 1977, p. 123.

10. D. Menzel and R. Gomer, *J. Chem. Phys.,* 41, 3311 (1964).

11. P. Redhead, *Can. J. Phys.,* 42, 886 (1964).

12. M. L. Knotek and P. J. Feibelman, *Phys. Rev. Lett.,* 40, 964 (1978).

13. P. A. Redhead, *Vacuum,* 12, 267 (1962).

14. T. E. Hartman, *Rev. Sci. Instr.,* 34, 1190 (1963).

15. P. Marmet and J. D. Morrison, *J. Chem. Phys.,* 36, 1238 (1962).

16. H. F. Winters and P. Sigmund, *J. Appl. Phys.,* 45, 4760 (1974).

17. E. Taglauer and W. Heiland, *J. Appl. Phys.,* 9, 261 (1976).

18. D. Edwards Jr., *J. Vac. Sci. Technol.,* 16, 758 (1979).

19. G. W. Fabel, S. M. Cox and D. Lichtman, *Surface Science,* 40, 571 (1973).

20. O. Gröbner, A. G. Mathewson, H Störi and P. Strubin, *Vacuum,* 33, 397 (1983).

21. E. M. Williams, F. Le Normand, N. Hilleret and G. Dominichini, *Vacuum,* 35, 141 (1985).

22. J. R. Vig, *J. Vac. Sci. Technol. A,* 3, 1027 (1985).

23. M. J. Drinkwine and D. Lichtman, *J. Vac. Sci. Technol.,* 15, 74 (1978).

PROBLEMS

4.1 †Define equilibrium vapor pressure.

4.2 †Mercury is used in certain diffusion pumps and pressure gauges. What is the vapor pressure of mercury at 25°C? What precautions should be taken when using a pump or gauge containing Hg? What safety items should be available?

4.3 Nitrogen has been collected in a beaker inverted over a water bath held at 25°C. The nitrogen has been added until the water inside the beaker has been displaced so that its level equals that of the water outside the beaker. What is the pressure of the nitrogen in the bath if the external pressure is 101,323.2 Pa?

4.4 Aluminum is deposited at a rate of 2×10^{-9} m/s (1.2×10^{20} at./(s-m^2)). Background gas, assumed to be air, also strikes the substrate. At 22°C calculate the ratio of Al atoms to air molecules incident on the surface for (a) $P = 10^{-3}$ Pa (decorative metallurgy), (b) $P = 10^{-5}$ Pa (metallization of a silicon wafer), and (c) $P = 10^{-7}$ Pa (surface studies).

4.5 A 1-L cubical chamber has been pumped (ideally) of all the volume gas. However, the internal surfaces are covered with one monolayer of nitrogen, corresponding to about 10^{19} molecules/m^2. Assume all the monolayer instantly desorbs into the empty volume. (a) Calculate the resulting pressure in the chamber. (b) What fraction of atmospheric pressure is this? (c) In what pressure region are cleaning techniques most important?

4.6 A gas display panel is constructed from two large, flat glass plates spaced 7.6×10^{-3} cm apart. The space between the plates is filled with a gas mixture comprised of 99.9 (vol%) neon, 0.1% argon, with a total pressure of 6×10^4 Pa. (a) What is the partial pressure of argon? (b) Assuming one monolayer of argon can be ion implanted over one-half of the glass surface during display operation, how much argon remains in the gas phase?

4.7 At what temperature is the residence time of H_2O on steel equal to 1 s?

4.8 †(a) Define diffusion. †(b) By what two mechanisms does gas dissolved in a solid enter a vacuum chamber?

4.9 †In what way do diffusion and permeation of molecular gases differ from atomic gases in a metal?

4.10 Why does the permeation rate of a gas through an elastomer increase when the elastomer swells?

Measurement

In these five chapters we discuss the tools used to measure pressure in a vacuum system: the pressure gauge, the flow meter and the residual gas analyzer (RGA). With the pressure gauge routine system performance is monitored and many problems are discovered. The flow meter is now a necessary part of any deposition or etching system. The residual gas analyzer adds a degree of sophistication to our analytical skills. Its ability to single out the gas or vapor that is limiting the system pressure or causing a process problem greatly reduces the difficulty in troubleshooting large and complex systems. Chapter 5 is devoted to a discussion of the commonly used pressure gauges. Chapter 6 describes the operation of common flow meters. In Chapter 7 we describe pumping speed measurements, which require the use of both pressure and flow measuring instruments. Chapter 8 discusses the operation and installation of residual gas analyzers on vacuum systems, while Chapter 9 describes qualitative and quantitative methods of interpreting the data obtained from an RGA spectrum.

CHAPTER 5

Pressure Gauges

In this chapter we discuss common pressure gauges. To discuss in detail each of the 20 or more pressure measuring instruments would result in a reduced presentation of those gauges that are the most important. Descriptions of the less frequently used gauges are contained in other texts [1-4].

Over the last 50 years many techniques have been developed for the measurement of reduced pressures. Gauges are either direct or indirect reading. Those that measure pressure by calculating the force exerted on the surface by incident particle flux are called direct-reading gauges. Indirect gauges record the pressure by measuring a gas property which changes in a predictable manner with gas density. Figure 5.1 sketches a way of classifying many pressure gauges. Their operating ranges are illustrated in Fig. 5.2.

5.1 DIRECT-READING GAUGES

The diaphragm, Bourdon, and capacitance manometers are the most common direct-reading gauges. Two rather well-known gauges which have a necessary place in pressure measurement, the U-tube manometer and the McLeod gauge, are not described in detail because they are not routinely used by the average vacuum-system operator. In its simplest form a manometer consists of a U-tube that contains a low vapor pressure fluid such as mercury or oil. One arm is evacuated and sealed, the other is connected to the unknown pressure. The unknown pressure is read as the difference in the two liquid levels. The McLeod gauge is a mercury manometer in which a volume of gas is compressed before measurement; for example, precompressing a small volume of gas at 10^{-2}

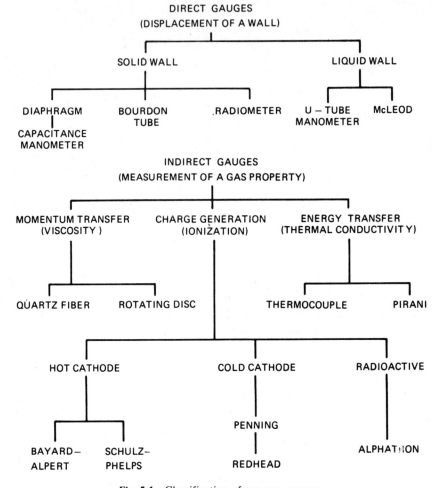

Fig. 5.1 Classification of pressure gauges.

Pa by 10,000 times results in a measurable pressure of 100 Pa. The U-tube manometer, which is used in the 10^2 to 10^5 Pa range, and the McLeod gauge, which is a primary standard in the 10^{-3} to 10^2 Pa pressure range, are described in many texts [1-4].

5.1.1 Diaphragm and Bourdon Gauges

The simplest mechanical gauges are the diaphragm and Bourdon gauges. Both are operated by a system of gears and levers to transmit the deflection of a solid wall to a pointer. The Bourdon tube (Fig. 5.3) is a coiled tube of elliptical cross section, fixed at one end and connected to the pointer mechanism at the other. Evacuation of the gas in the tube causes

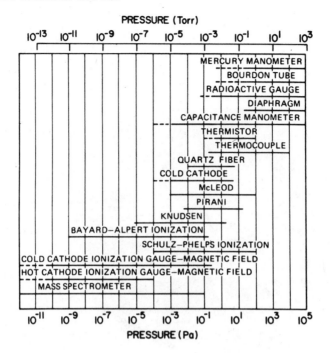

Fig. 5.2 Pressure ranges for various gauges. Adapted with permission from *Scientific Foundations of Vacuum Technique*, 2nd ed., J. M. Lafferty, Ed., p. 350. Copyright 1962, John Wiley & Sons.

rotation of the pointer. The diaphragm gauge contains a pressure-sensitive element from which the gas has been evacuated. By removing gas from the region surrounding the element, the wall is caused to deflect, and in a manner similar to the Bourdon tube the linear deflection of the wall is converted to angular deflection of the pointer.

Simple Bourdon or diaphragm gauges, for example those of the 50-mm-diameter variety, will read from atmospheric pressure to a minimum pressure of about 10^3 to 5×10^3 Pa. They are inaccurate and used only as a rough indication of pressure. They are commercially available with 316 stainless steel tubes by which they may be attached to clean systems.

Diaphragm and Bourdon gauges, which are more accurate than those described above, are available in a variety of ranges extending from 10^3 to 2×10^5 Pa and with sensitivities of an order of 25 Pa. Figures 5.3 and 5.4 illustrate two types of gauge. The gauge described in Fig. 5.3 is a diaphragm in which the entire instrument case is attached to the vacuum vessel and evacuated. The case is protected from possible overpressure damage by a blow-out plug. This gauge with its large internal volume, brass parts, and high vapor pressure lubricating materials is not the type to be appended to a clean system. The Bourdon gauge, especial-

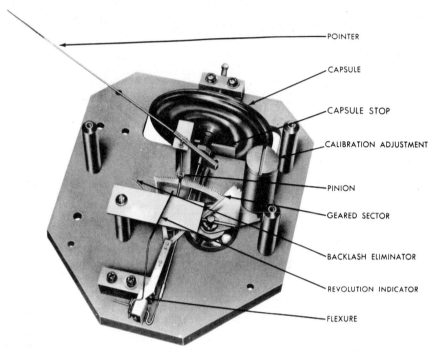

POINTER

CAPSULE

CAPSULE STOP

CALIBRATION ADJUSTMENT

PINION

GEARED SECTOR

BACKLASH ELIMINATOR

REVOLUTION INDICATOR

FLEXURE

Fig. 5.3 Diaphragm mechanism for absolute pressure measurement.   Reprinted with permission from the Wallace and Tiernan Division, Pennwalt Corp., Newark, NJ.

ly the differential tube type illustrated in Fig. 5.4, is quite suitable for attachment to clean systems;   only the small interior volume of the tube is added to the system.   Because tubes can be fabricated from many materials, the gauge can be designed to handle corrosive gases.   The differential gauge adds only a small surface area to the system and, when fabricated from 316 stainless steel, is excellent for use on clean chambers.

5.1.2   Capacitance Manometers

A capacitance manometer is simply a diaphragm gauge in which the deflection of the diaphragm is measured by observing the change in capacitance between it and a fixed counter electrode.   The first gauge was described in 1951 by Alpert, Matland, and McCoubrey [5].   They used a differential gauge head as a null reading instrument between the vessel of unknown pressure and another whose pressure was independently adjustable and monitored by a U-tube manometer.

The capacitance of the diaphragm-counter electrode structure is proportional to geometry (area/gap) and to the dielectric constant of the gas

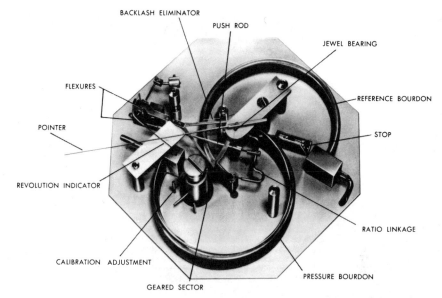

Fig. 5.4   Bourdon tube mechanism for absolute pressure measurement. Reprinted with permission from the Wallace and Tierman Division, Pennwalt Corp., Newark, NJ.

being measured in relation to that of air. Except for the few gases that have relative dielectric constants significantly different from air, for example, certain conductive and heavy organic vapors or gases ionized by radioactivity, the use of capacitance change to measure pressure represents a true, absolute-pressure measurement: that is, the pressure may be calculated from the geometry and the observed capacitance change. A 1% difference in the dielectric constant of the measured gas and air will result in an error of 1/2% of reading. A single-sided structure is not dependent on the dielectric constant of the measuring gas because both electrodes are in the vacuum, or reference side.

Modern capacitance manometers consist of two components, a transducer and an electronic sense unit that converts the membrane position to a signal linearly proportional to the pressure. A common design for a transducer is shown in Fig. 5.5. The flexible metal diaphragm, which has been stretched and welded in place, is located between two fixed electrodes. The differential transducer shown in Fig. 5.5 may be a null detector or a direct-reading gauge. When used as a null detector, the pressure at the reference side P_r is adjusted until the diaphragm deflection is zero. In this mode a second gauge is necessary to read pressure. To use as a direct-reading gauge the reference side must be pumped to about 10^{-5} Pa. After calibration the instrument may be used directly over the pressure range for which it was designed. Transducers are available with the reference side open for evacuation or evacuated and permanent-

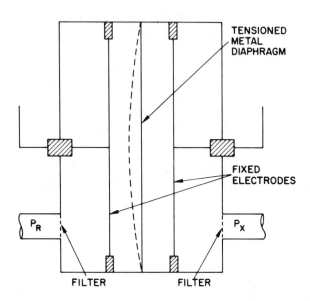

Fig. 5.5 Double-sided capacitance manometer head assembly. Reprinted with permission from *Industrial Research/Development*, January 1976, p. 41. Copyright 1976, Technical Publishing Co.

ly sealed. The permanent seal is usually a copper pinch seal; a getter is activated inside the tube at the time of manufacture.

Care should be taken in attaching the transducer to a system. It is generally advisable to have a bellows section in series with one tubulation if the transducer is to be permanently welded to a system with only short tube extensions. Even though some transducers contain filters to prevent particulates from entering the space between the diaphragm and the electrodes, it is advisable to force argon through a small diameter tube placed inside the tubulation or bellows extension to be welded on the sensor. The end of this small tube should be pushed in to a point beyond the weld location to allow the flow of argon to flush particulates out of the tube and away from the sensor during the welding operation. This procedure also stops the formation of oxide on the tube's interior walls.

Because the transducer contains ceramic insulators, cleanliness is in order; a contaminated head is hard to clean. Cleaning solvents are difficult to remove from the ceramic and may cause contamination of the system at a later time. To avoid this problem one transducer has been designed with a single-sided sensor. Both electrodes are on the reference side (see Fig. 5.6). One electrode is placed at the center of the diaphragm and the second is an annular ring located around the center electrode. For zero deflection of the membrane the circuit is adjusted for zero output signal. The deflection or bowing of the diaphragm causes a

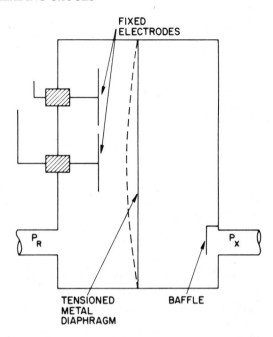

Fig. 5.6   Single-sided capacitance manometer head assembly.   The outer electrode is an annular ring.   Reprinted with permission from *Industrial Research/Development*, January 1976, p. 41. Copyright 1976, Technical Publishing Co.

capacitive imbalance, which is converted to a voltage proportional to pressure.   A proper choice of materials results in a transducer suitable for service in corrosive environments without head damage or in extremely clean environments without contamination by the head.

The electronic sensing unit applies an ac signal to the electrodes.   The changes in signal strength produced by the diaphragm are amplified and demodulated in phase to minimize the noise level.   The dc output is then used to drive an analog or digital read-out.   Because the resolution of the instrument is limited by system noise, the system bandwidth must be stated when specifying resolution; noise is proportional to the half-power of the bandwidth.   Typical capacitance manometer systems have a resolution of 1 part in 10^6 full scale at a bandwidth of 2 Hz.

Just as low electronic noise is of prime importance in obtaining high resolution, thermal stability of the head is necessary for stable, accurate, and drift-free operation.   The diaphragm deflection in the transducer can be as low as 10^{-9} cm; therefore motion of parts due to temperature change becomes a large source of error.   Transducers are available with

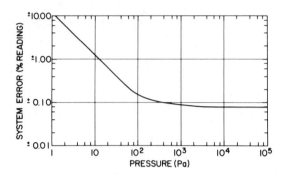

Fig. 5.7 Performance curve for a capacitance manometer with a 10^5-Pa full-scale transducer. Reprinted with permission from *Industrial Research/Development*, January 1976.
Copyright 1976, Technical Publishing Co.

heaters that maintain the ambient temperature at about 50°C and avoid
some of the problems of ambient temperature change. Many transducers
can be operated at temperatures as high as 250°C. The readings, however, must be corrected for thermal transpiration (see Section 2.3.4).
Stable operation of a transducer requires that the thermal expansion
coefficients of the diaphragm and electrode assemblies be well matched,
but in practice designs must make a trade-off between expansion coefficient and corrosion resistance. Without proper temperature regulation a
transducer may have zero and span coefficients of 5 to 50 ppm full scale
and 0.004 to 0.04% of reading per degree celsius, respectively, at ambient temperature [6]. Proper temperature regulation can result in an
order of magnitude improvement in the zero and span coefficients.

Capacitance manometers can be operated over a large dynamic range, a
factor of 10^4 to 10^5 for most instruments, but the overall system accuracy
deteriorates at small fractions of full head range, as illustrated in Fig. 5.7
for the 1.3×10^5-Pa head. Transducers with a full-scale deflection of
130 Pa have been checked in the 2.5×10^{-2} to 6.5×10^{-4}-Pa pressure
range by volumetric division and have been found to be linear to the
lowest pressure and in agreement within 0.6% plus 5.3×10^{-5} Pa [7].

5.2 INDIRECT-READING GAUGES

In this section the most familiar indirect-reading gauges are discussed.
Indirect gauges calculate pressure by measuring a pressure dependent
property of the gas. In the pressure range above 0.1 Pa, energy and
momentum transfer techniques can be used for pressure measurement.
The spinning rotor gauge [8,9] operates on this principle. It is a research
instrument suitable as a secondary standard [10,11]. Thermal conductivi-

ty gauges measure the heat transfer between two surfaces at different temperatures. A Pirani or a thermocouple gauge is found on every vacuum system for measuring pressure in the medium vacuum region. Ionization gauges, which measure gas density, have found wide acceptance. Hot cathode gauges are used in the Schulz-Phelps and the Bayard-Alpert geometries; together they span the pressure range 100 Pa to 10^{-9} Pa. Systems operating in the 10^{-3} to 1 Pa range often use the simpler Penning cold cathode gauge. Hot and cold cathode magnetron gauges which are capable of operation at pressures as low as 10^{-11} Pa are found on some ultrahigh vacuum systems but are not used on ordinary high vacuum systems.

5.2.1 Thermal Conductivity Gauges

Thermal conductivity gauges are a class of pressure measuring instruments that operate by measuring in some way the rate of heat transfer between a heated wire and its surroundings. The heat transfer between a heated wire and a nearby wall is pressure dependent in the $0.01 <$ Kn $<$ 10 range, where Kn is Knudsen's number. For $d = 10$ mm the pressure range is about 66 to 0.06 Pa, although the sensitivity of heat transfer with pressure is highly non-linear at each end of the scale. The heat transfer regimes in a thermal conductivity gauge are illustrated in Fig. 5.8. At high pressures where Kn < 0.01 the heat flow is given by (2.27) and is independent of pressure except for a small convection effect. In the $0.01 <$ Kn < 10 region the *free molecular* heat flow is given by (2.28). In this region the heat flow is linearly proportional to the pressure, provided that the accommodation coefficient and the temperature difference between the heated wire and the case remain constant. In the lowest pressure region the heat flow is predominantly accounted for by radiation and conduction through the wire to the supports:

$$H = A\sigma\varepsilon_1(T_2^4 - T_1^4) + \text{end losses} \qquad (5.1)$$

To extend the range of a gauge to its lowest possible pressure limit it is necessary to reduce the radiation and end conduction losses. The end losses are predominant only when the length of the wire is short. The radiant heat losses can be minimized by reducing the diameter and the emissivity of the hot wire. The emissivity of a clean tungsten wire is about 0.1, but in practice most are not clean. The upper pressure limit of a thermal conductivity gauge is determined by the saturation pressure of the thermal conductivity. This occurs at a Knudsen number of about 0.01. The two most commonly found gauges have upper pressure limits in the 15 to 150 Pa range, but tubes which read to 10^5 by taking advantage of pressure dependent convection losses [12] are available.

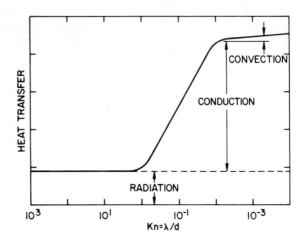

Fig. 5.8   Heat transfer regimes in a thermal conductivity gauge.

The sensitivity of the gauge is determined by tube construction and the gas as well as by the technique for sensing the change in heat flow with pressure. Tungsten is commonly used for the heater wire because it has a large thermal resistance coefficient. (When a semiconductor is the heat-sensitive element, the device is referred to as a thermistor gauge, even though it is strictly speaking a Pirani gauge.) Equation (2.28) describes the sensitivity of heat flow to a change in hot-wire temperature. The ratio of specific heats and thermal velocity in (2.23) depend on the gas species and in combination can produce as much as a fivefold difference in sensitivity between two gases. The accommodation coefficient α for clean materials can be of an order 0.1, but for contaminated surfaces it can be as high as unity. For most cases α is stable but not known. With all other factors well-controlled changes in emissivity and accommodation coefficient are large enough to allow thermal conductivity gauges to be used as only rough indicators of vacuum.

The change in temperature can be detected by monitoring the resistance of the heated wire. When a Wheatstone bridge circuit is used to measure the resistance change, the device is termed a Pirani gauge. Alternatively, the temperature change can be measured directly with a thermocouple, in which case it is called a thermocouple gauge.

Pirani Gauges

The term *Pirani gauge* is given to any type of thermal conductivity gauge in which the heated wire forms one arm of a Wheatstone bridge. A simple form of this circuit is shown in Fig. 5.9. The gauge tube is first activated to a suitably low pressure, say 10^{-4} Pa, and R_1 is adjusted for balance. A pressure increase in the gauge tube will unbalance the bridge because the increased heat loss lowers the resistance of the hot wire. By increasing the voltage, more power is dissipated in the hot wire, which causes it to heat, increase its resistance, and move the bridge toward balance.   In this method of gauge operation, called the constant-temperature method and the most sensitive and accurate technique for operating the bridge, each pressure reading is taken at a constant wire temperature.   To correct for changes that ambient temperature would have on the zero adjustment, an evacuated and sealed compensating gauge tube is used adjacent to the active gauge tube in another arm of the bridge. Bridges with a compensating tube can be used to 10^{-3} Pa.

The constant-voltage and constant-current techniques were devised to simplify the operation of the Pirani gauge. In each case the total bridge voltage or current is kept constant. The constant-voltage method is widely used in modern instruments because no additional adjustments need to be made after the bridge is nulled at lower pressures. The out-of-balance current meter is simply calibrated to read the pressure.

The constant-temperature method is the most sensitive and accurate because at constant temperature the radiation and end losses are constant.   Because the wire temperature is constant, the sensitivity is not diminished in the high pressure region.   This method does not lend itself to easy operation; balancing is required before each measurement.   A sudden drop in pressure can also cause overheating of the wire if the bridge is not immediately rebalanced.   Direct-reading, constant-temperature bridges that need only a zero adjustment are now commercially available, although at somewhat greater expense than a constant-

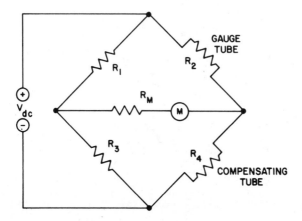

Fig. 5.9 Basic Pirani gauge circuit. Adapted with permission from *Vacuum Technology*, A. Guthrie, p. 163. Copyright 1963, John Wiley & Sons.

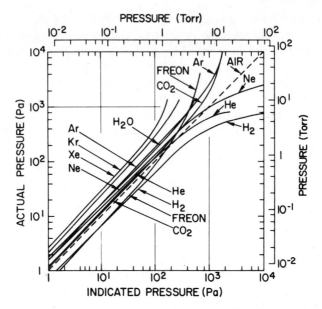

Fig. 5.10 Calibration curves for the Leybold TR201 Pirani gauge tube. Reprinted with permission from Leybold-Heraeus G.m.b.H., Postfach 51 07 60, 5000 Köln, West Germany.

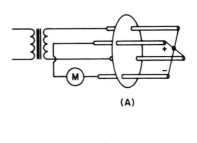

(A)

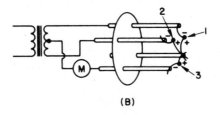

(B)

Fig. 5.11 Thermocouple gauge tubes for the 0-100 Pa range. (a) uncompensated gauge tube, (b) compensated gauge tube, (no. 3 is the compensating couple).

voltage or -current bridge. Modern circuitry has eliminated tedious bridge balancing. Because the heat conductivity varies considerably among gases and vapors, the calibration of the gauge is dependent on the nature of the gas. Most instruments are calibrated for air; therefore a chart like the one shown in Fig. 5.10 is needed when the pressure of other gases is measured.

Thermocouple Gauges

The thermocouple gauge measures pressure dependent heat flow. Constant current is delivered to the heated wire and a tiny thermocouple, perhaps iron- or copper-constantan, is carefully spot welded to its midpoint. As the pressure increases, heat flows to the walls and the temperature of the wire decreases. A low resistance dc microammeter is connected to the thermocouple and its scale is calibrated in pressure units.

Figure 5.11 shows the four-wire and three-wire versions of the gauge tubes. The four-wire gauge tube uses a dc meter to read the temperature of the thermocouple, while the power supply is regulated to deliver a

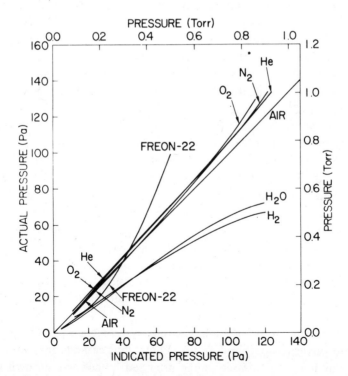

Fig. 5.12 Calibration curves for the Hastings DV-6M thermocouple gauge tube. Reprinted with permission from Hastings Instruments Co., Hampton, VA.

constant current to the wire. The current can be ac or dc. The three-wire gauge circuit reduces the number of leads between the gauge tube and controller and the number of vacuum feedthroughs by using ac to heat the wires and a dc microammeter to read the voltage between one thermocouple wire and the center tap of the transformer, which is a dc connection to the other junction. In both tubes the power delivered is not constant; instead the wire current is constant. Because the resistance of the wire is temperature dependent, the actual power delivered decreases slightly at high pressures. Both gauge forms are rugged and reliable but inaccurate. Calibration curves for one thermocouple gauge are given in Fig. 5.12.

5.2.2 Ionization Gauges

In the high and ultrahigh vacuum region where the particle density is extremely small, it is not possible, except in specialized laboratory situations, to detect the minute forces that result from the direct transfer of momentum or energy between the gas and a solid wall; for example, at a pressure of 10^{-8} Pa the particle density is only $2.4 \times 10^{12}/m^3$. This may be compared with a density of $3 \times 10^{22}/m^3$ at 300 K which is required to raise a column of mercury 1 mm. Even a capacitance manometer cannot detect pressures lower than 10^{-4} Pa. The basic principle used for the measurement of pressures lower than 10^{-3} Pa is the ionization of gas molecules and the collection of the ions and their subsequent amplification by sensitive and stable circuitry.

Each ionization gauge has its own lower pressure limit at which the ionized particle current is equal to a residual or background current. The best of these gauges have lower limits of an order of 10^{-11} to 10^{-12} Pa. In special research environments, where pressures far below 10^{-12} Pa may be encountered, the pressure is considerably below the limit of current ionization gauge technology. At a pressure of 5×10^{-15} Pa and a temperature of 4.2 K there are only 100 (nitrogen) molecules per cubic centimeter. Even with the most efficient ionization schemes available the ion current would be lost in the system noise. In those situations adsorbed gas can be collected on a particular surface for an extremely long time, after which the pressure pulse that results from flash desorption of the surface can be recorded [13].

In routine operation of high vacuum systems in the 10^{-1} to 10^{-7} Pa range the Bayard-Alpert and Schulz-Phelps hot cathode ionization gauges or the Penning cold cathode gauge are used. Each has its own pressure range, advantages, and disadvantages.

Hot Cathode Gauges

The operation of the ion gauge is based on ionization of gas molecules by electron impact and the subsequent collection of these ions by an ion collector.   This positive ion current is proportional to pressure, provided that all other parameters, including temperature, are held constant.   The number of positive ions formed is actually proportional to the number density, not the pressure; the ion gauge is not a true pressure measuring instrument but rather a particle density gauge.   It is proportional to pressure only if the temperature is constant.

The earliest form of ion gauge, the triode gauge, consisted of a filament surrounded by a grid wire helix and a large diameter, solid cylindrical ion collector.   This gauge, which is not illustrated here, looks a lot like a triode vacuum tube.   Electrons emitted by the heated filament were accelerated toward the grid wire which was held at a positive potential of about 150 V.   The external collector was biased about -30 V with respect to the filament and could collect the positive ions generated in the space between the filament and the ion collector.   This gauge measured pressures as low as 10^{-6} Pa but would not give a lower reading even if indirect experimental evidence indicated the existence of lower pressures. Further progress was not made until after 1947, when Nottingham [14] suggested that the cause of this effect was an x-ray-generated photocurrent.   Nottingham proposed that soft x-rays generated by the electrons striking the grid wire collided with the ion collector cylinder and caused photoelectrons to flow from the collector to the grid.   Some photoemis-

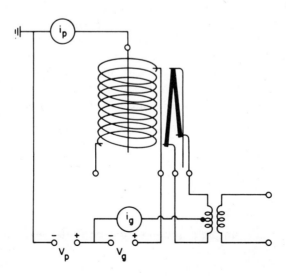

Fig. 5.13   Control circuit for a Bayard-Alpert ionization gauge tube.

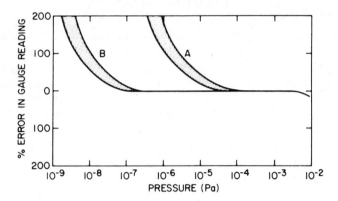

Fig. 5.14 Qualitative x-ray-generated error in ion gauge tube readings: (a) triode gauge tube, (b) efficient Bayard-Alpert tube.

sion is also caused by ultraviolet radiation from the heated filament. As they leave the collector these photo-electrons produce a current in the external circuit which is not distinguishable from the positive ion flow toward the ion collector and mask the measurement of reduced pressures.

In 1950 Bayard and Alpert [15] designed a gauge in which the large area collector was replaced with a fine wire located in the center of the grid (Fig. 5.13). Because of its smaller area of interception of x-rays, this gauge could measure pressures as low as 10^{-8} Pa. Today this gauge is the most popular design for the measurement of high vacuum pressures. It is available in a tubulated glass envelope (tubulated gauge) or mounted on a metal base (nude gauge). A more efficient design increased the sensitivity of the gauge tube by capping the end of the grid to prevent electron escape (Nottingham [16]) and reduced the x-ray limit even more by use of a fine collector wire. This efficient design can measure pressures as low as 2×10^{-9} Pa. Figure 5.14 qualitatively illustrates the x-ray limits of the triode gauge and the efficient Bayard-Alpert gauge.

The proportionality between the plate current and pressure is given by

$$i_p = S' i_e P$$

or

$$P = \frac{1}{S'} \frac{i_p}{i_e} \qquad \blacktriangleright (5.2)$$

where i_p and i_e are the plate and emission currents, respectively, and S' is the sensitivity of the gauge tube. This sensitivity has dimensions of reciprocal pressure, which in SI is Pa^{-1}, and is dependent on the tube geometry, grid and plate voltages, type of control circuitry, and nature of the gas being measured. For the standard-design Bayard-Alpert tube with external control circuitry, a plate voltage of +150 V, and a grid voltage of 30 V the sensitivity for nitrogen is typically 0.07/Pa. Variations in tube design, voltage, and control circuitry can cause it to range from 0.05 to 0.15/Pa. The tube's sensitivity for other gases varies with the ionization probability. Alpert [17] has suggested that the relative sensitivity (e.g., the ratio of the absolute sensitivity of a gas to that of nitrogen) should be independent of structural and electronic variations and thus be more meaningful to tabulate.

The relationship between the gauge pressure and the unknown pressure is

$$P(x) = \frac{S(N_2)}{S(x)} P(N_2) \qquad (5.3)$$

or because the sensitivity has been normalized to nitrogen, $S(N_2) = 1$,

$$P(x) = \frac{P(\text{meter reading})}{\text{Relative sensitivity of gas}(x)} \qquad \blacktriangleright (5.4)$$

With the help of (5.3) and Table 5.1 [18] the pressure of gases other than nitrogen can be measured with an ion gauge, even though all ion gauges are calibrated for nitrogen. This is done by dividing the gauge reading by the relative sensitivity of the gas of interest.

Gauge sensitivity is often given in units of microamperes of plate current per unit of pressure per manufacturer's specified emission current; for example, a typical nitrogen sensitivity is (100 μA/mTorr)/10 mA. This is a confusing way of saying the sensitivity is 10/Torr, but it does illustrate an important point; not all gauge controllers have the same calibration value of emission current, and not all gauge tubes have the same sensitivity. Checking the instruction manual can avoid potential embarrassment.

The classical control circuit is designed to stabilize the potentials and emission current while measuring the plate current. The plate current meter is then calibrated in appropriate ranges and units of pressure. The accuracy of the gauge is dependent in part on moderately costly, high quality emission current regulation. One gauge controller [19] avoids the problem of close regulation of the emission current by use of an integrated circuit to take the ratio of plate to emission current. Examination of (5.2) shows that except for a constant scale factor this current ratio is indeed proportional to pressure.

Tungsten and thoriated iridium (ThO$_2$ on iridium) are two commonly used filament materials. Thoriated iridium filaments are not destroyed when accidentally subjected to high pressures—an impossible feat with fine tungsten wires—but they do poison in the presence of some hydrocarbon vapors. The remarks in Section 8.2 about filament reactivity with

Table 5.1   Approximate Relative Sensitivity
of Bayard–Alpert Gauge Tubes to
Different Gases[a]

Gas	Relative Sensitivity
H_2	0.42 - 0.53
He	0.18
H_2O	0.9
Ne	0.25
N_2	1.00
CO	1.05 - 1.1
O_2	0.8 - 0.9
Ar	1.2
Hg	3.5
Acetone	5

Source: Adapted with permission from *J. Vac. Sci. Technol.*, **8**, p 661. T. A. Flaim and P. D. Owenby. Copyright 1971, The American Vacuum Society.

[a] The pressure of any gas is found by dividing the gauge reading by the relative sensitivity.

gases in the ionizer of a residual gas analyzer also pertain to the ion gauge.

Ion gauge outgassing is accomplished by direct or electron bombardment heating. Either the grid wire is heated directly by connecting it to a low voltage high current transformer or the grid and plate wire are connected to a high voltage transformer and heated by electron bombardment. It is best to wait until the pressure is on a suitably low scale ($<10^{-4}$ Pa) before outgassing. An unbaked tubulated gauge should be outgassed until the walls have desorbed. (The pressure may be monitored during outgassing on gauges that use resistance-heated grids.) The time for this initial outgassing is variable but 15 to 20 min is typical. After the initial outgassing the tube should be left on. Subsequently only short outgassing times, say 15 s, are periodically needed to clean the electrodes. It is useful to operate the gauge at reduced emission (0.1 mA) because it will pump the least when the emission current is the lowest.

At pressures greater than 10^{-2} Pa space charge reduces the number of electrons capable of producing ionizing collisions and the apparent sensi-

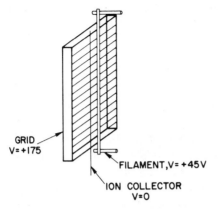

GRID
V= +175

FILAMENT,V= +45V

ION COLLECTOR
V=0

Fig. 5.15 Schulz-Phelps type ion gauge tube for operation at high pressures. Reprinted with permission from Varian Associates, 611 Hansen Way, Palo Alto, CA.

tivity is reduced. In addition, the mean free path becomes small and ions are scattered before reaching the collector. A high pressure gauge has been designed by Schulz and Phelps [20], versions of which are marketed by several manufacturers (Fig. 5.15). The close spacing of the electrodes allows this tube to be used at high pressures. Ion generation however, is reduced because the chance for an ionizing collision is proportional to the path length. A typical sensitivity for a Schulz-Phelps tube is 4×10^{-3} /Pa and a typical pressure range is 10^{-4} to 100 Pa. The ability to read lower pressures is again limited by x-ray-generated electrons. These tubes are excellent for monitoring chamber pressure during sputtering, reactive-ion etching, and other plasma processes. It is necessary to mount the Schulz-Phelps tube in a way that will prevent it from being affected by optical or other electromagnetic energy radiating from the plasma. This is accomplished by mounting it on an elbow and placing a piece of stainless screen over the end of the elbow at its entrance to the process chamber.

Cold Cathode Gauge

The cold cathode gauge developed by Penning [21] about 50 years ago provides an alternative to the hot cathode gauge which in some respects is superior but in other respects more limited. The gauge tube illustrated in Fig. 5.16 uses a wire anode loop maintained at a potential of 2 to 10 kV and grounded cathode electrodes. Surrounding the tube is a permanent magnet of about 0.1 to 0.2 T.

The arrangement of the electric and magnetic fields causes electrons to travel long distances in spiral paths before finally colliding with the anode. These long trajectories considerably enhance the ionization probability and result in a gauge with a much higher ionization efficiency than the hot cathode gauge. The total current, which is the sum of the electron and positive ion currents, is so much greater than in a hot cathode gauge that a current amplifier is not needed. Output currents of 10 to 50 mA/Pa are typical.

The range of operation of the cold cathode gauge is 1 to 10^{-4} Pa; gauge operation becomes erratic at low pressures because of the difficulty of maintaining the discharge. Penning and Nienhuis [22] were able to overcome some of the problems in this design by using a cylindrical anode with cathode plates at each end and a cylindrical magnet. Cold cathode gauges like hot cathode gauges have sensitivities which vary with gas species and in a similar manner. One advantage of the cold cathode gauge is that it overlaps the range of the hot cathode and thermal conductivity gauges but has the disadvantage of not operating below 10^{-4} Pa. Because of sputtering, the nitrogen pumping speed of these tubes is typically in the 0.1 to 0.5-L/s range—a factor of 10 to 100 times greater than a hot cathode. Because of this high pumping speed, gauges are fabricated with a large diameter entrance tubulation, typically 30 mm; they should not be connected to a system with tubulation of a smaller diameter or a considerable pressure drop will result. Cold cathode gauges should be mounted in a way that will not allow metal particles to fall inside the tube.

5.2.3 Ultrahigh Vacuum Gauges

The realization that photoelectrons emitted from the ion collector were indistinguishable from the positive ions incident on the ion collector led to the development of the Bayard-Alpert ionization gauge and a reduction of almost three orders of magnitude in the minimum detectable pressure. This, in turn, led to the development of a number of gauges with even lower x-ray, field emission, or photon limiting currents. Both hot and cold cathode gauges have been developed. Each gauge was

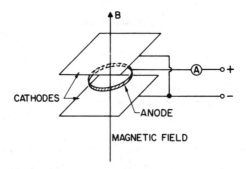

Fig. 5.16 Penning type cold cathode discharge tube.

designed to screen the ion collector from the pressure independent current or to increase the electron ionization path length.

Redhead investigated the possibility of using a Penning-type cold cathode gauge for pressures below 10^{-8} Pa because the electron current that produced x-rays was proportional to pressure (no x-ray limit). He developed the inverted magnetron gauge [23,24] which contained auxiliary cathode electrodes. These electrodes shielded the ion collector from the field emission current that limited the minimum detectable pressure of the Penning gauge. The inverted magnetron gauge and the magnetron gauge developed by Redhead [25] have low pressure limits of 10^{-11} Pa. Young and Hession [26] developed the trigger gauge, a cold cathode gauge that is started by momentary operation of a hot cathode. The trigger gauge has a low pressure limit of 5×10^{-10} Pa.

Several hot cathode gauges have also been developed. The modulated Bayard-Alpert gauge designed by Redhead [27] contains an extra electrode wire adjacent to the ion collector. By measuring the ion current at two modulator potentials the effects of the x-ray current may be subtracted. With careful calibration this gauge can measure pressures as low as 10^{-12} Pa. Lafferty [28] devised a high sensitivity hot cathode magnetron gauge with an x-ray limit of 5×10^{-12} Pa. Other hot cathode gauges, which shield the ion current from the x-ray-generated photocurrent with supressor grids [29], extracting electrodes [30], and bent beams, have also been developed.

The ultrahigh vacuum gauges we have discussed are usually calibrated or referenced to a mass spectrometer which, in turn, is calibrated by a Bayard-Alpert gauge.   The Bayard-Alpert gauge is referenced to the primary standard, the McLeod gauge.   Not all the ultrahigh vacuum gauges we have described are in widespread use today.   Many workers choose to use the nude, enclosed grid ion gauge to its limiting pressure of 2×10^{-9} Pa and a residual gas analyzer for pressures below that value because a knowledge of the spectrum is usually desired.

5.2.4   Tubulated versus Nude Gauges

Ion gauges are particle density gauges.   They measure the density of molecules in the ionizing chamber.   This is true for both nude and tubulated gauges.   Condensable vapors and directional gas effects each can cause tubulated gauges to read differently than nude gauges.

Blears [31] observed that a nude gauge gave pressure readings much higher than a tubulated gauge when the dominant species was a pump oil vapor.   He found that the vapors pumped by the walls of the tube caused the vapor pressure to be much lower inside the tubulated gauge than in the vicinity of the nude gauge.   Some of the apparent difference in pressure at the two gauge locations was due to the higher sensitivity for hydrocarbons ($S' = 5$–10).   Not only are hydrocarbon molecules prevented from reaching the electrodes of a tubulated gauge but those that reach either gauge cause a deflection 5 to 10 times greater than an equal number of nitrogen molecules.

The particle density inside a tubulated gauge may differ from its surroundings because of directional gas flow effects. The inside density may be different because gas enters the gauge only through the tube opening. In this sense a tubulated gauge is a flux gauge and not a particle density gauge. The density inside the gauge tube is determined from a balance between the rate at which particles enter the tube and the rate at which they leave. In equilibrium $\Gamma_{exit} = \Gamma_{enter}$, where Γ_{enter} is the rate at which particles bombard the entrance, and Γ_{exit} is the rate at which particles leave. The density inside the tube is related to the exit flux by (2.9). If the gas surrounding the glass envelope obeys a cosine distribution, the density inside will equal the density outside and the gauge will read the chamber pressure for any gauge position or orientation. If the gas around the tube exhibits directional effects, the gauge reading will be highly dependent on gauge orientation. This can happen when a nearby surface is heated or when it is cooled to a temperature where it is a good cryogenic pump. It can also happen when the arrival rates are not symmetrical because of the location of gas sources and pumps. See Chapter 7.

Tubulated and nude gauges are both widely available. In UHV systems a nude gauge may be desirable because its enclosed grid gives it a higher sensitivity and a lower x-ray limit than a tubulated gauge. Nude gauges should not be used for applications where the electrodes can be exposed to contaminating deposits. Tubulated gauges should be used with care near the pump entrances, especially cryogenic pumps. For most vacuum work tubulated gauges are used. (A nude gauge located in a small diameter pipe is effectively a tubulated gauge.) They will give the same readings (except for the Blears effect) and the tubulated structures are more rugged.

5.2.5 Accuracy of Indirect–Reading Gauges

It has already been noted that changes in emissivity and accommodation coefficient in thermal conductivity gauges are large enough to allow these gauges to be used only to indicate the degree of vacuum. Ionization gauges also suffer from inaccuracy but for different reasons.

Redhead [32] has shown that the sensitivity of a Bayard-Alpert gauge varied as much as a factor of 2.5 when the filament-to-grid spacing was varied from 0.5 to 6 mm. This variation resulted from the changing electron orbit length; hence the total ionization was a function of the filament-to-grid spacing. Redhead also observed that a number of nominally identical Bayard-Alpert gauges had sensitivity variations as large as ±15%. The grid-to-filament spacing appeared to be the most inadequately controlled dimension in most Bayard-Alpert structures. For this reason a sagging filament induces error. Electron-induced ion desorption is another common source of error in pressure measurement. The high collector current observed with a Bayard-Alpert gauge after it had been exposed to a relatively high gas pressure results from desorption of gases chemisorbed on the grid. Some gas molecules are desorbed as ions that strike the collector and result in artificially high collector currents [33-37]. Ion gauges may often be a dominant source of gas in ultrahigh vacuum systems. Ion desorption causes major errors in pressure measurement and residual gas analysis because the conditions of measurement reflect the conditions in the ion gauge or RGA and not the system as a whole. In spite of the problems that relate to construction tolerances, pumping, dissociation, gas generation, and temperature effects, the Bayard-Alpert gauge can give readings that are accurate within 25%; the various ultrahigh vacuum versions are accurate within an order of magnitude. Under suitable laboratory conditions, in which a gauge has been calibrated for a known gas composition, accuracies of a few percent can be obtained [38]. The average gauge is performing well, however, if it reads within 25%. The calibration of high and ultrahigh vacuum gauges has been reviewed by Sellenger [39] and Fowler and Bock [40]. Procedures for hot-filament gauge calibration are given in AVS Tentative Standard 6.4 [41].

REFERENCES

1. J. H. Leck, *Pressure Measurement in Vacuum Systems*, 2nd ed. Chapman and Hall, London, 1964.

2. J. P. Roth, *Vacuum Technology*, North Holland, Amsterdam, 1982.

3. S. Dushman, *Scientific Foundations of Vacuum Technique*, 2nd ed., J. M. Lafferty, Ed., Wiley , New York, 1962, p. 220.

4. A. Guthrie, *Vacuum Technology*, Wiley, New York, 1963, p. 150.

5. D. Alpert, C. G. Matland, and A. C. McCoubrey, *Rev. Sci. Instrum.*, **22**, 370 (1951).

6. J. J. Sullivan, *Ind. Res. Dev.*, January 1976, p. 41.

7. G. Loriot and T. Moran, *Rev. Sci. Instrum.*, **46**, 140 (1975).

8. J. W. Beams, D. M. Spitzer, Jr., and J. P. Wade, Jr., *Rev. Sci. Instrum.*, **33**, 151 (1962).

9. J. K. Fremerey, *J. Vac. Sci. Technol.*, **9**, 108 (1972).

10. G. Comsa, J. K. Fremerey, B. Lindenau, G. Messer and P. Röhl, *J. Vac. Sci. Technol.*, **17**, 642 (1980).

11. G. Reich, *J. Vac. Sci. Technol.*, **20**, 1148 (1982).

12. For example, the Granville-Phillips Convectron Gauge, Series 275, Granville Phillips Co, Boulder CO., or Leybold-Heraeus TR201 gauge, Leybold-Heraeus G.m.b.H., Köln, West Germany.

13. W. Thompson and S. Hanrahan, *J. Vac. Sci. Technol.*, **14**, 643 (1977).

14. W. B. Nottingham, *7th Ann. Conf. on Phys. Electron.*, M.I.T., 1947.

15. R. T. Bayard and D. A. Alpert, *Rev. Sci. Instrum.*, **21**, 571, (1950).

16. W. B. Nottingham, *1954 Vacuum Symp. Trans.*, Comm. Vacuum Techniques, Boston, 1955, p. 76.

17. D. Alpert, *J. Appl. Phys.*, **24**, 7 (1953).

18. T. A. Flaim and P. D. Owenby, *J. Vac. Sci. Technol.*, **8**, 661 (1971).

19. Ratio-matic Gauge,[®] Varian Associates, 611 Hansen Way, Palo Alto, CA.

20. G. J. Schulz and A. V. Phelps, *Rev. Sci. Instrum.*, **28**, 1051 (1957).

21. F. M. Penning, *Physica*, **4**, 71 (1937).

22. F. M. Penning and K. Nienhuis, *Philips Tech. Rev.*, **11**, 116 (1949).

23. P. A. Redhead, *Can. J. Phys.*, **36**, 255 (1958).

24. J. P. Hobson and P. A. Redhead, *Can. J. Phys.*, **36**, 271 (1958).

25. P. A. Redhead, *Can. J. Phys.*, **37**, 1260 (1959).

26. J. R. Young and F. P. Hession, *Trans. 9th Nat. Vacuum Symp.*, Macmillan, New York, 1963, p. 234.

27. P. A. Redhead, *Rev. Sci. Instrum.*, **31**, 343 (1960).

28. J. M. Lafferty, *J. Appl. Phys.*, **32**, 424 (1961).

29. W. L. Schuemann, *Rev. Sci. Instrum.*, **34**, 700 (1963) .

30. P. A. Redhead, *J. Vac. Sci. Technol.*, **3**, 173 (1966).

31. J. Blears, *Proc. R. Soc. London*, **188A**, 62 (1946).

32. P. A. Redhead, *J. Vac. Sci. Technol.*, **6**, 848 (1969).

33. P. A. Redhead, *Vacuum*, **12**, 267 (1962).

34. P. A. Redhead, *Vacuum*, **13**, 253, (1963).

35. T. E. Hartman, *Rev. Sci. Instrum.*, **34**, 1190 (1963).

36. G. Rettinghaus and W. K. Huber, *J. Vac. Sci. Technol.*, **6**, 89 (1969).

37. H. F. Winters, *J. Vac. Sci. Technol.*, **7**, 262 (1970).

38. J. M. Lafferty, *J. Vac. Sci. Technol.*, **9**, 101 (1971).

39. F. R. Sellenger, *Vacuum*, **18**, 645 (1969).

40. P. Fowler and F. J. Bock, *J. Vac. Sci. Technol.*, **7**, 507 (1970).

41. AVS Tentative Standard 6.4, *J. Vac. Sci. Technol.*, **7**, (1970).

PROBLEMS

5.1 †Indicate whether the following gauges will or will not be damaged if they are turned on at atmospheric pressure: (a) thermocouple gauge, (b) Pirani gauge, (c) cold cathode gauge, (d) hot cathode ion gauge, (e) capacitance manometer.

5.2 The following relation has been given for the capacitance between the tensioned diaphragm and the fixed electrode of a capacitance manometer (*Methods of Experimental Physics* **14**, *Vacuum Physics and Technology*, G. W. Wessler and R. W. Carlson, Eds., Academic Press, New York, 1979, p. 50).

$$C = \frac{\varepsilon_o A E t^3}{D E t^3 - K r^2 P}$$

where A is the area of the diaphragm of radius r, thickness t, and elastic constant E. D is the diaphragm-counterelectrode spacing at $P = P_{ref} = 0$. Show that, to first order, the change in capacitance is not dependent on a small change in either disk thickness or elastic constant at low pressures.

5.3 The pressure in a very large chamber is monitored by a capacitance manometer and a McLeod gauge. The McLeod gauge uses mercury and therefore is trapped with liquid nitrogen. The McLeod gauge reads 1000 Pa, and the capacitance manometer reads 1100 Pa. What can you tell about the molecules in the chamber?

5.4 A capacitance manometer head is heated to 50°C with a small oven. What thermal transpiration correction is required for P < 0.2 Pa (air) in a head with 0.03-m-diameter tubing? Assume ambient temperature is 20°C.

5.5 What thermal transpiration correction is required for a differential capacitance manometer heated to 50°C in which one side is referenced to 100 Pa and the gauge reads 0.2 Pa? Assume the same geometry and ambient as the last problem.

5.6 †What mechanisms account for the inaccuracy of thermocouple gauges?

5.7 An ion gauge reads a pressure of 5×10^{-4} Pa in an argon-filled chamber at 20°C. What is the true pressure?

5.8 An ion gauge that is calibrated at 10 mA emission current is operated at 100 μA emission current. The gauge reads 10^{-5} Pa (nitrogen). (a) What is the true pressure? (b) Why are ion gauges operated at low emission currents?

5.9 A large, clean, leak-free chamber initially at 20°C contains nitrogen. A nude ion gauge located within the chamber reads 5×10^{-3} Pa. The chamber is heated to 200°C. What does the ion gauge now read? Neglect any pumping effects of the gauge.

5.10 A small chamber containing an ion gauge is connected to a vacuum system with an intervening cold trap. The tubulation connecting the main chamber and the cold trap and that connecting the cold trap to the gauge are each 2 cm diameter. The temperature in the main chamber is 200°C, in the cold trap 77 K, and in the gauge chamber 25°C. Assuming the gases in the system don't condense, does the presence of the cold trap affect the pressure reading?

CHAPTER 6

Flow Meters

Flow measurements are performed to characterize components and monitor system operating conditions. Pump manufacturers and some users measure the speed of pumps. Pumping speed is not directly measured. It is calculated from experimental measurements of pressure and gas throughput. Process engineers control the gas flow in systems for plasma deposition or etching, chemical vapor deposition, reactive sputtering or ion milling. Researchers measure the gas flow into a chamber to calibrate systems used for studying gas desorption kinetics. For some applications accuracy is important, while for others only repeatability is necessary.

At atmospheric pressure a moderately large 50-L/s (100-cfm) mechanical pump has a gas throughput of 5×10^6 Pa-L/s; a small 100-L/s ion pump operating at 10^{-5} Pa pumps 10^{-3} Pa-L/s—a range of more than nine orders of magnitude. Figure 6.1 illustrates the flow ranges of several pumps and processes, and the capabilities of some gauges. No one gauge covers the entire range. Techniques such as moving oil or mercury pellets, and time to exhaust a reservoir [1-3] are still used to measure the very low flows used in pumping speed measurements.

In this chapter we define molar flow and mass flow, relate them to throughput, and review several techniques and instruments used for flow measurement.

6.1 MOLAR FLOW, MASS FLOW AND THROUGHPUT

Gas flow can be expressed in two ways: It is frequently expressed in units of throughput, such as Pa-m³/s or Torr-L/s. It may also be expressed in terms of the conservable quantities moles/s or kg/s. Confu-

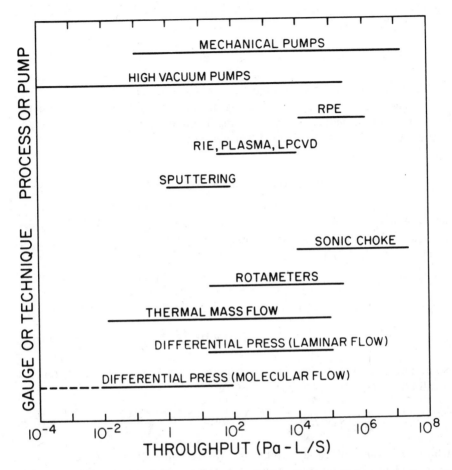

Fig. 6.1 Gas flow requirements and instrument ranges: (top) gas flow ranges of pumps and requirements of several processes; (bottom) gas flow ranges of flow measuring instruments and devices.

sion arises because the two ways of expressing flow are not dimensionally the same, and throughput does not conserve energy. In SI throughput has units of Pa-m³/s. Throughput has the dimensions of power and 1 Pa-m³/s = 1 J/s = 1 W. This is the power required to transport the gas. One Pa-m³ is the quantity of gas in 1 m³ at a pressure of 1 Pa. Molar flow or mass flow have dimensions of moles or mass per unit time. These can be related to throughput only if the temperature of the gas is known.

The second distinction between mass flow and throughput involves conservation of mass. A throughput measurement does not conserve mass. The numerical value of Pa-m³ does not uniquely define the number of molecules. For example 1 Pa-m³ of air could contain 2.45×10^{20} molecules/m³ at 300 K or it could contain 1.225×10^{20} molecules/m³ at 600 K. Moles and mass are conserved quantities. Knowledge of the number of moles/s or kg/s flowing through a system allows us to perform calculations when parts of the system are at different temperatures. We need to know when it is appropriate to use molar flow or throughput, and how to convert from throughput to mass or molar flow.

The molar flow rate N' has SI units of (kg-moles)/s and represents the total number of moles of gas passing a plane in 1 s. Molar flow and throughput are related by the ideal gas law. If we replace n in (2.13) with N/V we get

$$PV = NkT = \frac{N}{N_o}(N_o k)T = NRT \tag{6.1}$$

where N is the number of kg-moles of gas and $R = N_o k = 8314.3$ kJ/ (K − kgmole). The molar flow rate is obtained by taking the time derivative of (6.1):

$$\frac{d}{dt}(PV) = Q = N'RT$$

$$N' = \frac{Q}{RT} = 1.21 \times 10^{-4}\left(\frac{Q}{T}\right) \qquad \blacktriangleright (6.2)$$

Sometimes we wish to express the flow as mass flow, kg/s in SI, and we recall each kg-mole has a mass of M kg:

$$m'(\text{kg/s}) = \left(\frac{MQ}{RT}\right) = 1.21 \times 10^{-4}\left(\frac{MQ}{T}\right) \qquad \blacktriangleright (6.3)$$

where M is the molecular weight, and Q has units of Pa-m³/s at temperature T. The flow may be expressed alternatively as the number of molecules per second passing a plane $\Gamma(\text{molecules/s}) = N_o N'$. With these relationships we can convert from throughput to molar flow rate, mass flow rate or molecular flow rate. Erlich [4] notes that it is customary to label flow in standard leaks with units of "atm-cc/s at T". He reminds us that the value given is numerically equal to Q and not to kg-moles/s. The statement "at T" is included to allow conversion from throughput to molar flow. Some of the equations given in this chapter have flow given with dimensions of throughput and others with molar or mass flow.

Both throughput and mass or molar flow can be used in calculations. Throughput is a convenient term to use when the system is at a constant temperature for measurements such as pumping speed calculations. Molar flow is best used for studying reaction kinetics and for calculations which would otherwise have to be referred to several temperatures. For example, a calibrated leak labeled in "Pa-m³/s at 23°C" is connected to a system whose chamber is at 35°C and pump is at 50°C. In this example it would be much easier to have the leak labeled in kg-mol/s. The important concept to remember is the fundamental difference between throughput Q and molar flow. Throughput is dimensionally different from molar flow, and is not a quantity which is conserved. A kg-mole of gas is a fixed, known number of molecules and does not change with temperature.

6.2   ROTAMETERS AND CHOKES

A rotameter [5] is a flow measuring device constructed from a precision-tapered bore which contains a ball of accurately ground diameter and known mass. See Fig. 6.2a. The gas flow through the tube raises the ball to a height proportional to the throughput or mass flow. The general equation for continuum flow in an orifice is

$$Q = AP_1C'\left(\frac{2\gamma}{\gamma-1}\frac{kT}{m}\right)^{1/2}\left(\frac{P_2}{P_1}\right)^{1/\gamma}\left[1-\left(\frac{P_2}{P_1}\right)^{(\gamma-1)/\gamma}\right]^{1/2} \quad (6.4)$$

The equation for continuum flow in a rotameter with a low pressure drop can be obtained from this by letting $P_2 = P_1-\Delta P$. When ΔP is small compared to P_1 or P_2, we get

$$Q = \left(\frac{2kT\Delta PP_1}{m}\right)^{1/2} A \quad (6.5)$$

Except for a geometrical factor which accounts for the non-zero thickness of the orifice, (6.5) describes gas flow through a rotameter. The gas flow is a function of the inlet pressure P_1, gas temperature, molecular weight, height h, and mass of the ball. The mass of the ball is constant

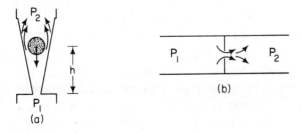

Fig. 6.2 Rotameter (a), and (b) choke flow elements.

and creates a constant pressure difference ΔP.

$$Q \propto \left(\frac{T}{M} \Delta P P_1 \right)^{1/2} f(h) \tag{6.6}$$

The mass flow rate can be expressed as

$$m' \propto \left(\frac{M}{T} \Delta P P_1 \right)^{1/2} f(h) \tag{6.7}$$

From (6.6) and (6.7) we see that the inlet pressure and temperature must be known to calibrate the throughput and mass flow. Rotameters are initially calibrated for one gas and must be recalibrated for use with any other gas. Rotameters are made for flows ranging from 5×10^3 to 5×10^6 Pa-L/s. The accuracy of these instruments is of order 10 to 20% full scale, while repeatability is about 2 to 3%.

Chokes [6,7] (Fig. 6.2b) are used to measure or more commonly set the throughput in the range 5×10^4 to 10^7 Pa-L/s. After the flow through an orifice reaches its sonic limit ($P_2/P_1 < 0.52$), it is practically independent of outlet pressure and is expressed as

$$Q = AP_1 C' \left(\frac{kT}{m} \frac{2\gamma}{\gamma+1} \right)^{1/2} \left(\frac{2}{\gamma+1} \right)^{1/(\gamma-1)} \tag{6.8}$$

For air at 22°C,

$$Q(\text{Pa} - \text{L/s}) = 2 \times 10^5 P_1 C' A(\text{m}^2) \tag{6.9}$$

The mass flow rate is given by

$$m' = AP_1 \left(\frac{m}{kT} \frac{2\gamma}{\gamma+1} \right)^{1/2} \left(\frac{2}{\gamma+1} \right)^{1/(\gamma-1)} \tag{6.10}$$

For air at 22°C,

$$m'(\text{kg/s}) = 2.58 \times 10^{-3} P_1 C' A(\text{m}^2) \qquad \blacktriangleright (6.11)$$

Again the throughput is dependent on the area, inlet pressure, temperature and gas species. These devices are not accurate in small sizes, less than 1 mm diameter, because the nature of the choke is critically dependent on the length of the hole as well as the radius and shape of the entrance edge. Van Atta [7] discusses the large radius orifices which are designed to make the flow more uniform and repeatable. Chokes are useful as flow restricting devices where accuracy and repeatability are not necessary. Chokes are used to limit flow from gas cylinders and to control turbulence during venting and rough pumping.

6.3 DIFFERENTIAL PRESSURE TECHNIQUES

Rotameters and chokes measure gas flow at a known inlet pressure and an essentially constant pressure drop. They do not operate in the flow ranges below 5×10^3 Pa-L/s. Low flow rates are easily measured in the molecular or laminar viscous flow region by measuring the pressure drop across a known conductance. The concept is the same for low, medium and high vacuum. Only the form of the conductance and the pressure gauge differ. High and ultrahigh vacuum flow measurements are almost always limited to pumping speed measurements and are treated separately in Chapter 7.

A molecular or laminar viscous element is used in combination with a capacitance manometer to measure the gas flow in the low and medium vacuum range. A laminar flow element [8] is incorporated into a flow meter as sketched in Fig. 6.3a. It is simply a capillary tube long enough to satisfy the Poiseuille equation. See Section 3.3:

$$Q = \frac{\pi d^4}{128 \eta l} P_{ave} \Delta P \qquad (6.12)$$

The flow is proportional to ΔP and the average pressure in the tube. Flow measurement with a long capillary requires two pressure gauges and knowledge of the temperature as well as the gas species. This is an accurate technique, but not the most convenient. It is most often used for calibration of thermal mass flow meters.

Molecular flow elements [9] are constructed from a parallel bundle of capillaries, v-grooves or similar shapes. See Fig. 6.3b. The diameter of each channel must be kept small for the flow to remain molecular at usably high pressures, say 100 Pa. Typically a single channel will have a diameter of a fraction of a millimeter. The conductance of one channel is of order 10^{-4} L/s so that a large number of parallel channels (>10,000) are necessary to achieve a practical device. The flow through such an element can be expressed as

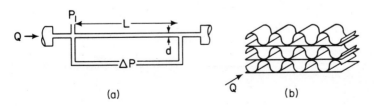

Fig. 6.3 Differential pressure flow elements, (a) laminar and (b) molecular.

$$Q = C\Delta P = Na'A\frac{v}{4}\Delta P \tag{6.13}$$

where N is the number of channels and a' is the transmission probability of one channel. We see the flow to be proportional to $(T/M)^{1/2}$. As long as the line pressure is less than 100 Pa it is not necessary to know its value. These devices are available commercially for use in the range 0.01 to 100 Pa-L/s. They are attitude insensitive, stable and easy to use. However the holes can become clogged. They have a slow response time (5 to 60 s), can have a high pressure drop, and cannot be used at inlet pressures greater than 100 Pa. The output of the capacitance manometer that measures ΔP can be used to control the opening and closing of an adjacent valve and achieve closed loop flow control.

6.4   THERMAL MASS FLOW TECHNIQUE

Mass flow can be calculated from the quantity of heat per unit time required to raise the temperature of a gas stream a known amount. Flow meters have been constructed which are sensitive to either thermal conductivity [10-12] or heat capacity [13,14]. Devices sensitive to heat capacity have become widely accepted because they are accurate and can measure large gas flows with a low power input.

The concept of the Thomas [15] thermal mass flow meter is illustrated in Fig. 6.4. We can measure the gas flow by applying constant power to the uniformly spaced grid and observing the temperature rise of the gas on the downstream side of the grid. The amount of heat required to warm the gas stream is linearly dependent on the mass flow and the specific heat

$$m' = \frac{H}{C_p(T_2 - T_1)} \tag{6.14}$$

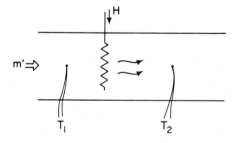

Fig. 6.4   Principle of thermal mass flow measurement.

For example, if we apply a nitrogen mass flow of 0.001 kg/s to this device, we will observe a temperature rise of 10°C for each 8.1 J/s of heat input to the grid. The change in thermal capacity with temperature is small and has only a slight effect on the measurement of *mass* flow. Typical thermal coefficients due to heat capacity variations range from +0.075%/°C (CO_2) to +0.0025%/°C (Ar).

Operation of a thermal mass flow meter is based on (6.14). Modern versions of this meter use a small bore heated tube to reduce the amount of heater power. One form of the device is sketched in Fig. 6.5. The meter must be mounted in the position shown as the heat distribution is attitude sensitive. The two thermocouples measure the change in temperature profile between the no flow and flow condition. Another form of the device uses a bridge circuit to keep the temperature profile constant. With this technique the mass flow is proportional to the amount of power required to maintain a constant profile [16]. Flow meters are constructed with full-scale deflections of 2 to 150 Pa-L/s using tubes of 0.2 to 0.8 mm internal diameter. The range of measurement may be increased to 10^5 Pa-L/s by use of a laminar flow bypass with a fixed divider ratio. A flow divider allows a known fraction of the flow to pass through the flow measuring tube.

A thermal mass flow meter directly *measures* the amount of heat absorbed by the gas stream, and therefore indirectly measures the mass flow. However, the gauge scales are normally *calibrated* in units of throughput - usually air. We can convert throughput to mass flow with the aid of (6.3). A gauge calibrated in units of throughput has a different temperature coefficient than one calibrated in units of mass flow, because throughput is density dependent. A mass flow of 10^{-6} kg/s (air) at a temperature of 20°C, according to (6.3), corresponds to a throughput of 83.64 Pa-L/s. If the gas were heated to 30°C, a mass flow of 10^{-6} kg/s would correspond to a throughput of 86.5 Pa-L/s because the heated air is less dense than the room temperature air. Near room temperature the temperature coefficient due to density changes is about −0.33%/°C. The temperature coefficient for throughput is therefore slightly less than this because of the small positive temperature coefficient of the heat capacity previously discussed. Temperature stability is improved by use of an insulation layer. Adding insulation increases the response time from 1 or 2 s to 6 or 10 s.

The thermal mass flow gauge will have to be readjusted for gases other than air. We can purchase a gauge especially calibrated for one gas, or we can multiply the meter reading by a correction factor. The meter deflection is proportional to the gas density and heat capacity, so we can

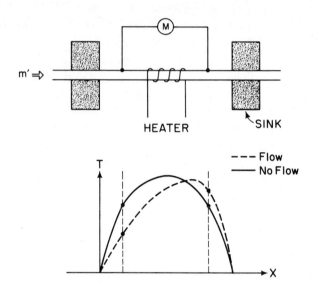

Fig. 6.5 Thermal mass flowmeter.

correct with the aid of (6.15).

$$Q_x = \left(\frac{\rho(\text{air})C_p(\text{air})}{\rho(x)C_p(x)} \right) Q_{\text{meter}} = f_x Q_m \qquad \blacktriangleright (6.15)$$

The factor in parentheses is known as the meter correction factor f. Some examples are given in Table 6.1. The mass flow of the uncalibrated gas is found from (6.3).

Thermal mass flow meters have the advantages of convenience, accuracy (<1 %) and moderately short response time. These meters are attitude sensitive and have a high temperature coefficient. Designs using very small diameter tubes can clog. They are used to monitor or measure gas flow and, when used in combination with a servo-controlled valve, can be used to control the flow. The tubes are usually constructed from stainless steel but are available in Monel for corrosive applications.

The maximum flow and the pressure drop across the flow meter and valve must be known to choose the proper size flow meter and valve. We first determine the equivalent air throughput from $Q_{\text{meter}} = Q_x/f_x$. We then choose a flow meter with the next largest full-scale meter deflection. From the manufacturer's data we determine the pressure drop ΔP across the meter and valve combination at maximum flow. The value of ΔP is little concern if the delivery pressure is above atmosphere. Vapors such as CCl_4 which are liquid at room temperature have vapor pressures below atmospheric pressure. If the gas or vapor source is at a

Table 6.1   Thermal Mass Flow Meter Correction Factors[a]

Gas	Heat Capacity J/(kg-°C)	Density (kg/m³) at 0°C	Correction Factor f[b]
Air	1004.2	1.293	1.00
NH_3	2058.5	0.760	0.73
Ar	520.5	1.782	1.45
AsH_3	488.3	3.478	0.67
BCl_3	535.1	5.227	0.41
CCl_4	692.5	6.86	0.31
Cl_2	478.7	3.163	0.86
CF_4	692.0	3.926	0.42
B_2H_6	2125.5	1.235	0.44
SiH_2Cl_2	627.6	4.506	0.40
C_2H_6	1714.2	1.342	0.50
He	5192.3	0.1786	1.45
H_2	14,305.1	0.0899	1.01
HBr	360.0	3.61	1.00
HF	1455.6	0.893	1.00
Kr	2481.5	3.793	1.543
CH_4	2229.2	0.715	0.72
Ne	1029.3	0.9	1.46
NO	974.0	1.339	0.99
N_2	1039.7	1.250	1.00
NO_2	808.8	2.052	0.74
NF_3	751.9	3.168	0.48
O_2	917.6	1.427	1.00
PH_3	993.3	1.517	0.76
SiH_4	1334.3	1.433	0.60
$SiCl_4$	531.4	7.58	0.60
SF_6	666.1	6.516	0.26
CCl_2FCClF_2	673.6	8.360	0.20[c]
WF_6	338.9	13.28	0.25
Xe	158.2	5.858	1.32

a Reproduced with permission from MKS Instruments, 6 Shattuck Road, Andover, MA 01810

b $Q_x = fQ_{meter}$

c at 60°C

reduced pressure, we must size the meter and valve so that the pressure drop is less than the difference between the delivery pressure P_1, and the chamber pressure P_2. $\Delta P < (P_1 - P_2)$. The pressure drop can be reduced by choosing a low conductance valve or if necessary as somewhat larger flow meter than might otherwise be desired.

6.5   GAUGE PROPERTIES

The gauges we have discussed differ in significant ways. Not only do they function in different pressure and flow ranges, but they sense flow in unique ways. All of the meters, with the exception of the thermal mass flow meter measure Q, and that meter is commonly calibrated in units of throughput. The calibration of a sonic choke and rotameter are valid only when the inlet pressure is known. The average pressure and the pressure drop are required to calculate flow in a laminar element, while only the pressure drop is necessary to read the molecular gauge. The thermal mass flow meter requires us to know only the temperature.

REFERENCES

1. C. E. Normand, *1961 Trans. 8th. Nat'l Vac. Symp.,* L. E. Preuss, ed., Pergamon, New York, 1962, p. 534.

2. D. J. Stevenson, *1961 Trans. 8th. Vac. Symp.,* L. E. Preuss, ed., Pergamon, New York, 1962, p. 555.

3. AVS Standard 4.1, *J. Vac. Sci. Technol.,* **8**, 664 (1971).

4. C. Erlich, *J. Vac. Sci. Technol. A,* **4**, 2384 (1986).

5. C. M. Van Atta, *Vacuum Science & Engineering*, McGraw-Hill, New York, 1965, Chapter 7.

6. R. W. Kuzara, in *Flow - Its Measurement and Control in Science and Industry*, Vol. 2, W. W. Durgin, Ed., Instrument Society of America, Research Triangle Park, 1981, p. 741.

7. Flow Measurement, PTC 19.5.4 *American Society of Mechanical Engineers*, New York, 1959, Chaper 4.

8. D. A. Todd, Jr., in *Flow - Its Measurement and Control in Science and Industry*, **2**, W. W. Durgin, ed., Instrument Society of America, Research Triangle Park, 1981, p. 695.

9. R. M. Kiesling, J. J. Sullivan, and D. J. Santeler, *J. Vac. Sci. Technol.*, **15**, 771 (1978).

10. C. E. Hastings and C. R. Wcislo, *AIEE*, March, 1951.

11. F. MacDonald, *Instruments and Control Systems*, October, 1969.

12. J. H. Laub, *Electrical Engineering*, December, 1947.

13. J. M. Benson, W. C. Baker, and E. Easter, *Instrum. and Control Systems*, p. 85, February, 1970.

14. C. E. Hawk and W. C. Baker, *J. Vac. Sci. Technol.*, **6**, 255 (1969).

15. C. C. Thomas, *J. Franklin Institute*, **152**, 411 (1911).

16. MKS Instruments Inc. Burlington, MA, 01803.

PROBLEMS

6.1 †What is the fundamental difference between mass flow and throughput?

6.2 (a) What is the mass flow rate in kg/s of 1000 Pa-L/s of air at 20°C? Express this as (b) a molecular flow rate and (c) a molar flow rate.

6.3 What is the limiting flow of 20°C atmospheric pressure air through a sonic choke whose diameter is 3.5 mm?

6.4 †A 0.5-L/s molecular flow element is used with a capacitance manometer with a range of 0.01 to 100 Pa. What range of flow can it measure (a) for air and (b) for argon?

6.5 A laminar flow tube is being used to calibrate a 0 to 200 SCCM thermal mass flow meter for air. The tube is 1 mm in diameter and 5 cm long. The room-temperature air supply has a maximum pressure of 10^5 Pa and the differential capacitance manometer has a full scale reading of 200 Pa. Does this calibrated source provide enough air flow to calibrate the flow meter?

6.6 †What are two advantages and disadvantages of a thermal mass flow meter?

6.7 How much power is absorbed from a heater which causes air [heat capacity of 1004.16 J/(kg-K)] flowing at the rate of 0.06 kg/min to rise 2°C?

6.8 The specific heat of neon is 20.85 kJ/(kg mole-K). What is its heat capacity in kJ/(kg-K)?

6.9 A thermal mass flow meter calibrated for air is used to measure a neon flow. The meter reads 10 SCCM. (a) What is the indicated meter reading in Pa-L/s? (b) What is the actual neon flow in Pa-L/s?

6.10 1,1,2-Trichloro-1,2,2-trifluoroethane (CCl_2FCClF_2) is a liquid with a vapor pressure of 37,730 Pa (288 Torr) at 20°C. (a) What is the maximum flow that can be read on a thermal flow meter that is calibrated for a full scale nitrogen flow of 500 SCCM (844 Pa-L/s)? (b) This flow meter is used in series with a control valve to regulate the flow into a vacuum chamber held at a pressure of 10 Pa. For trichlorotrifluoroethane the flow in the meter-valve combination can be expressed as $Q(Pa - L/s) = 2 \times 10^{-7} P_{ave} \Delta P$. Can this meter-valve combination provide a trichlorotrifluoroethane flow equal to the full-scale value of the flow meter?

CHAPTER 7

Pumping Speed

Pumping speed is a quantity few of us measure. We are usually content with published data; however, we do need to understand how speed is measured. Pumping speed is measured by well-established, but sometimes inaccurate, standard techniques. We need to know the accuracy of the available data. We may also wish to measure the pumping speed to check for proper pump operation or to provide specialized information. For example, the pumping speed may be needed for a specific gas, pump temperature or pump fluid. If we know the correct pumping speed at the inlet of a pump and the correct line conductance, we can calculate the pumping speed at the chamber entrance with reasonable accuracy.

We begin by defining pumping speed and then describe its measurement in mechanical and high vacuum pumps. We include a simplified technique for checking an operating system and discuss the measurement of water pumping speed. We conclude with a discussion of the errors inherent in standard test fixtures.

7.1 PUMPING SPEED

Pumping speed is the volumetric rate at which gas is transported across a plane. In mathematical terms speed is the gas throughput divided by the pressure at the plane of the pressure gauge:

$$S = \frac{Q}{P} \qquad \blacktriangleright (7.1)$$

Like conductance, it has dimensions of volume per unit time, and in SI it is expressed in units of m^3/s. Units of L/s or m^3/h are also used. Unlike conductance, pumping speed is not a property of a passive component like a length of pipe or a baffle. Recall the definition of conductance from (3.3). It is the gas throughput divided by the pressure drop across a component. Pumping speed is defined at a plane and not across a component.

Equation (7.1) implies independent measurements for the throughput Q and pressure P. The vacuum gauges used for pressure measurement are described in Chapter 5. The ionization gauge and capacitance manometer are used for routine laboratory measurements, while Spinning rotor and calibrated ionization gauges are used in exacting standards work. Devices such as a thermocouple gauge or a small Bourdon tube will not be used because their accuracy is poor. Several flow measuring devices are needed to span the range of flows necessary to characterize low, medium and high-vacuum pumps. Many of the flow methods have been replaced by the thermal mass flow meter or a combination of a capacitance manometer and a molecular or laminar flow orifice for routine measurements of pumping speed. All of these with the exception of high vacuum flow techniques are described in Chapter 6. High vacuum flow measuring techniques are described in Section 7.3.1.

7.2 MECHANICAL PUMPS

The AVS standard test dome [1] used for mechanical pump speed measurements is shown in Fig. 7.1. For inlet pressures below 10^{-2} Pa the AVS Standard specifies calibrated, trapped ionization gauges for measuring the pressure, McLeod gauges in the 10^{-2} to 100 Pa range, and mercury manometers for pressures above 100 Pa. A capacitance manometer will measure the pressure over this range and is more convenient for general laboratory use than the gauges described in the standard. A

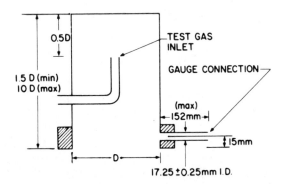

Fig. 7.1 Test dome for the measurement of mechanical vacuum pumping speed. Reprinted with permission from *J. Vac. Sci. Technol.*, **5**, p. 92. Copyright 1968, The American Vacuum Society.

thermal mass flow meter is the most convenient, but not necessarily the most accurate, instrument for gas flow measurement.

After the gas has been flowing into the test dome for at least 3 min, the equilibrium pressure is recorded and the speed is calculated from $S = Q/P$. The pumping speed is measured over the entire operating pressure range of the pump. The ambient temperature, barometric pressure, rotation speed, and type of oil in the pump should be recorded.

7.3   HIGH VACUUM PUMPS

Measurement of high vacuum pumping speed requires flow measurements considerably below the range of the techniques discussed in Chapter 6. Speed measurements are further complicated by the anisotropic gas flow patterns in the test dome and a flowing gas stream. Here we examine one technique for measuring the speed of a pump and one technique for estimating speed at the entrance to the working chamber.

7.3.1   Measurement Techniques

Volumetric pumping speed is measured with a standard metering dome. There were several standards [2-4]; however the majority of standards organizations have settled on a design similar to that described in the new AVS Recommended Practice [5] and described in Fig. 7.2.

In the AVS orifice method the gas flow is obtained from the pressure drop across an orifice of known dimensions $C(P_1 - P_2)$ and the resulting speed is calculated from

$$S = C\left(\frac{P_1 - P_{01}}{P_2 - P_{02}} - 1\right) \qquad (7.2)$$

In the AVS flowmeter method the flow is measured by an accurate flowmeter and $S = Q/P - P_0$

Pump Dependence

The technique described above is applicable to turbomolecular, ion, getter and cryogenic pumps. These pumps behave differently from diffusion pumps, so we need to observe certain procedures when measuring their speed. Air is the gas specified in the AVS standard for characterizing a diffusion pump, but speed can be and is measured for other gases. The International Standards Organization (ISO) standard specifies nitrogen. Diffusion pumps pump all gases and ideally their speed should vary as $1/m^{1/2}$. In Chapter 12 we show the light gas pumping speeds to be somewhat higher than the speeds for air, or nitrogen, but not as great as

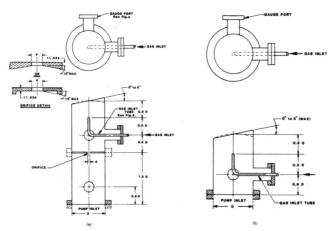

Fig. 7.2 AVS high vacuum test domes: (a) orifice method; (b) flowmeter method. Reprinted with permission from *J. Vac. Sci. Technol. A*, **5**, p. 2552, M. Hablanian, Copyright 1987, The American Vacuum Society.

predicted, $1/m^{1/2}$. Ion, getter, and cryogenic pumps do not pump all gases. Each pump will not pump certain gases. As a result, capture pumps are characterized with one gas, say nitrogen, because their speed is a function of gas composition, as well as prior history.

Some time may be required for a capture pump to reach equilibrium after admitting gas at a fixed flow rate [5]. This is particularly true for ion pumps at low pressures [6]. If pumping speed data are recorded too quickly, data taken in order of decreasing pressure will yield an incorrectly small value of speed, while those taken in order of increasing pressure will yield an incorrectly large value of speed. It may also be necessary to erase the pump's memory for one or more gases pumped before measurement. This can be accomplished by pumping for 1 h at a pressure of 10^{-3} Pa with the gas under study [5].

The operating characteristics of each pump should be recorded with the speed measurements. The type of fluid, the size of forepump and foreline pressure in compression pumps, the boiler power in a diffusion pump, the rotating frequency of a turbomolecular pump, and the refrigeration capacity of cryogenic pumps are some of the factors which need to be known.

Measurement of Water Vapor Pumping Speed

Few measurements of water vapor pumping speed have been made because of the experimental problems. Water is difficult to degas; it can freeze on evaporation, boil at room temperature and plug valves. It will sorb on the test dome at very low pressures. Landfors et al. [7] have measured the water pumping speed of a diffusion pump with and without a liquid nitrogen cold trap and the speed of a cryogenic pump. They constructed a chamber for admitting known amounts of degassed water vapor at constant pressure into an ISO test dome. Speed was measured at pressures greater than 10^{-2} Pa to avoid sorption effects. They found the water pumping speed to be the essentially the same as air for a diffusion pump without a liquid nitrogen trap. The water pumping speed for a liquid-nitrogen-trapped diffusion pump was approximately the same as a cryogenic pump of the same throat area. They concluded the pumping speed for water should be given by the projected area of the cold surface at the inlet of the trap as reduced by the conductance of the intervening tubing.

Pumping Speed at the Chamber

Only pump manufacturers and large projects will have the necessary test domes for measuring pumping speed, so it is necessary to have an approximate method anyone can use on an existing system. The approximate pumping speed at the entrance to the chamber can be measured without the trouble of attaching an elaborate dome and metering system. The speed can be deduced approximately if it is assumed to be independent of pressure in the region of interest. If this is true, then it follows that

$$S = \frac{(Q_2 - Q_1)}{(P_2 - P_1)} \tag{7.3}$$

where Q_1 is the flow that results in P_1, and so on. It can be shown that this flow is equivalent to

$$S = V\frac{\left(\dfrac{dP_2}{dt}\right) - \left(\dfrac{dP_1}{dt}\right)}{(P_2 - P_1)} \qquad \blacktriangleright (7.4)$$

To measure the pumping speed the system is first pumped to its base pressure P_1 and the high vacuum valve is closed. At this time the pressure rise dP_1/dt is plotted over at least one decade pressure increase. The high vacuum value is opened and the system pumped to its original base pressure. A gas is then admitted through a leak valve until the pressure rises to a value P_2 which is several times that of the base pressure. The high vacuum value is closed and dP_2/dt is recorded. The system volume is then estimated and the speed is calculated by use of (7.4). This method, called the rate of rise or constant volume method, is only approximate, because gas flow at the base pressure Q_1 is, in general, background desorption and not the same gas species as admitted through the leak. Fixing the starting pressure P_2 at a value of at least 10 times P_1 will ensure reasonable accuracy.

The pressure at the pump throat can be estimated by subtracting the impedance drop of the conducting pipe and trap located between the gate valve and the pump.

7.3.2   Measurement Error

High vacuum pumping speed measurement error arises from three sources. First, the definition of speed given by (7.1) requires Q and P to be measured at the *same surface*. This is not true for the dome described in Fig. 7.2. Second, the definitions of Q and P assume cosine distribution of molecular arrival, and that is not the case when gas enters from a pipe. Third, the gas flow is sometimes measured inaccurately. All three sources of measurement error exist because we do not completely compensate for the non-uniform gas distribution in the system. The non-uniform gas distribution affects both the capture coefficient or Ho [8] coefficient of the pump and the measurement of pressure. The Ho coefficient is the ratio of actual pumping speed to the maximum pumping speed of an aperture of the same size in which no molecules are reflected.

In Chapter 3 we defined the intrinsic conductance of a tube for cosine distribution of incoming molecules and noted this definition did not apply to beamed flow. The same effect occurs at the inlet of a pump. The capture probability is a function of arrival angle. When we place a pipe at the inlet of a pump the capture coefficient will change, and the pump no longer has its "intrinsic" speed, that is, the speed it would have if appended to an extremely large chamber. Its speed may increase because molecules shot straight into the pump will have a greater probability of being captured than will those arriving in a cosine distribution [9], and because molecules bounce around—the "maze" effect [10]. Test domes constructed from pipes of the same diameter as the pump are criticized because the inlet flux is not cosine and the resultant speed is not what is measured on a large chamber. Proponents of the AVS or ISO type test dome argue that the measured speed is meaningful because pumps are almost always connected to a pipe.

Pressure measurement is our second consideration. The pressure in the dome is not isotropic or uniform. Mathematically the pressure in the dome can be described by a tensor [11]. The pressure is a function of gauge location and orientation, and the surface temperature may not be uniform. Examine the pump and pipe sketched in Fig. 7.3. The pump is an ideal pump from which no molecules return ($a = 1$). Gas is flowing to the pump inlet. Gauge 1 reads the pressure at the pump entrance, while gauge 2 reads a pressure corresponding to the flux incident on its opening. One-half the gas arriving at the entrance to gauge 2 will come

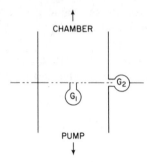

Fig. 7.3 The effect of orientation on pressure gauge readings.

from the top and one-half from the bottom. Since there is no flow from the bottom, gauge 2 will read $P_1/2$. The speed we would calculate from the side-mounted gauge located in the plane of the pump inlet would be two times the actual speed. Side reading gauge error has been understood for quite some time and has been discussed by many authors [9,11-14]. Projecting parts and non-unity sticking coefficients will usually prevent a pump from capturing all entering molecules. A fraction $(1-a)$ of them will be returned and gauge 2 will read $P_2 = P(1-a/2)$, where a is the transmission coefficient. The error resulting from the use of a side reading gauge is a function of the Ho coefficient. The relation between the *actual* Ho coefficient a and the *apparent* Ho coefficient a' due to the side reading gauge is found to be [13]

$$a = \left(\frac{a'}{1 + \dfrac{a'}{2}} \right) \qquad \blacktriangleright (7.5)$$

The apparent speed of a perfect pump ($a = 1$) is $2S$. The error will become less as the Ho coefficient goes to zero. Equation (7.5) is valid provided the reflected gas obeys a cosine distribution. In general this is not true, but objects such as inlet screens and chevrons randomize the flow and minimize the deviation from a cosine distribution.

Feng and Xu [15] calculated the pumping speed measurement error of the new AVS or ISO test dome to be within 4% for Ho coefficients in the range 0.2 to 1.0. They found the error in the older AVS dome [3] to be greater than 10%. Measurement error was the reason for changing the gauge location in the new AVS Recommended Practice. This is sufficient accuracy for the applications envisioned. Measuring pumping speed in the high vacuum region is complicated by the inhomogenous gas flow patterns. But with the choice of of a suitable dome we can measure a speed which we can use to calculate the pumping rate of a chamber to which the pump is appended. It is not necessary to know the pumping speed to a fraction of a percent because we will use formulas for joining conductances which are not corrected for beaming.

REFERENCES

1. Apparatus of AVS Tentative Standard 5.3, *J. Vac. Sci. Technol.*, **5**, 1968.

2. A. Venema, *Vacuum*, **4**, 272 (1954).

3. Apparatus of AVS Tentative Standard 4.1, 4.7, and 4.8, *J. Vac. Sci. Technol.*, **8**, 664 (1971).

4. D. R. Denison and E. S. McKee, *J. Vac. Sci. Technol.*, **11**, 337 (1974).

5. M. Hablanian, *J. Vac. Sci. Technol. A*, **5**, 2552 (1987).

6. D. Andrew, *Vacuum*, **16**, 653 (1966).

7. A. A. Landfors, M. H. Hablanian, R. F. Herirck, and D. M. Vaccarello, *J. Vac. Sci. Technol. A*, **1**, 150 (1983).

8. T. L. Ho, *Physics*, **2**, 386 (1932).

9. W. Steckelmacher, *Vacuum*, **15**, 249 and 503 (1965).

10. R. Buhl and E. A. Trendelenburg, *Vacuum*, **15**, 231 (1965).

11. B. B. Dayton, *Ind. Eng. Chem.*, **40**, 795 (1948).

12. D. J. Santeler et al., *Vacuum Technology and Space Simulation*, NASA SP-105, National Aeronautics and Space Administration, Washington, D.C., 1966, p. 119.

13. B. B. Dayton, *Vacuum*, **15**, 53 (1965).

14. D. R. Denison, *J. Vac. Sci. Technol.*, **12**, 548 (1975).

15. Y Feng and T. Xu, *Vacuum*, **30**, 377 (1980).

PROBLEMS

7.1   †Define volumetric pumping speed.

7.2   †What is the distinction between volumetric pumping speed and conductance?

7.3   A 500-L/s high vacuum pump is operating at an inlet pressure of 2 × 10⁻³ Pa. What is the throughput of the pump? The pressure at the top of the cold trap above the pump is 2.1 × 10⁻³ Pa. What is the conductance of the trap?

7.4   †A chamber containing air at reduced pressure is connected to a perfect vacuum by a thin circular aperture of area A. What is the pumping speed of the aperture when the mean free path of the gas is much larger than the diameter of the aperture?

7.5   Sketch the speed and conductance of an aperture in molecular flow as a function of the ratio of the two pressures P_1/P_2 across the aperture.

7.6   †For what reasons are pumping speed measurements more complex in high vacuum than in low and medium vacuum regions?

7.7 The pumping speed of an oil-sealed rotary vane mechanical pump filled with mineral oil was measured using an AVS test dome. The flow and pressure data were:

Gas Flow (T-L/min)	Gauge Reading (Torr)
3.8×10^1	1.0×10^{-1}
3.3×10^0	1.0×10^{-2}
3.0×10^{-1}	1.0×10^{-3}
6.4×10^{-2}	2.0×10^{-4}
3.1×10^{-2}	1.0×10^{-4}
5.3×10^{-3}	4.0×10^{-5}
7.5×10^{-4}	2.0×10^{-5}

Plot the inlet speed versus the inlet pressure. Estimate the ultimate pressure.

7.8 A 0.11-m³ chamber is connected by way of an adjacent gate valve, duct and cold trap to a 6-in. diffusion pump. The pump contains polyphenyl ether pump fluid. A rate of rise measurement is performed on the system for the purpose of measuring the net system pumping speed. Figure 7.4a shows the rate of rise of the background following valveclosure. Residual gas analysis shows it to be mainly water vapor. At the end of this measurement the system was pumped to the base pressure and a constant but unknown argon flow was admitted to the chamber. The base pressure rose to 5×10^{-6} Torr. Figure 7.4b shows the rate of rise after the gate valve was closed. Calculate the pumping speed for argon at the entrance to the chamber.

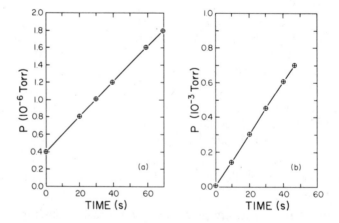

Fig. 7.4

7.9 A 6-in. (0.197 m) diameter diffusion pump has been appended to the dome like that in Fig. 7.2 for measurement of the air pumping speed. The diameter of the connecting orifice is 1.97 cm. The lower ion gauge reads 8.3×10^{-5} Pa, while the upper gauge reads 4.8×10^{-3} Pa. (a) What is the pumping speed for air? (b) What is the Ho coefficient?

7.10 Calculate the pumping speed at the chamber for the pumping system sketched in Fig. 7.5. Assume the real pumping speed of the diffusion pump is 2100 L/s at the pump throat.

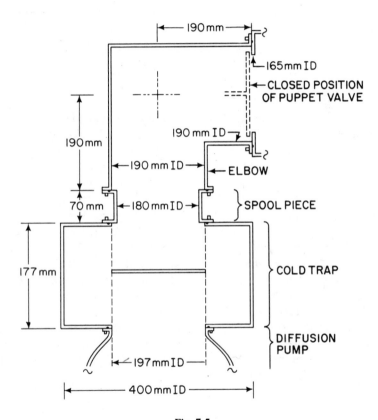

Fig. 7.5

CHAPTER 8

Residual Gas Analyzers

Residual gas analyzers have become so common that almost every vacuum technologist has seen one in the laboratory or attached to an occasionally troublesome system on the manufacturing line. In 1960 Caswell [1] used the RGA to study the residual gases in vacuum evaporators and demonstrated that the performance of a conventional system could be improved by Viton gaskets, Meissner traps, and getters. For thin-film deposition the importance of thoroughly outgassing the source material was stressed. Caswell also used the RGA to study the effects of residual gases on the properties of tin and indium films [2,3]. Since that time the instrument has proved its value in the solution of device fabrication problems by measuring other properties such as gas purity, gas loads evolved from films, and background gas composition during sputtering and other plasma processes.

This chapter is concerned with the theory of the operation of magnetic sectors and RF quadrupoles and the methods of installation of and data collection on vacuum and sputtering chambers. The interpretation of data is discussed in Chapter 9.

8.1   INSTRUMENT DESCRIPTION

RGAs and mass spectrometers are used to measure the ratio of mass to electric charge of a molecule or atom. First the molecules are ionized, then directed through a mass separator, and finally detected. See Fig. 8.1. A variety of methods has been developed for each of the three stages of particle identification. Some approaches are sophisticated and are applicable to analytical laboratory mass spectrometers that are capable of differentiating small fractional mass differences; for example,

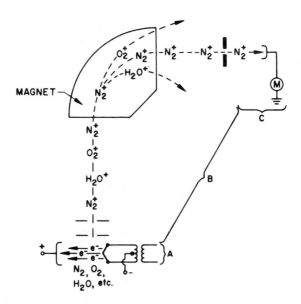

Fig. 8.1 Three stages of partial pressure analysis: (*a*) Ionization—hot filament illustrated; (*b*) mass separation—magnetic sector illustrated; and (*c*) detection—Faraday cup illustrated.

carbon monoxide ions (M = 27.9949 AMU) and molecular nitrogen ions (M = 28.0061 AMU) are easily separable on a double-focusing magnetic sector. (The weight of ^{12}C is 12 AMU.) Other methods have been developed into portable instruments that scan the mass range 1 to 50 AMU or perhaps 1 to 300 AMU and are able to resolve adjacent peaks 1 AMU apart. The latter instruments are collectively referred to as residual gas analyzers. This section reviews briefly the most commonly used ionization, separation, and detection methods in RGAs.

8.1.1 Ionization Sources

Virtually the only technique applicable to the production of positive ions in commercial residual gas analyzers is electron impact ionization. Other techniques such as field ionization and chemical ionization are useful in some research applications. Figure 8.2 is an ionization chamber that might be used in a residual gas analyzer. The electrons from the filament are drawn across the chamber to the anode. While crossing this space some of the electrons collide with gas molecules, strip off one or more of their electrons, and create positive ions. Not all ionization chambers are geometrically similar to the one sketched in Fig. 8.2. One instrument looks very much like a Bayard-Alpert ionization gauge except for the absence of the wire collector and the addition of an electron reflector

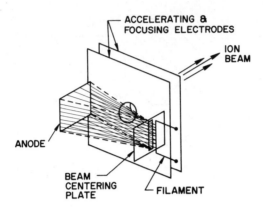

ION
BEAM

ACCELERATING &
FOCUSING ELECTRODES

ANODE

BEAM
CENTERING
PLATE   FILAMENT

Fig. 8.2   One form of an ionizing source used in a residual gas analyzer.

outside the filament.   These and other ionizers were designed to maximize ion production and instrument sensitivity.

As in the ion gauge, positive ion production is not the same for all gases.   The RGA differs from the ion gauge in that it sorts ions by their mass-to-charge ratio and counts each ratio separately.   Thus for nitrogen the ion gauge makes no distinction between a current due to N^+ ($M/z = 14$) or N_2^+ ($M/z = 28$), while the RGA does distinguish the two ion currents.   Table 8.1 gives the total positive ion cross sections relative to N_2 for several common gases at an ionizing energy of 70 eV [4].   Although the ionization cross section does not peak at the same energy for all gases, it is generally greatest for most gases somewhere in the 50-to-150-eV range.   For this reason most ionizers operate at a potential of 70 V.   Some instruments make provision for the adjustment of the ionizing voltage because it is sometimes desirable to reduce the potential in order to reduce the dissociation of complex molecules.   This is essential in qualitative analysis.

The ion production of each species is proportional to its density or partial pressure.   Consider a sample of a gas mixture containing only equal portions of nitrogen, oxygen and hydrogen whose total pressure is 3×10^{-5} Pa.   A mass scan of this mixture would show three main peaks of unequal amplitudes.   All other factors being equal, the main oxygen peak would be slightly larger and the hydrogen peak about half as large as the nitrogen peak because of differences in relative sensitivity or ionizer yield.   If, however, the total pressure of the gas mixture were increased to 6×10^{-5} Pa, the amplitudes of each of the three main peaks would double.   In other words, the instrument is linear with pressure.   Linearity of the ionizer extends to a maximum total pressure of order 10^{-3} Pa.   At higher pressures space charge effects and gas collisions become important.   The ions produced in the space between the filament

Table 8.1 Experimental Total Ionization
Cross Sections (70 V) for Selected
Gases Normalized to Nitrogen

Gas	Relative Cross Section
H_2	0.42
He	0.14
CH_4	1.57
Ne	0.22
N_2	1.00
CO	1.07
C_2H_4	2.44
NO	1.25
O_2	1.02
Ar	1.19
CO_2	1.36
N_2O	1.48
Kr	1.81
Xe	2.20
SF_6	2.42

Source. Reprinted with permission from *J. Chem. Phys.*, **43**, p. 1464, D. Rapp and P. Englander-Golden. Copyright 1965, The American Institute of Physics.

and anode are drawn out of that region, focused, and accelerated toward the mass separation stage. The acceleration energy depends on the type of mass analyzer that follows, and in a magnetic sector instrument ion acceleration is really a part of the mass separation stage.

8.1.2 Mass Separation

Almost a dozen techniques have been developed for mass separation of ions generated by a method like the one just described. For various reasons only two techniques, the magnetic sector and the RF quadrupole, have survived the test of widespread commercial development. Those who are interested in a thorough discussion of all types of mass separation schemes are referred to other sources [5]. The common methods are outlined in Fig. 8.3. This section discusses the quadrupole and magnetic

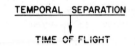

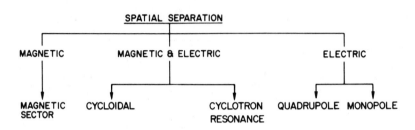

Fig. 8.3 Mass separation methods.

sector mass separation methods as they are commonly used in residual gas analyzers.

Magnetic Sector

The magnetic sector analyzer, which was developed 60 years ago [6], separates ions of different mass-to-charge ratios by first accelerating the ions through a potential V_a and then directing them into a uniform magnetic field perpendicular to the direction of the ion motion. While under the influence of this magnetic field the ions are deflected in circular orbits of radii r given by

$$r = \frac{1}{B}\left(\frac{2mV_a}{ze}\right)^{1/2} \qquad\qquad \blacktriangleright(8.1)$$

If B is given in units of teslas, the ion accelerating energy V_a, in volts, the mass M, in mass units, and z is the degree of ionization, the radius of curvature will be

$$r = \frac{1.44 \times 10^{-4}}{B}\left(\frac{MV_a}{z}\right)^{1/2} \qquad\qquad (8.2)$$

A practical mass analyzer that uses magnetic separation is shown in Fig. 8.4 for a 60° magnetic sector. In principle, any angle will work, but angles of 180°, 90°, and 60° are common. The 60° sector is a common filter for RGA applications. It provides sufficient separation between source and collector, and good focusing for divergent ions and requires a minimum amount of magnetic material.

As illustrated in Fig. 8.4, the location of the exit and entrance slits determine the radius r at which the beam will be properly focused. With

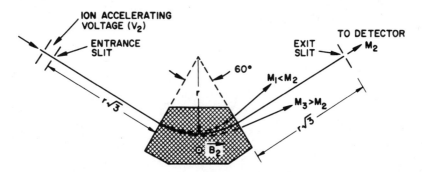

Fig. 8.4 A magnetic sector mass separator (60°) with symmetrical entrance and exit slits. Adapted with permission from *Mass Spectroscopy in Science and Technology*, p. 18, F. A. White. Copyright 1968, John Wiley & Sons.

the radius so specified, the mass-to-charge ratio M/z of the beam in focus is determined by the accelerating potential and the magnetic field strength. In the example shown a singly ionized molecule $z = 1$ of mass M_2 is focused on the exit slit for $V_a = V_2$ and $B = B_2$. Masses $M_1 < M_2$ and $M_3 > M_2$ will be deflected through greater and lesser angles, respectively, than M_2. To focus mass M_3 on the detector, B must be increased or V_a must be decreased. Commercial RGAs generally use permanent magnets and a variable acceleration voltage. Electromagnets are available to extend the range and provide magnetic scanning.

Equation (8.1) states that for constant r and B, the quantity MV_a/z is a constant. From this it can be seen that sweeping a large mass range, say, $1 \leq M \leq 300$, requires prohibitively large linear sweep voltages. Because of this limitation, permanent magnetic sector analyzers divide the instrument range into at least two scales by changing the magnet; for example, one scale might cover the mass range $2 \leq M < 50$ with a magnet of about 0.1 T and a second mass range of $12 \leq M \leq 300$ with a magnet of 0.25 T. Traditionally, but not exclusively, such instruments sweep the voltage linearly with time. Because MV_a/z is a constant, the resulting mass scan is not linear with time; as the mass number increases, the peak separation decreases. Somewhat more expensive instruments allow the accelerating potential to be held constant while an electromagnet of 0 to 0.25 T sweeps the range $1 \leq M \leq 100$.

Differentiation of (8.1) reveals that the mass dispersion Δx of the instrument is mass dependent. For ions of equal energy traversing a uniform magnetic sector the mass dispersion, or spatial separation between adjacent peaks of mass m and $m + 1$, has been found to be [5]

$$\Delta x \propto \frac{r}{m} \qquad \blacktriangleright (8.3)$$

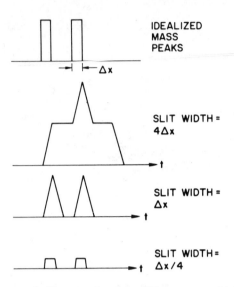

Fig. 8.5   Idealized mass peaks illustrate the trade-off between sensitivity and resolution.

Equation (8.3) illustrates why instruments of small radii cannot effective-ly separate adjacent heavy mass peaks. The resolution and sensitivity of a magnetic sector are dependent on mass and exit slit width.

Figure 8.5 shows two idealized mass peaks being scanned with slits of different widths and indicates that wide slits are efficient collectors of ions (high sensitivity) but poor resolvers of adjacent mass peaks (low resolving power). To first order there are no mass dependent transmis-sion losses in a fixed-radius, magnetic sector instrument. This means that if equal numbers of, say, hydrogen and xenon molecules pass through the entrance slit, equal numbers will pass through the exit slit. It will be noted later that this is not always true of the quadrupole. Figure 8.6 shows an RGA trace taken with a small sector instrument [7] on a 35-in. oil diffusion pumped system. This trace clearly illustrates the non-linear nature of the sweep and the mass dependent resolution. Because the slit width at the detector is fixed and the distance between adjacent mass peaks varies as $1/m$, the valley between two peaks of unit mass differ-ence is not so pronounced at high mass numbers as at low mass numbers. Figure 8.7 was taken on an instrument in which the magnetic field was varied to produce a linear scan. The same kind of mass dependent resolution is also evident here. More detailed discussions of magnetic sectors are found in a number of texts [8,9].

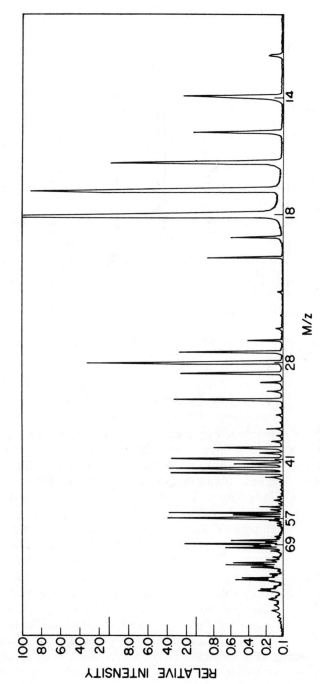

Fig. 8.6 Mass scan from a large diffusion pumped evaporator taken with a magnetic sector instrument with a permanent magnet and voltage sweep.

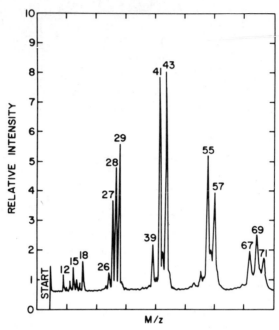

Fig. 8.7 Linear mass scan taken with an Aero Vac 700 series magnetic sector instrument. Reprinted with permission from High Voltage Engineering Corp., Burlington, MA.

RF Quadrupole

The RF quadrupole, developed by Paul [10] and co-workers, is the most popular non-magnetic mass filter in modern RGAs. Its acceptance has been due, in part, to the development of the necessary stable, high power quadrupole power supplies. Figure 8.8 illustrates the mass filter geometry and the path of a filtered ion. The ideal electrodes are hyperbolic in cross section. In practice they are realized by four rods of cylindrical cross section located to provide the optimum approximation to the hyperbolic fields. Each of the rods is spaced a distance r_o from the central axis. Mosharrafa [11] has provided a non-mathematical explanation of quadrupole operation. The two rods with positive dc potential, $+U$ in Fig. 8.9, create a potential valley near the axis in which positive ions are conditionally stable. The potential is zero along the axis of symmetry, shown in the dotted curve of Fig. 8.9. This field is zero only if the potential $-V$ is simultaneously applied to the other pair of quadupole rods. It is a property of a quadrupole, not a dipole, field. The addition of an RF field of magnitude greater than the dc field ($U + V \cos \omega t$) creates a situation in which positive ions are on a potential "hill" for a small portion of the cycle. Heavy ions have too much inertia to be affected by this short period of instability, but light ions are quickly collected by the rods after a few cycles. The lighter the ion, the fewer number of cycles required before ejection from the stable region. This rod pair acts as a "high pass" filter.

The rod pair with the negative dc potential $-U$ creates a potential "hill" that is unstable for positive ions. However, the addition of the RF field creates a field $-(U + V \cos \omega t)$ which allows a potential "valley" to exist along the axis of the quadrupole for a small portion of the cycle provided $V > U$. In this field light ions are conditionally stable and heavy ions drift toward the electrodes because the potential hill exists for most of the cycle. This half of the quadrupole forms a "low pass" filter.

Together the high and low pass filters form a band pass filter that allows ions of a particular mass range to go through a large number of stable, periodic oscillations while traveling in the z-direction. The width of the pass band, or resolution, is a function of the ratio of dc to RF potential amplitudes U/V, while the "sharpness" of the pass band is determined by the electrode uniformity, electrical stability, and ion entrance velocity and angle. A detector is mounted on the z-axis at the filter's exit to count the transmitted ions. Ions of all other M/z ratios will follow unstable orbits and be collected by the rods before exiting the filter. The stability limits for a particular M/z ratio are determined from the solutions of the equations of motion of an ion through the combined RF and dc fields and involve ratios of ω, M/z, r_o^2 and the potentials U and V. A thorough discussion of the RF quadrupole has been given by Dawson [12]. By sweeping the RF and dc potentials linearly in time the instrument can be made to scan a mass range. Scan times as slow as 10

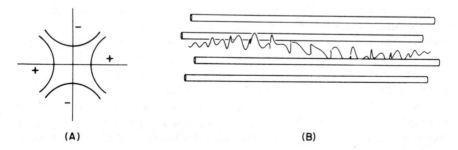

(A)   **(B)**

Fig. 8.8 Quadrupole mass filter: (*a*) idealized hyperbolic electrode cross section; (*b*) three-dimensional computer-generated representation of a stable ion path. Courtesy of A. Appel, IBM T. J. Watson Research Center.

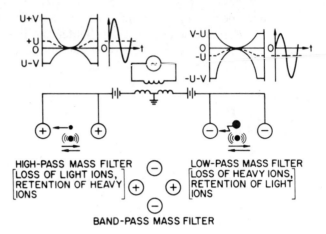

Fig. 8.9 Electric fields in a quadrupole mass filter. Reprinted with permission from *Industrial Research/Development* (March 1970), p. 24, M. Mosharrafa. Copyright 1970, Technical Publishing Co.

to 20 min and as fast as 80 ms are typically attainable in commercial instruments with a range of $1 \leq M/z \leq 300$. One noticeable distinguishing feature of the quadrupole is that no additional restriction other than linear sweeping of the RF and dc potentials is needed to obtain a graphical display that is linear in mass scan.

Although the stability of the trajectory of an ion may be calculated without consideration of the z-component of the ion velocity or the beam divergence, experimentally the situation is more complicated. There is a reasonable range of velocities and entrance angles that does yield stable trajectories. In the magnetic sector both the ion energy and the magnetic field determine the focus point of an ion. One of the advantages of the quadrupole is that ions with a range of energies or entrance velocities will focus even though not with the same resolution. The slow ions are resident in the filter for a longer time and therefore are subjected to a greater number of oscillations in the RF field than are those ions with larger z-components of velocity. As a result the slow ions are more finely resolved but suffer more transmission losses than the light ions. For this reason quadrupole transmission usually decreases with increasing mass. In a typical instrument adjusted for unity resolution the gain is constant to about $20 < M/z < 50$, after which it decays at the rate of approximately a decade per 150 AMU. See Fig. 8.10. This is only typical; there is considerable instrumental and manufacturer variation. By proper choice of the potentials U and V the mass dependence of the transmission can be considerably improved at the expense of resolution. Experimentally mass independent transmission can be achieved by adjusting the sensitivity for two gases, say, argon and xenon, to be the same. If accu-

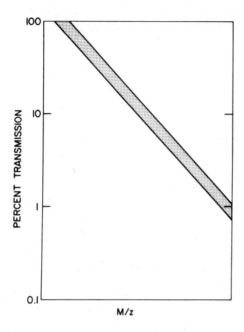

Fig. 8.10 Relative transmission of a typical RF quadrupole as a function of charge-to-mass ratio when adjusted for unity absolute resolution. This may be varied by changing the sensitivity.

rate knowledge of the transmission versus mass is desired, it must be measured for the particular filter and potentials in question. A typical mass scan taken on a small oil diffusion pumped system with a quadrupole adjusted for constant absolute resolution is displayed in Fig. 8.11.

Resolution and Resolving Power

The absolute resolution of an RGA is a measure of the ion-separating ability of the instrument of a given mass M and is given by the peak width ΔM. The American Vacuum Society's tentative standard for absolute resolution [13] specifies that the peak width shall be measured at a point equal to 10% of the peak height. See Fig. 8.12a. The resolving power of an RGA is the ratio of mass to resolution, $M/\Delta M$. In addition to the AVS definition of ΔM, Fig. 8.12 provides some other common definitions. It is evident that the numerical value of the resolving power of a given instrument will vary with the definition. A difference of a factor of 2 in numerical values between these definitions is not uncommon.

The reason that myriad definitions exist is partly historical and partly because no one definition seems to be suitable for all situations. Definition (*b*) in Fig. 8.12 is acceptable for two peaks of equal size but does not cover the trace peak next to a main peak. Definition (*c*) in the same figure is adequate for peaks of unequal magnitude, while definitions (*a*) and (*d*) do not require adjacent mass peaks to compute resolving power. In RGA work it is necessary to resolve adjacent peaks separated by one mass unit so that the minimum absolute resolution needed is unity. Analytical spectroscopy necessitates the discrimination of mass peaks separated by small fractional mass units. According to definition (*b*) in Fig. 8.12 a resolving power of 2000 is needed to distinguish $^{32}S^+$ ($M/z =$ 31.9720) from $^{16}O_2^+$ ($M/z =$ 31.9898). One important aspect of the definitions of resolving power and resolution has not been adequately emphasized; that is, the definitions are incomplete unless the sensitivity is specified. See (8.4). As illustrated in Fig. 8.5, the sensitivity and resolution of a magnetic sector vary with the slit width, while the same parameters are electronically controlled in the RF quadrupole. Because the resolving power can be adjusted over a wide range at the expense of sensitivity, it is misleading to quote the value of only one parameter.

8.1.3 Detection

The ion current detector located at the exit of the mass filter stage must be sensitive to small ion fluxes. The ion current at mass *n* is related to

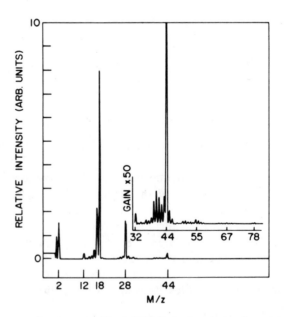

Fig. 8.11 Mass scan taken on a small oil diffusion pumped chamber with an RF quadrupole instrument.

the pressure in the linear region by

$$i_n = S'_n P_n \qquad (8.4)$$

where i_n, S'_n, and P_n are, respectively, the ion current, sensitivity of the ionizer and filter, and partial pressure of the nth gas. A typical sensitivity for nitrogen is 5×10^{-6} A/Pa to 2×10^{-5} A/Pa. We might ask why the sensitivity is defined with dimensions of current per unit pressure instead of reciprocal pressure as in ion gauge tubes. The answer is that the ion sources used in RGAs are sometimes space charge controlled, and their ion current is not linearly proportional to their emission current. Some instruments with high sensitivity use very high emission currents, up to 50 mA, but a typical ionizer with a nitrogen sensitivity of 7×10^{-6} A/Pa will have an emission current of 1 to 5 mA. For an emission current of 1 mA the sensitivity, defined as an ion gauge, would be 7×10^{-3} Pa^{-1}, which is an order of magnitude smaller than that of an ion gauge. The design of the mass analyzer is responsible for the low "ion

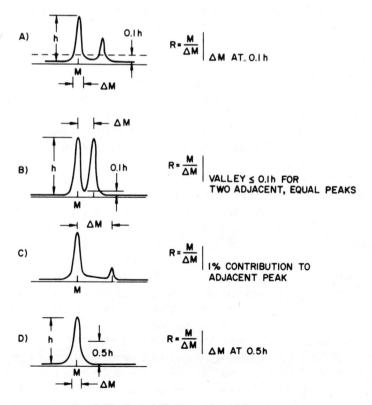

Fig. 8.12 Four definitions of resolving power.

gauge" sensitivity of the mass analyzer. Not all the ions generated in the ionizer are extracted through the drawing-out electrode nor do all the extracted ions traverse the mass filter.

If we assume an average sensitivity of 10^{-5} A/Pa and a dynamic pressure range of 10^{-1} to 10^{-12} Pa, the ion current at the entrance to the detector can range from 10^{-6} to 10^{-17} A. For the upper half of this range a simple Faraday cup detector followed by a stable, low noise, high gain FET amplifier will suffice, but below 10^{-12} A an electron multiplier is needed. Figure 8.13 illustrates a typical installation in which high gain is required—a combination Faraday cup–electron multiplier. When the Faraday cup is in operation, the first dynode is grounded to avoid interference. When the electron multiplier is used, the Faraday cup is grounded or connected to a small negative potential to improve the focus of the ions as they make a 90° bend toward the first dynode. In quadrupole analyzers the first dynode is generally located off-axis to avoid x-ray and photon bombardment.

Amplification in an electron multiplier is achieved when positive ions incident on the first dynode generate secondary electrons. The secondary electrons are amplified as they collide with each succeeding dynode. The multipliers are usually operated with a large negative voltage (−1000 to −3000 V) on the first dynode. The gain of the multiplier is given by

$$G = G_1 \cdot G_2^n \qquad\qquad \blacktriangleright (8.5)$$

where G_1 is the number of secondary electrons generated on the first dynode per incident ion and G_2 is the number of secondary electrons per incident electron generated on each of the n succeeding dynodes. The values of G_1 and G_2 depend on the material, energy, and nature of the incident ion. For a 16-stage multiplier whose overall gain G is 10^6, $G_2 = 2.37$; for $G = 10^5$, G_2 would be 2.05.

The electron multiplier sketched in Fig. 8.13 typically uses a Cu-2 to -4% Be alloy as the dynode material and when suitably heat treated to form a beryllium oxide surface will have an initial gain as high as 5 × 10^5. These tubes should be stored under vacuum at all times because a continued accumulation of contamination will cause the gain to decrease slowly. If the tube has been contaminated with vapors, such as halogens or silicone-based pump fluids, the gain will drop below 10^3; at this point the multiplier must be cleaned or replaced. If the multiplier has had only occasional exposure to air or water vapor and has not been operated at high output currents near saturation for prolonged times, its gain can usually be restored with successive ultrasonic cleanings in toluene and acetone followed by an ethyl alcohol rinse, air drying, and baking in air or oxygen for 30 min at 300°C. Tubes contaminated with silicones cannot be reactivated. Tubes contaminated with chemisorbed hydrocarbons or fluorocarbons often form polymer films on the final stages in

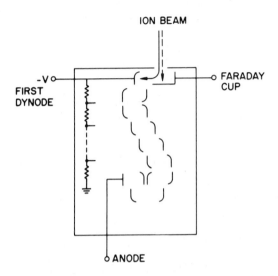

Fig. 8.13   Combination Faraday cup–electron multiplier detector.   Reprinted with permission from Uthe Technology Inc, 325 N. Mathilda Avenue, Sunnyvale, CA, 94086.

which the electron current is great.   Such contamination can be quickly removed by plasma ashing.

A channel electron multiplier is illustrated in Fig. 8.14.   The structure consists of a finely drawn $PbO–Bi_2O_3$ [14] tube typically 5 mm OD, and 1 mm ID.   A high voltage is applied between the ends of the tube.   The high resistivity of the glass makes it act like a resistor chain which causes secondary electrons generated by incoming ions to be deflected in the direction of the voltage gradient.   The tubes are curved to prevent positive ions generated near the end of the tube from traveling long distances in the reverse direction, gaining a large energy, colliding with the wall, and releasing spurious, out-of-phase secondary electrons [15,16].   If the tube is curved, the ions will collide with a wall before becoming energetic enough to release unwanted secondary electrons.   An entrance horn can be provided if a larger entrance aperture is needed.   A channel electron multiplier can be operated at pressures up to about 10^{-2} Pa and at temperatures up to 150°C; it has a distinct advantage over the Be-Cu type multiplier in that it is relatively unaffected by long or repeated exposure to atmosphere.   Because the secondary emission material is a suboxide, its emission is adversely affected by prolonged operation at high oxygen pressures.   It seems to be less affected by some contaminants than a Be-Cu multiplier.   Under normal operating conditions the gain of either multiplier will degrade in one or two years to a point at which replacement or rejuvination is necessary.   Little increase in gain will be realized by increasing the voltage on a multiplier whose gain is less than 1000.

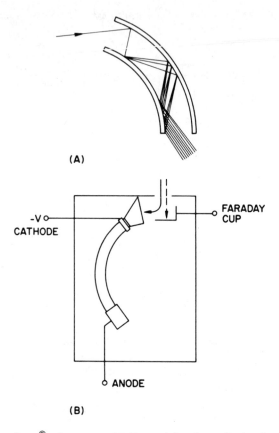

(A)

(B)

Fig. 8.14 Channeltron® electron multiplier: (*a*) schematic detail of capillary; (*b*) incorporation into mass analyzer.   Reprinted with permission from Galileo Electro-Optics Corp., Galileo Park, Sturbridge, MA 01518.

Although channel electron multipliers offer a significant advantage in their ability to accept air exposure, they saturate at a lower current than Be-Cu multipliers and cannot be operated with a linear output at high pressures unless the operating voltage or the emission current is reduced. The manufacturer's literature should be consulted for the precise values of saturation current and range of linearity before any electron multiplier is used.   The gain of a multiplier is not a single-valued function.   Ions illuminate the first dynode and electrons collide with each succeeding dynode.   The gain of the latter dynodes G_2 is dependent on the interstage voltage and the dynode material, while the gain of the first dynode G_1 is more complex.   The gain of the first stage is primarily dependent on the material, mass and the energy of the impinging ion, as illustrated in Fig. 8.15 for an AgMg dynode [17].   Ideally, the gain is proportional to $m^{1/2}$,

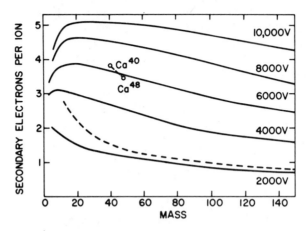

Fig. 8.15 Secondary electron yield of an AgMg dynode as a function of the mass and energy of the impinging ion.   Reprinted from *Mass Spectroscopy* (1954), p. 43, with permission of the National Academy of Sciences, Washington, DC.

the dotted curve in Fig. 8.15, but this is valid only for ions of low energy and high mass.   The gain of any multiplier is dependent on several other factors. Stray magnetic fields can distort the path of the electrons, and complex molecules may dissociate on impact to produce more electrons than a simple compound or element of the same mass.   Isotopes and doubly ionized molecules also react somewhat differently than the curve in Fig. 8.15; for example, the gains of ^{40}Ca and ^{48}Ca are different because of the variations in binding energy of the outer electron [17].

If the gain of a specific electron multiplier needs to be accurately known, a calibration curve of gain versus mass must be experimentally measured for all gases and vapors of interest.   The gain of an electron multiplier is never constant.   It may be measured by taking the ratio of the currents from the Faraday cup and the electron multiplier output, a tedious process, done only when semiquantitative analysis is required. Even then the gain of the multiplier must be periodically checked with one major gas to account for day-to-day aging of the tube.   Only periodic checking of the gain at mass 28 with N_2 is necessary for ordinary residual gas analysis work.

8.2   INSTALLATION AND OPERATION

8.2.1   High Vacuum

For routine gas analysis work the RGA head is mounted on a work chamber port. A valve is usually placed between the head and the chamber to keep the electron multiplier clean when the chamber is frequently vented to air. This is especially important with a Be-Cu electron multiplier. The head should be positioned to achieve maximum sensitivity in the volume being monitored. If the instrument is to monitor beams or evaporant streams, a line-of-sight view is necessary. It is often worthwhile to mount the head inside the chamber and shield its entrance so that the beam impinges on a small portion of the ionizer. This shielding will reduce the contamination of the ionizer and associated ceramic insulators.

Bayard-Alpert ion gauge tubes should be turned off when the RGA is operated. In the high vacuum region surface desorption by ion bombardment near the ion gauge will cause increases in all background gas levels. In the ultrahigh vacuum region inert gases are ionically pumped, while hydrogen and mass 28 are desorbed by electron bombardment [18]. If the system is to be baked or operated at an elevated temperature, the mass head should be baked at the same temperature or higher to avoid contamination from the condensable vapors that collect on the coolest surfaces. The necessity of baking will affect the choice of electron multiplier. Mass heads that contain channel electron multipliers can be baked to 320°C and operated at temperatures up to 150°C, while those with a Be-Cu multiplier can usually be baked to about 400°C and operated at temperatures up to 185°C. These temperature limitations are due to the materials used in channel electron structures and the glass encapsulated resistors in Be-Cu multiplier chains. Further contamination of the head and multiplier can be prevented in diffusion pumped systems by avoiding silicone-based pump fluids in favor of polyphenylethers or perfluoropolyethers. Silicone-based fluids will polymerize under electron bombardment to form insulating layers; therefore the absence of fluid fragments from the spectrum does not imply the absence of polymer films on a surface. During the initial operation of an RGA in an unbaked system heat dissipation by the filament will warm nearby surfaces and cause them to outgas. After 30 to 60 min of operation these surfaces will equilibrate thermally. Ceramic insulators will often exhibit a "memory effect" in which they continuously evolve fragments of hydrocarbons, fluorides, or chlorides after having been exposed to their vapors. This memory effect often confuses analysis when a mass head is moved from one system to another. Ion sources are easily cleaned by vacuum firing at 1100°C.

Filaments made from tungsten, thoriated iridium, and rhenium are available for RGA use. Tungsten filaments are recommended for general purpose work, although they generate copious amounts of CO and CO_2. Rhenium filaments do not consume hydrocarbons. Thoriated iridium filaments are used for oxygen environments although their emission characteristics are easily changed after contamination by hydrocarbons or halocarbons. As in the ion gauge, the thoriated iridium or "non-burnout"

Table 8.2   Some Properties of RGA Filament Materials

Property	ThO$_2$	W/3% Re	Re
CO production	Unknown	High	High
CO$_2$ production.	High	Mod.	Mod
O$_2$ consumption	Low	High	High[a]
H-C consumption	Unknown	High	Low
Water entrapment	Maybe high	Low	Low
Volatility in O$_2$	Low	High	V. high[b]
Good filament for ...	Nitrogen oxides	Hydrogen	Hydrogen
	Oxygen	Hydrogen halides	Hydrocarbons
	Sulfur oxides	Halogens	Hydrogen halides
		Halocarbons[c]	Halocarbons[c]
			Halogens

Source. Reprinted with permission from Uthe Technology Inc., 325 N. Mathilda Ave., Sunnyvale, CA 94086.

[a] Loss of one filament caused O$_2$/N$_2$ ratio to increase 17.6%.

[b] Exposure to air at 10^{-4} Pa caused failures of the filament pair at 30 and 95 h.

[c] Freons, etc.

filaments are advantageous when a momentary vacuum loss occurs. Table 8.2 gives some of the properties of three filament materials as tabulated [Raby [19]]. Molecular hydrogen dissociates at a temperature of 1100 K [20]. This be avoided by the use of a lanthanum hexaboride filament operating at 1000 K [21].

An oscilloscope and a chart recorder simplify data taking. The oscilloscope is handy for observing the desired mass range and gain before plotting and necessary for using the instrument to detect transient leaks. It is important to remember that the electron multiplier may saturate at pressures above 10^{-3} Pa if the applied voltage is too large. The gain of the multiplier should be reduced by decreasing the voltage so that the largest ion current is in the multiplier's linear range. Because of the long time constants used in low current, high gain amplifiers, measurements of extremely small signals must be made slowly on a chart recorder at speeds of about 1 AMU/s or less. The turbomolecular pump should be included in the usual grounding procedures used with all high gain amplifiers.

8.2.2 Differentially Pumped

It is sometimes desired to sample the gas in a system operating at a pressure higher than can be tolerated by the RGA, one in which sputtering or ion etching is performed. This can be done by differentially pumping the RGA. Honig [22] has given three primary conditions for gas analysis with a differentially pumped RGA: (1) The beam intensity should be directly proportional to the pressure of the gas in the sample chamber but should be independent of its molecular weight. (2) In a gas mixture the presence of one component should not affect the peaks due to another component. (3) The gas flow through the orifice should be constant during the scan. These conditions are satisfied if there is good mixing in the chamber, if the leaks are in molecular flow, and if the pump is throttled.

Examine the model of the differentially pumped system sketched in Fig. 8.16. In this model P_c is the pressure in the sputtering system, P_s is the pressure in the spectrometer chamber, and P_p is the pressure in the auxiliary pump whose speed is S_p. The conductances C_1 and C_2 which connect these chambers are schematically shown as capillary tubes, but in practice they may be tubes or small diameter holes in thin plates. The gas flow through the auxiliary system which is everywhere the same leads to the following equation:

$$C_1(P_c - P_s) = C_2(P_s - P_p) = S_p P_p \qquad (8.6)$$

When P_p is eliminated, we obtain

$$P_s = \frac{P_c}{1 + \left(\dfrac{C_2}{C_1}\right)\left(\dfrac{S_p}{S_p + C_2}\right)} \qquad (8.7)$$

Now consider two cases: Case 1: let the conductance C_2 be much larger than the speed of the pump. The auxiliary pump is located in or immediately adjacent to the spectrometer chamber with no interconnecting

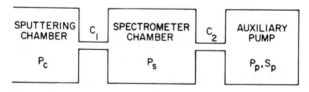

Fig. 8.16 Analysis of a differentially pumped residual gas analyzer.

conductance. Equation (8.7) then reduces to

$$P_s = \frac{P_c}{1 + (S_p/C_1)} \qquad \blacktriangleright (8.8)$$

The conductance C_1 has a mass dependence which varies as $m^{1/2}$, while the pumping speed has a mass dependence that is a function of the pump type; for example, an ion pump has different pumping speeds for noble and reactive gases and diffusion pump speeds increase somewhat at low mass numbers but not as $m^{1/2}$. The important result of this calculation is that the ratio of the gas pressure in the spectrometer to the pressure in the chamber, $P_s/P_{c,}$ is mass dependent when the auxiliary pump is appended directly onto the spectrometer chamber. Gases will not exist in the spectrometer region in the same proportion as in the chamber. Honig's first criterion will therefore not be satisfied.

Case 2 considers the situation in which a small conductance C_2 is placed between the spectrometer and pump such that $C_2 \ll S_p$. For this condition (8.7) becomes

$$P_s = \frac{P_c}{1 + (C_2/C_1)} \qquad \blacktriangleright (8.9)$$

Light gases will still pass from the sputtering chamber to the spectrometer chamber more rapidly than heavy gases, but so will they exit to the pump. Stated another way, the mass dependences of C_1 and C_2 are the same and negate each other so that (8.9) is mass independent. Therefore to sample the ratios of gases in the sputter chamber accurately, the auxiliary pump should be throttled to about 1/10 of its speed. If this is done, the pressure ratio of two gases in the chamber (P_{AC}/P_{BC}) will be the same as their ratio in the spectrometer (P_{AS}/P_{BS}).

The orifice molecular flow ratio is given by

$$\frac{N_A}{N_B} = \left(\frac{M_B}{M_A}\right)^{1/2} \frac{P_A}{P_B} \qquad (8.10)$$

Therefore the ratio of the two gases in the sputtering chamber will eventually change as the lighter gas is selectively removed at a faster rate unless the gas in the sputtering chamber is in dynamic equilibrium. If the gas throughput to the main sputtering pump is much larger than the flow through the aperture, the steady-state composition of the gas in the sputtering chamber will not be affected by the auxiliary pump. Sullivan and Busser [23] note that the time constant of the spectrometer chamber (V_s/C_2) must be equal to or less that the reaction time of the process to obtain time dependent information concerning the process.

The RGA may be connected to the working chamber with a valve that will permit monitoring the background when the system is evacuated to

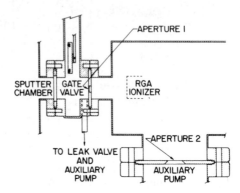

Fig. 8.17   Installation of a differentially pumped RGA on a sputtering chamber.

the high vacuum range and a parallel leak valve for use when the chamber is being operated at sputtering pressures. In this manner the RGA can be used to monitor rough gas composition in the chamber during the process. For those situations in which high sensitivity is desired the leak valve may take the form of a capillary tube or a small hole in a plate located in front of a clear aperture gate valve, (Fig. 8.17). The size of the aperture should be small enough to keep the gas in molecular flow: for example, at a work chamber pressure of 2 Pa (15 mTorr), the mean free path is 3 mm. If the aperture diameter is made to be equal to or less than 0.1λ (0.3 mm), the flow of gas will be completely molecular. To maintain the spectrometer chamber at a pressure of 1.3×10^{-3} Pa (10^{-5} Torr), and the pump at a pressure of 1.3×10^{-4} Pa (10^{-6} Torr) the conductance C_2 must be 17 L/s, which is equivalent to a 13.5-mm-diameter aperture; the pump speed required is 150 L/s. For these values (8.7) is essentially independent of mass. The flow from the sputtering chamber may be increased by drilling several holes to allow the flow to remain molecular in each hole. A reduced flow is possible for use with small auxiliary pumps by choosing an orifice or a capillary tube for C_1 that is smaller than that in the above example. The capillary tube has the advantage of beaming the sample gas to the ionizer and increasing the signal-to-noise ratio and the disadvantage of promoting wall collisions. For high chamber pressures (atmospheric) the aperture is replaced by a porous plug.

Small triode ion pumps are sometimes used for differentially pumping the RGA. They seem convenient but not suitable. Do not use an ion pump on an RGA. The amount and kind of gas desorbed from the walls of an ion pump is a function of the quantity of sputtering gas being pumped. For example, the amount of H_2 backstreaming is a function of sputter chamber pressure. This creates a high background for hydrogen and some other gases. Diffusion pumps suffer from the problem that some gases, caught and slowly desorbed on the cold trap, yield a steady-state background level that is neither constant nor reliable. Argon is one of the gases subject to this memory effect, but turbomolecular pumps show no evidence of it and are recommended for this application.

Residual gas analysis is then performed by recording a process and a background spectrum. The background spectrum is obtained under the conditions of gas flow in the sputtering chamber and into the RGA at the desired sputtering pressure. The process spectrum is taken with the discharge operating. Because the pressures in the RGA are lower than in the work chamber, the sensitivity of the instrument is similarly reduced. The large pressure reduction factor (1500 for the above example) and the ever-present background gases reduce the detection limit of the RGA. If, in this example, the minimum detectable partial pressure of nitrogen were 10^{-6} Pa in the spectrometer after subtracting the background, the minimum detectable pressure of nitrogen in the sputter chamber would be 1.5×10^{-3} Pa, or 0.1% of the sputter gas pressure. The minimum detectable partial pressure is usually limited to ~ 0.2 to 0.5% of the sputter gas pressure for an unbaked auxiliary chamber pumped by an ion pump. The ultimate detection sensitivity is determined by the pressure reduction factor and the minimum detectable pressure in the analyzer region. If the RGA were capable of detecting a partial pressure of 10^{-9} Pa and the pressure reduction factor between the sputtering chamber and the ionizing chamber were 1500, a partial pressure of 1.5×10^{-7} Pa (1 part in 10^6) would be detectable in the sputtering chamber. The unbaked ion-pumped system described above cannot reach this limt for several reasons.   Visser [24] showed that vapors such as pump fluids or water cannot be detected easily because they sorb on the walls of the leak valve and analyzer chamber. Heating the walls will stop this sorption but will increase the background pressures of other gases. Gases that do not sorb also present problems. Argon ionized in the analyzer will collide with nearby surfaces and release gases and vapors—even those surfaces that have been thoroughly baked. These problems can be avoided by using a turbomolecular pump, an aperture, a chopper, a synchronous detector or energy filter, and a line-of-sight path to the ionizer. If necessary, a hole may be cut in the ionizer grid.   In this manner only those molecules traversing the space between the aperture and ionizer without collision are counted. With a phase-sensitive or energy-selective technique such as this it is possible to achieve a detection sensitivity of $1:10^4$ for a gas like hydrogen and a sensitivity of $1:10^7$ for a gas or vapor with no background contribution.

Differentially pumped gas analysis is a useful technique for determining the time constants of contaminant decay during presputtering and for sampling the background gases during sputtering. Coburn [25] has used a carefully constructed differentially pumped mass spectrometer to sample the ionization in a sputtering environment. A small hole in the anode allowed study of the energy distribution of ionized plasma particles and the thickness profile of multilayered films located on the cathode. Visser [24] demonstrated an alternative to differentially pumped gas analysis. A mini-quadrupole was operated directly in the glow discharge; this technique however is applicable only at pressures of $\leq$ 0.5 Pa.

8.3 INSTRUMENT SELECTION

The instrument chosen for residual gas analysis work must be simple and reliable and have adequate resolution, sensitivity, and mass range. A variety of instruments is available from the elementary sector or quadrupole, with a mass range of 1 to 50 AMU, a Faraday cup detector, and a resolving power of 1 to 3 mass units. Other instruments have mass ranges up to 800 AMU. The only differences are complexity and cost.

For simple monitoring of background gases in regurlarly cycled, non-bakable high vacuum systems the simplest of instruments with a mass range of 1 to 50 AMU and a resolving power of 1 mass unit is adequate. With such an instrument the dominant fixed gases up to mass 44 and hydrocarbons at mass numbers 39, 41, and 43 may be monitored. Units with Faraday cup detectors and a sensitivity of 10^{-8} Pa are commercially available. Some instruments have an alarm that alerts the operator when a particular mass number has exceeded a preset ion current.

More detailed residual gas analysis with a sector or a quadrupole requires both more sensitivity and a greater mass range. Partial pressures of 10^{-12} Pa are detectable with an electron multiplier. A mass range of at least 1 to 80 AMU is necessary to distinguish most pump oils, but some solvents such as xylene have major peaks in the mass range 104 to 106 AMU. A mass range of 1 to 200 AMU permits identification of many heavy solvents. A resolving power of 1 mass unit is desired. The ability to detect partial pressures in the 10^{-10} to 10^{-12} Pa range is needed to leak-check and analyze the background of ultrahigh vacuum systems and to do serious semiquantitative analysis. An electron multiplier is necessary for differentially pumped systems. Instruments for these purposes need both the Faraday cup and an electron multiplier to measure the gain and to work over a wide pressure range.

The recent microprocessor revolution has resulted in a variety of instruments that provides automatic control and information display. The operator convenience is a definite asset, but the accuracy with which data are often displayed can seduce the user into believing that the data are indeed that well known. An RGA is limited by the stability of the ionizer, mass filter, and detector, as well as the history and cleanliness of the instrument, and *not* by the accuracy to which data can be presented.

REFERENCES

1. H. L. Caswell, *J. Appl. Phys.*, **32**, 105 (1961).
2. H. L. Caswell, *J. Appl. Phys.*, **32**, 2641 (1961).
3. H. L. Caswell, *IBM J. Res. Dev.*, **4**, 130 (1960).
4. D. Rapp, and P. Englander-Golden. *J. Chem. Phys.*, **43**, 1464 (1965).
5. F. A. White, *Mass Spectrometry in Science and Technology*, Wiley, New York, 1968, p. 13.
6. A. J. Dempster, *Phys. Rev.*, **11**, 316 (1918).
7. Aero Vac Model 610 Magnetic Sector, Aero Vac Products, High Voltage Engineering Corp., Burlington, MA.
8. G. P. Barnard, *Modern Mass Spectrometry*, The Institute of Physics, London, 1953.
9. C. A. McDowell, *Mass Spectrometry*, McGraw-Hill, New York, 1963.
10. W. Paul and H. Steinwedel, *Naturforsch*, **8A**, 448 (1953).
11. M. Mosharrafa, *Industrial Research/Development*, March, 24 (1970).
12. P. H. Dawson, Ed., *Quadrupole Mass Spectrometry and its Applications*, Elsevier,n New York, 1976.
13. American Vacuum Society Standard (tentative), AVS 2.3-1972, *J. Vac. Sci. Technol.*, **9**, 1260 (1972).
14. U.S. Patent #3, 492,523.
15. J. G. Timothy and R. L. Bybee, *Rev. Sci. Instrum.*, **48**, 292 (1977).
16. E. A. Kurz, *Am. Lab.*, March, 67 (1979).
17. M. G. Inghram and R. J. Hayden, National Academy of Science, National Research Council, Report #14, Washington, D.C., 1954.
18. G. A. Rozgonyi and J. Sosniak, *Vacuum*, **18**, 1 (1968).
19. B. Raby, UTI Technical Note 7301, May 18, 1977.
20. T. W. Hickmott, *J. Vac. Sci. Technol.*, **2**, 257 (1965).
21. T. W. Hickmott, *J. Chem. Phys.*, **32**, 810 (1960).
22. R. E. Honig, *J. Appl. Phys.*, **16**, 646 (1945).
23. J. J. Sullivan and R. G. Busser, *J. Vac. Sci. Technol.*, **6**, 103 (1969).
24. J. Visser, *J. Vac. Sci. Technol.*, **10**, 464 (1973).
25. J. Coburn, *Rev. Sci. Instrum.*, **41**, 1219 (1970).

PROBLEMS

8.1 †What are the functions of the ionizer, mass filter and detector in a residual gas analyzer?

8.2 †Can a residual gas ionizer indicate the mass of an atom or molecule?

8.3 For what reason would you wish to reduce the ionization energy in a residual gas analyzer?

8.4 Why must we determine the acceleration energy before injecting ions into a magnetic sector?

8.5 Describe how to measure the sensitivity of an electron multiplier at $M/z = 28$.

8.6 Give two advantages of a Be-Cu multiplier, and of a channel electron multiplier.

8.7 A gas at a pressure of 100,000 Pa is to be sampled by an RGA with a maximum operating pressure of 0.001 Pa. An orifice designed by the method of Section 8.2.2 is too small to work. How would you do it?

8.8 Describe how gas sampling with a reducing orifice and a big pump (Eq. 8.8) affects the analysis of a mixture of argon, nitrogen and hydrogen when pumped by (a) a diffusion pump, (b) a turbomolecular pump, and (c) an ion pump.

8.9 What limits the sensitivity of a differentially pumped analysis system?

8.10 Why should we use a metal-gasketed valve instead of an elastomer-gasketed valve when differentially pumping with the arrangement shown in Fig. 8.17?

CHAPTER 9

Interpretation of RGA Data

The method of interpreting mass scans like those in Figs. 8.6, 8.7, and 8.11 is based on detailed knowledge of the cracking or fragmentation patterns of gases and vapors found in the system. The complexity depends on the type and quality of information sought. After becoming familiar with an RGA, qualitative analysis of many major constituents is rather straightforward, whereas precise quantitative analysis requires careful calibration and complex analysis techniques. This section discusses cracking patterns, some rules of qualitative analysis, and methods of determining the partial pressures of the gases and vapors in the spectrum.

9.1 CRACKING PATTERNS

When molecules of a gas or vapor are struck by electrons whose energy can cause ionization, fragments of several mass-to-charge ratios are created. The mass-to-charge values are unique for each gas species, while the peak amplitudes are dependent on the gas and instrumental conditions. This pattern of fragments, called a cracking pattern, forms a fingerprint that may be used for absolute identification of a gas or vapor. The various peaks are primarily created by dissociative ionization, isotopic mass differences, and multiple ionization.

9.1.1 Dissociative Ionization

The cracking pattern of methane CH_4, illustrated in Fig. 9.1, shows, in addition to the parent ion CH_4^+, the electron dissociation of the molecule into lighter fragments, CH_3^+, CH_2^+, CH^+, C^+, H_2^+, H^+. Fragments

149

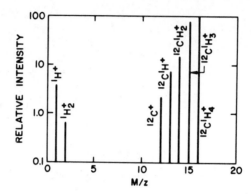

Fig. 9.1 This cracking pattern of methane illustrates the five largest dissociative ionization peaks.

containing ^{13}C are not shown. The fragments shown in this cracking pattern were produced by dissociative ionization.

9.1.2 Isotopes

Figure 9.2 illustrates several peaks in the mass spectrum of singly ionized argon due to isotopes. By comparing the peak heights of the isotopes with the relative isotopic abundances given in Appendix D, we see that this is a spectrum of naturally occurring argon. The relative isotopic peak heights observed on the RGA will generally mirror those given in Appen-

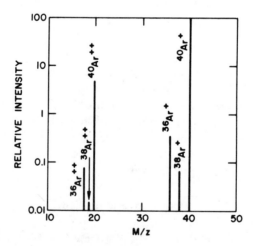

Fig. 9.2 The argon cracking pattern illustrates isotopic mass differences (three isotopes) and two degrees of ionization (single and double).

dix D unless the source was enriched or the sensitivity of the RGA was not constant over the isotopic mass range. Some compressed gas cylinders may contain gas in which a rare component has been selectively removed.

9.1.3 Multiple Ionization

Higher degrees of ionization are also visible in the argon spectrum. Argon has three isotopes of masses 36, 38, and 40; the doubly ionized peaks Ar^{++} show up at $M/z = 18$, 19, and 20. The cracking patterns of heavy metals may show triply ionized states.

9.1.4 Combined Effects

The cracking pattern of a somewhat more complex molecule CO, given in Appendix E.2, is illustrated in Fig. 9.3 to show the combined effects of isotopes, dissociation, and double ionization. The amplitude of the largest line $^{12}C^{16}O^{+}$ has been normalized to 100, while the amplitude of the weakest line shown, $^{16}O^{++}$, is 10^6 times smaller. The spectrum, as sketched in Fig. 9.3, can be observed only under carefully controlled laboratory conditions. During normal operation of an RGA the spectrum will be cluttered with other gases and only the four or five most intense

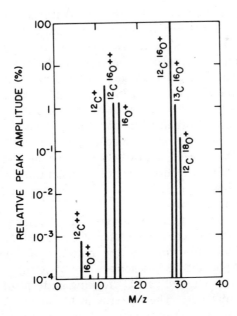

Fig. 9.3 Cracking pattern of CO illustrates dissociative ionization, isotopic and doubly ionized peaks.

peaks will be identifiable as a part of the CO spectrum. Even then some of the peaks will overlap the cracking patterns of other gases. Under the assumption that this clean spectrum has been obtained, it is instructive to classify the eight lines according to their origin. The main peak at mass 28 is due to the single ionization of the dominant isotope $^{12}C^{16}O$. Energetic electrons decompose some of these molecules into two other fragments, $^{12}C^+$ and $^{16}O^+$. Because carbon has two isotopes and oxygen three, several isotopic combinations are possible. The two most intense lines are $^{13}C^{16}O^+$ and $^{12}C^{18}O^+$. Isotopic fragments such as $^{13}C^+$ and $^{18}O^+$ are too weak to be seen here. Other complexes, for example, $^{12}C^{16}O^{++}$ ($M/z = 14$), $^{16}O^{++}$ ($M/z = 8$), and $^{12}C^{++}$ ($M/z = 6$), have high enough concentrations to be seen.

The relative amplitudes of the eight lines of CO are determined by the source gas and instrument. If isotopically pure source gas were used (^{12}C and ^{16}O), only six peaks would be found. Changes in the operating conditions of the instrument also affect the relative peak heights. The ion temperature and electron energy affect the probability of dissociation and the formation of higher ionization states. Dissociation of a molecule into fragments also changes the kinetic energy of the ion fragment. This can be a serious problem in a magnetic sector instrument because the kinetic energy of the ion directly affects the focusing and dispersion of the instrument. Quadrupoles can focus ions with a greater range of initial energies and therefore are less sensitive to transmission losses of fragment ions than sector instruments.

Electron- and ion-stimulated desorption of atoms from surfaces [1,2] can also add to the complexity of the cracking pattern. Oxygen, fluorine, chlorine, sodium, and potassium are some of the atoms that can be released from surfaces by energetic electron bombardment. In a magnetic sector instrument energetic oxygen or fluorine will often occur at fractional mass numbers of mass 16⅓ and 19¼, respectively [3]. These peaks are not representative of gaseous oxygen and fluorine in the chamber but of molecules desorbed from the walls. They occur at fractional mass numbers because they are formed at energies different from those corresponding to the ionized states of free molecules [1] and leave the surface with some kinetic energy [3]. The generation of gases resulting from the decarburization of tungsten filaments can add other spurious peaks to the cracking pattern.

Representative cracking patterns are given in the appendix. Appendixes E.2 and E.3 give the cracking patterns of some common gases and vapors, and Appendix E.4 contains the patterns of frequently encountered solvents. Appendix E.5 describes the patterns of gases used in semiconductor processing, and partial cracking patterns of six pump fluids are given in Appendix E.1. The patterns in the appendixes are intended to be representative of the substances and are not unique. It cannot be emphasized too strongly that each pattern is quantitatively meaningful only to those who use the same instrument under identical operating conditions.   Nonetheless, there are enough similarities between patterns to warrant their tabulation.

9.2   QUALITATIVE ANALYSIS

Perhaps the most important aspect of the analysis of spectra for the typical user of an RGA is qualitative analysis; that is, the determination of the types of gas and vapor in the vacuum system.   In many cases the existence of a particular molecule points the way to fixing a leak or correcting a process step.   The quantitative value or partial pressure of the molecular species in question usually does not need to be known because industrial process control is frequently done empirically.   The level of a contaminant, for example, water vapor, which will cause the process to fail, is determined experimentally by monitoring the quality of the product.   An inexpensive RGA tuned to the mass of the offending vapor is then used to indicate when the vapor has exceeded a predetermined partial pressure. With experience many gases, vapors, residues of cleaning solvents, and traces of pumping fluids will be recognizable without much difficulty.

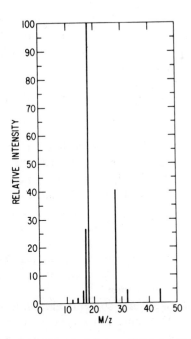

Fig. 9.4 Background spectrum constructed from a (20:4:4:1:1) mixture of H_2O, N_2, CO, O_2, and CO_2.

The mass spectra shown in Figs. 8.6, 8.7, and 8.11 contain considerable information about the present condition as well as the history of the systems on which they were recorded. To help in their interpretation examine the hypothetical spectrum in Fig 9.4   which was constructed from five gases, H_2O, N_2, CO, O_2, and CO_2, in the ratio (20:4:4:1:1) from the cracking patterns in Appendixes E.2 and E.3. Let us study this pattern under the assumption that its origin and composition are not known.   Examination of the cracking patterns of common gases, like those tabulated in Appendixes E.2 and E.3, quickly verifies the presence of carbon dioxide, oxygen, and water vapor.   Notice that the mass 32 peak is not due to the dissociation of carbon dioxide.   The presence of oxygen at 32 AMU usually indicates an air leak unless it is being intentionally introduced.   Analysis of the mass 28 peak is not so clear.   Some of this peak is certainly due to the dissociation of CO_2; however, if we assume that the sensitivity of the instrument is reasonably constant over the mass range in question, that contribution cannot be very great.   The majority of the peak amplitude would then be attributable to N_2 or CO or both.   To distinguish these gases further, the amplitudes of the peaks at mass 16 (O^+), mass 14 (N^+, N_2^{++}, CO^{++}), and mass 12 (C^+) are examined.   In practice it is difficult to conclude much from the presence of carbon because it originates from so many sources, both organic and inorganic. The mass 14 peak is largely due to N^+; therefore nitrogen is definitely present.   Analysis of the mass 16 peak is complicated by the fact that there are other sources of atomic oxygen besides CO, namely O_2 and CO_2, as well as electron-stimulated desorption of O^+ from the walls. Oxygen desorption is a common phenomenon.   Referring to the cracking pattern tables, we see that the mass 16 peak looks too large to be accounted for totally by the dissociation of CO_2 and O_2, while the mass 14 peak looks too small to be only a fragment of nitrogen.   We then conclude that both nitrogen and carbon monoxide are present but in undetermined amounts.

After some familiarity with the combined effects of these common background gases they should be easily identifiable in the spectra shown in Figs. 8.6, 8.7, and 8.11.   The large peaks at mass 32, 28, and 14 seen in Fig. 8.6 are an indication of an air leak, and the fact that water vapor is the dominant gas load in an unbaked vacuum system is verified by the large mass 18 peak shown in Figs. 8.6 and 8.11. These spectra also show fragments that are characteristic of organics as well as other fixed gases. One way to determine the nature of the organics in the system qualitatively is to become familiar with the cracking patterns of commonly used solvents, pump fluids, and elastomers.

The cracking patterns of several solvents are listed in Appendix E.4. A common characteristic of organic molecules is their high probability of fragmentation in a 70-eV ionizer. It is so great that the parent peak is rarely the most intense peak in the spectrum.   The lighter solvents such as ethanol, isopropyl and acetone have fragment peaks bracketing nitrogen and carbon monoxide at mass 27 and 29, but each has a prominent peak at a different mass number. Methanol and ethanol both have a major fragment at mass 31; they can be distinguished by the methanol's absence of a fragment at mass number 27. Solvents that contain fluorine or chlorine have characteristic fragments at mass numbers 19, 20, and 35, 36, 37 and 38, respectively. The extra fragments at mass numbers 20 (HF) and 35 to 38 (HCl) seem to be present, whether ornot the solvent contains hydrogen. Fragments due to CF, CCl, CF_2, and CCl_2 are also characteristic of these compounds. As with all fragments, their relative amplitudes will vary with instrumental conditions. Although the fragments of mass number 27, 29, 31, 41, and 43 are prominent in these solvents, they are also common fragments of many pump fluids, which further complicates the interpretation of a spectrum.

Appendix E.1 lists the partial cracking patterns of six common pump fluids. All the fragment peaks up to mass number 135 are tabulated. For the sector instrument the largest peak (100%) often occurs at a higher mass number and therefore is not shown. The complete spectra of the four fluids which were taken on the sector instrument were tabulated by Wood and Roenigk [4]. Most organic pump fluid molecules are quite heavy, and the parent peaks are not often seen in the system because only the lighter fragments backstream through a properly operated trap. Saturated straight-chain hydrocarbon oils are characterized by groups of fragment peaks centered 14 mass units apart and coincide with the number of carbon atoms in the chain. Figure 9.1 shows the fragment peaks for group C_1 in the mass range 12 to 16. Higher carbon groups have similar characteristic arrangements; C_2 (mass numbers 24 to 30), C_3 (mass numbers 36 to 44), C_4 (mass numbers 48 to 58), and C_5 (mass numbers 60 to 72), and so on. The spectrum of Apiezon BW diffusion pump oil taken by Craig & Harden [5] and shown in Fig. 9.5 illustrates these fragment clusters. In most hydrocarbon oils the fragments at mass numbers 39, 41, 43, 55, 57, 67, and 69 are notably stable and their presence is a guarantee of hydrocarbon contamination. These odd-numbered peaks, which are more intense than the even-numbered peaks in straight-chain hydrocarbons, clearly stand out in the mass scans shown in Figs. 8.6 and 8.7. Some traces of mechanical pump oil are seen in Fig. 8.11. Hydrocarbon oils are used in most rotary mechanical pumps, except those for pumping oxygen or corrosive gases, and in diffusion pumps for many applications. These diffusion pump oils are not resistant to oxidation and will decompose when exposed to air while heated. Their continued popularity for certain applications is due to their low cost.

Silicones are an important class of diffusion pump fluids. They not only have low vapor pressures but also have extremely high oxidation resistance [6]. Cracking patterns for Dow Corning DC-704 and DC-705, given in Appendix E.1, show many of the characteristic fragments of

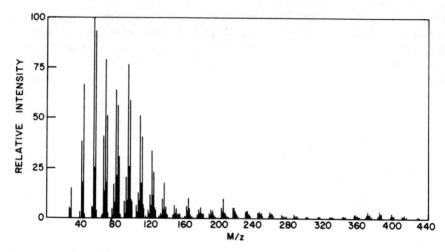

Fig. 9.5 Mass spectrum of Apiezon BW oil (obtained using MS9 sector spectrometer), source temperature 170°C. Reprinted with permission from *Vacuum*, **16**, 67, R. D. Craig and E. H. Harden. Copyright 1966, Pergamon Press, Ltd.

benzene. By way of illustration a partial spectrum taken in a contaminated diffusion-pumped system using DC-705 fluid is compared with the cracking pattern of benzene in Fig. 9.6. The peaks labeled M are due to mechanical pump oil. Systems that have been contaminated with a silicone pump fluid will always show the characteristic groups at $M/z = 77$ and 78 and usually those at $M/z = 50$, 51, and 52. Notice the lack of these peaks in Fig 8.6, which was taken on a system with a straight chain hydrocarbon oil in the diffusion pump. Wood and Roenigk [4] observed fragments of DC-705 at $M/z = 28$, 32, 40, and 44, all of which could naturally occur in a vacuum system. These peaks are due do dissolved gas.

Esters and polyphenylethers are also widely used pump fluids because they polymerize to form conducting layers. They find use in systems that contain mass analyzers, glow discharges, and electron beams. Octoil-S is characterized by its repeated C_mH_n groupings. Polyphenylether (Santovac 5, Convalex-10, and BL-10) also contains the characteristic fragments of the phenyl group which include the fragments at $M/z = 39$, 41, 43, 44, and 64. Cracking patterns will vary greatly from one instrument to another and the inability to match the data exactly to the patterns in any table should not be considered evidence of the absence or existence of a particular pumping fluid in the vacuum system. In fact, the cracking pattern for DC-705 given in Appendix E.1 does not show the same relative intensities at $M/z = 50$, 51, 52, and 78, as shown in Fig. 9.6a or seen by other workers. The spectrum taken from a gently heated liquid pump fluid source contains proportionally more high mass decomposition products than the spectrum of backstreamed vapors from a trapped diffusion pump because the trap is effective in retaining high molecular weight fragments. See Section 11.4. A liquid nitrogen trap is a mass-selective filter. These patterns cannot be used to differentiate between backstreamed DC-704 and DC-705. High mass ion currents are severely attenuated when a quadrupole instrument is operated at constant absolute resolution. This built-in attenuation, sketched in Fig. 8.10, can easily lead to the conclusion that the environment is free of heavy molecules.

Elastomers are found in all systems with demountable joints except those using metal gaskets. The most notable property of all elastomers is their ability to hold gas and release it when heated or squeezed. The mass spectra of Buna-N is shown in Fig. 9.7 during heating [7]. Also shown are the initial desorption of water followed by the dissociation of the compound at a higher temperature. The decomposition temperature is quite dependent on the material. Mass spectra obtained during the heating of Viton fluoroelastomer (Fig. 9.8 [7]) show the characteristic

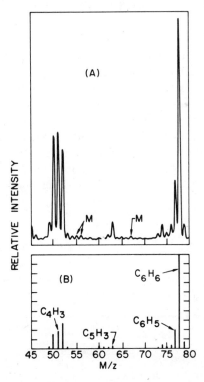

Fig. 9.6 Comparison of (a) residual gas background in a system contaminated with DC-705 fluid and (b) cracking pattern of benzene.

release of water at low temperatures and the release of carbon monoxide and carbon dioxide at higher temperatures. At a temperature of 300°C the Viton begins to decompose. Silicone rubbers have a polysiloxane structure. They are permeable, and their mass spectra usually show a large evolution of H_2O, CO, and CO_2. At high temperatures they begin to decompose. Their spectra show groups of peaks at $Si(CH_3)_n$, $Si_2O(CH_3)_n$, and $Si_3O_2(CH_3)_n$; $n = 3, 5, 6$, respectively, are the largest [8]. Polytetrafluoroethylene (Teflon) is suitable for use up to 300°C, although it outgasses considerably. A spectrum taken at 360°C shows major fragments at $M/z = 31(CF)$, $50(CF_2)$, $81(C_2F_3)$, and $100(C_2F_4)$ [8].

The potential limits of qualitative analysis become clear after some practice with an RGA. A knowledge of the instrument, the cleaning solvents, and the pumping fluids used in mechanical, diffusion, or turbo-molecular pumps combined with periodic background scans will result in the effective use of the RGA in the solution of equipment and process problems.

9.3 QUANTITATIVE ANALYSIS

The RGA, with a resolving power of 50 to 150, is not intended to be an analytical instrument. It cannot eliminate overlapping peaks, nor are its cracking patterns as stable as those of an analytical instrument. Quantitative analysis of a single gas or vapor or combination of gases and vapors with unique cracking patterns is a simpler task than the analysis of combinations that contain overlapping peaks. This section demonstrates approximation techniques for quickly obtaining quantitative partial pressures within a factor of 10 but points out that the acquisition of data accurate to, say, 10% requires careful calibration. Either crude data are obtained quickly or accurate data painstakingly—there is little middle

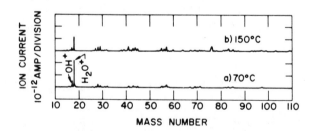

Fig. 9.7 Mass spectra obtained during the heating of Buna-N rubber. Reprinted with permission from *Trans. 7th Nat. Vacuum Symp. (1960)*, p. 39, R. R. Addis, L. Pensak and N. J. Scott. Copyright 1961, Pergamon Press, Ltd.

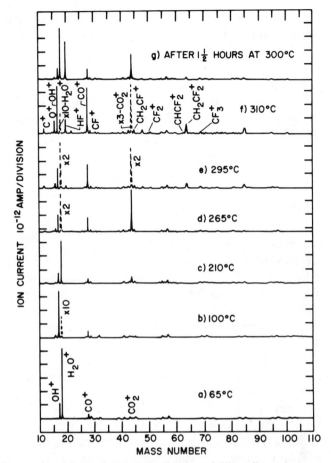

Fig. 9.8 Mass spectra obtained during the heating of Viton fluoroelastomer.   Reprinted with permission from *Trans. 7th Nat. Vacuum Symp. (1960)*, p. 39, R. R. Addis, L. Pensak, and N. J. Scott. Copyright 1961, Pergamon Press, Ltd.

ground.   We now consider techniques, both approximate, and precise, for gases with isolated and overlapping cracking patterns.

9.3.1   Isolated Spectra

Approximate analysis techniques for gases with isolated spectra will yield results of rough accuracy with minimal effort, for example, a mixture of Ar, O_2, N_2, and H_2O would be reasonably easy to examine because of the unique peaks at M/z = 40, 32, 28, and 18.   One technique is the summation of the heights of all peaks of any significant amplitude which are due to these gases followed by the division of that number into the total pressure.   The resulting sensitivity factor, expressed in units of pressure per unit scale division, is then applied to all the peaks without further correction. Some improvement in accuracy can be made if the ion currents are first corrected for the ionizer sensitivity by dividing by the values given in Table 8.1. There is an alternative technique that is equally accurate and does not require the knowledge of the total pressure. It relates the partial pressure of a given gas to the ion current, sensitivity of the mass analyzer, and gain of the electron multiplier according to (9.1):

$$P(x) = \frac{\text{total ion current}(x)}{GS(N_2)} \qquad \blacktriangleright (9.1)$$

The sensitivity is usually provided by the manufacturer for nitrogen and is typically of the order of 10^{-5} A/Pa. By taking the ratio of electron multiplier current to the Faraday cup current, we can determine the gain G of the multiplier. From this information the partial pressure of the gas is obtained. These two techniques are accurate to within a factor of 5.

A more accurate correction accounts for the gas ionization sensitivity and the mass dependencies of the multiplier gain and mass filter transmission. The mass dependence of the multiplier is often approximated as $M^{1/2}$, as shown on the dotted curve of Fig. 8.15, but this is not always valid. The transmission of a fixed-radius sector with variable magnetic field is independent of mass, but the most common RGA, the quadrupole, has a transmission that is dependent on the energy, focus, and resolution settings. The more accurate correction may turn out to be less accurate unless a significant amount of calibration is done to obtain the mass dependence of the mass filter and electron multiplier accurately. The time spent in applying corrections such as those shown in Figs. 8.10 and 8.15 is probably out of proportion to the information gained.

Accurate measurements of the partial pressures of gases with nonoverlapping spectra are best accomplished by calibrating the system for each gas of interest. The vacuum system must be thoroughly clean and baked if possible before the background spectrum is recorded. It is then backfilled with gas to a suitable pressure so that the cracking pattern can be recorded and the gas sensitivity measured. The values of all the ionizer potentials and currents, the gain of the electron multiplier, and the pressure should be recorded at that time. The system should be thoroughly pumped and cleaned between each successive background scan and gas admission. This is a laborious process and is only done when precise knowledge of the partial pressure of a particular species is required. Even then periodic checks of the multiplier gain, for example, at mass 28, are still necessary.

9.3.2   Overlapping Spectra

Analysis of overlapping spectra is made more difficult by the fact that the peak ratios for a given gas may not be stable with time because of electron multiplier contamination and because trace contaminants in the system will add unknown amounts to the minor peaks of gases under study. To gain an appreciation of the problems involved in determining partial pressures, two simplified numerical examples are worked out here.

A mass spectrum is taken on a system that contains peaks mainly attributable to N_2 and CO. Peaks due to nitrogen appear at mass numbers 28 and 14 and peaks due to CO appear at mass numbers 28, 16, 14, and 12. Trace amounts of carbon present in the system from other sources dictate that the amplitude of the mass 12 peak cannot be relied on for accurate determination of the CO concentration. In a similar manner the mass 16 peak is of questionable value because of the surface desorption of atomic oxygen, methane, or other hydrocarbon contamination. Therefore the analysis in this simplified example is weighted heavily in favor of using the peaks at mass numbers 14 and 28. From cleanly determined experimental cracking patterns the relative peak heights for nitrogen were found to be 0.09 and 1.00 for mass numbers 14 and 28, respectively, while values of 0.0154 and 1.10 were measured at the same mass numbers for CO. (The CO cracking patterns have been corrected for the difference in ionizer sensitivity; see Table 8.1.) The sensitivity and multiplier gain of the instrument were $S = 10^{-5}$ A/Pa and $G = 10^{+5}$. From the mass spectrum the ion currents were $i_{14} = 10.54$ μA and $i_{28} = 210$ μA. The individual partial pressures were then found by solving the following two equations simultaneously:

$$i_{28} = SG[a_{11}P(N_2) + a_{12}P(CO)]$$

$$i_{14} = SG[a_{21}P(N_2) + a_{22}P(CO)] \tag{9.2}$$

or

$$210 \ \mu A = 1 \ A/Pa[1.00P(N_2) + 1.10P(CO)]$$

$$10.54 \ \mu A = 1 \ A/Pa[0.09P(N_2) + 0.0154P(CO)]$$

which yielded $P(N_2) = P(CO) = 10^{-4}$ Pa.

Now consider how a change unaccounted for in the cracking pattern would affect the accuracy of this calculation. If the actual cracking pattern of nitrogen and consequently the measured ion currents were to change without the knowledge of the operator, because of contamination in the first dynode or a change in the temperature of the ion source, an error would be introduced into the calculation because the coefficients a_{mn} in the right-hand side of (9.2) were not altered simultaneously. Figure 9.9a shows the calculated values of $P(N_2)$ and $P(CO)$ that would be obtained for the example in (9.2) if the actual mass 14 fragment of nitrogen were changed from 9 to 5% of the mass number 28 peak without our knowledge and therefore without our having made the corresponding change in the coefficient a_{21}. It can be seen that even with moderate changes in the cracking pattern the partial pressures of the two gases can still be determined within 25% for this example in which the N_2 and CO are present in equal amounts. If greater accuracy is desired, the cracking pattern of the gases should be taken frequently and the coefficients a_{mn} adjusted to account for these instrumental changes.

For *unequal* concentrations of the two gases the errors are far greater than when the gases are present in equal proportions. Figure 9.9b illustrates the pressure measurement error as a function of the cracking pattern change for a 10:1 ratio of N_2 to CO. Again, this represents an actual change in cracking pattern which was not accounted for by a corresponding change in the coefficient a_{21}. This demonstrates that even modest changes in the cracking pattern ratios of the major constituent can cause the error in partial pressure calculation of the minor constituent to be as great as 200 to 400% when the major and minor constituents have overlapping cracking patterns.

These two illustrations demonstrate that quantitative analysis of overlapping spectra requires accurate and often frequent measurements of the cracking patterns and that accurate quantitative measurements of trace gases are difficult when the trace gas peaks overlap those of a major constituent.

The effects of certain cracking pattern errors in a residual gas spectrum have been illustrated by Dobrozemsky [9]. These data taken on an Orb-Ion pumped system are presented in Table 9.1. Column 1 shows the correctly analyzed partial pressures and their standard deviations, Column 2 and 3, respectively, show the effects of interchanging the peaks at mass numbers 17 and 18 and the effects of doubling the peak height at mass number 14. This analysis demonstrates vividly the effects of errors in the accuracy of calculating trace gas compositions. Column 1, the correct analysis, shows a standard deviation for H_2O of 0.3%, while for a trace gas such as hydrogen the standard deviations is large enough to render the measurement useless. The standard deviation for hydrogen is large because the signal at $M/z = 2$ arises from many sources. Literally any hydrocarbon that is ionized in the RGA has a fragment at $M/z = 2$ and the standard deviation of each fragment is additive. The result is that the hydrogen concentration, if any, is not known. It demonstrates the ease with which false data can be generated when cracking patterns are not accurately known. Note also that the incorrect analyses shown in Col-

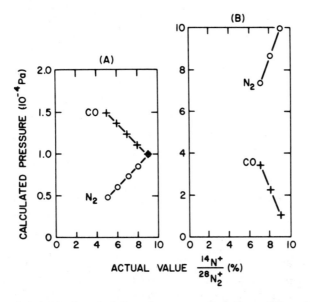

Fig. 9.9 Errors induced in the calculation of pressures of mixtures of N_2 and CO resulting from physical changes in the dissociation of nitrogen (ratio of N^{14} to N^{28}) which were not compensated for by appropriate changes in the coefficient a_{21} of the nitrogen cracking pattern. (*a*) 1:1 mixture of N_2 to CO, (*b*) 10:1 mixture of N_2 to CO.

umn 2 and Column 3 yield unreasonably large standard deviations for all gases even when only one ion current was in error.

Even though only a limited number of low molecular weight gases are present in a vacuum system, the analysis procedure is complicated by the fact that there are often several gases that produce peaks at the same mass numbers; for example, CO, N_2, C_2H_4, and CO_2 produce ion current at mass number 28 and CO_2, O_2, CH_4, and H_2O produce ion currents at mass number 16. For a system containing n gases and m ion current peaks, (9.2) becomes

$$
\begin{bmatrix} i_1 \\ i_2 \\ \cdot \\ \cdot \\ \cdot \\ \cdot \\ \cdot \\ i_m \end{bmatrix} = \begin{bmatrix} a_{11} & \cdots & a_{1n} \\ a_{21} & & a_{2n} \\ \cdot & & \cdot \\ \cdot & & \cdot \\ \cdot & & \cdot \\ \cdot & & \cdot \\ \cdot & & \cdot \\ a_{m1} & \cdots & a_{mn} \end{bmatrix} \begin{bmatrix} P_1 \\ P_2 \\ \cdot \\ \cdot \\ \cdot \\ \cdot \\ \cdot \\ P_n \end{bmatrix} \tag{9.3}
$$

where i_m is the ion current at mass m, a_{mn} are the components of the cracking pattern matrix, and P_n is the partial pressure of the nth gas. Most gases have more than one peak, so that $m > n$, and the system is

Table 9.1   Analysis of Background Gases in an Orb–Ion Pumped System[a]

Gas	Partial Pressures $\times$ 10^{-8} Torr		
	1	2	3
H_2	2.21 $\pm$ 3.15	2.85 $\pm$ 8.29	2.07 $\pm$ 2.67
He	15.54 $\pm$ 0.62	14.95 $\pm$ 22.6	14.94 $\pm$ 7.2
CH_4	5.37 $\pm$ 0.49	-13.5 $\pm$ 5.13	9.64 $\pm$ 1.63
NH_3	2.64 $\pm$ 0.89	49.56 $\pm$ 6.83	5.03 $\pm$ 2.17
H_2O	50.95 $\pm$ 0.15	15.4 $\pm$ 5.29	48.44 $\pm$ 1.68
Ne	4.54 $\pm$ 0.54	4.39 $\pm$ 16.5	4.38 $\pm$ 5.24
N_2	-1.99 $\pm$ 3.26	34.88 $\pm$ 57.6	-13.6 $\pm$ 18.3
CO	4.54 $\pm$ 2.91	-30.7 $\pm$ 51.4	15.38 $\pm$ 16.4
C_2H_6	3.42 $\pm$ 0.74	8.44 $\pm$ 18.4	1.89 $\pm$ 5.84
O_2	0.00 $\pm$ 0.13	-0.99 $\pm$ 4.87	0.3 $\pm$ 1.55
Cl	-0.02 $\pm$ 0.12	0.04 $\pm$ 4.64	0.02 $\pm$ 1.48
Ar	1.36 $\pm$ 0.09	1.29 $\pm$ 3.29	1.31 $\pm$ 1.04
CO_2	5.67 $\pm$ 0.14	5.3 $\pm$ 3.22	5.50 $\pm$ 1.02
C_3H_8	2.72 $\pm$ 0.56	0.27 $\pm$ 12.9	3.32 $\pm$ 4.11
Acetone	2.99 $\pm$ 0.31	7.94 $\pm$ 6.91	1.33 $\pm$ 2.2

Source. Reprinted with permission from *J. Vac. Sci. Technol.*, **9**, p. 220, R. Dobrozemsky. Copyright 1972, The American Vacuum Society.

[a] The total pressure is 1.3×10^{-7} Torr: (1) correct analysis; (2) incorrect spectrum obtained by interchanging ion currents at mass numbers 17 and 18; (3) incorrect spectrum obtained by doubling the ion current at mass number 14.

overspecified. A least mean squares or other smoothing criterion is then applied to the data to get the best fit. If the cracking patterns are carefully taken and if the standard deviations are measured as well, then accuracies of 1 to 3% may be obtained for major constituents [9,10]. The matrix for a real problem would contain about 10 $\times$ 50 elements and require the assistance of a computing machine in order to obtain a solution. In this case it is practical to use a computer for data acquisition and instrument control as well as for analysis [10]. These experiments are expensive and time consuming and are only performed in situations where such precision is required. Beware of instruments which directly convert ion currents to gas partial pressures and display them on a CRT. Since cracking patterns change with time and cleanliness and sensitivity is a function of gas mixture and pressure, simple algorithims cannot accurately determine pressure. **Beware.**

REFERENCES

1. P. Marmet and J. D. Morrison, *J. Chem. Phys.*, **36**, 1238 (1962).
2. P. A. Redhead, *Can. J. Phys.*, **42**, 886 (1964).
3. J. L.Robbins, *Can. J. Phys.*, **41**, 1383 (1963).
4. G. M. Wood, Jr. and R. J. Roenigk, Jr., *J. Vac. Sci. Technol.*, **6**, 871 (1969).
5. R. D. Craig, and E. H. Harden, *Vacuum*, **16**, 67 (1966).
6. C. W. Solbrig and W. E. Jamison, *J. Vac. Sci. Technol.*, **2**, 228 (1965).
7. R. R. Addis, Jr., L. Pensak, and N. J. Scott, *Trans. 7th A.V.S. Nat. Vacuum Symp. 1960*, Pergamon, Oxford, 1961, p. 39.
8. A. H. Beck, *Handbook of Vacuum Physics*, Vol. 3, Macmillan, New York, 1964, p. 243.
9. R. Dobrozemsky, *J. Vac. Sci. Technol.*, **9**, 220 (1972).
10. D. L. Ramondi, H. F. Winters, P. M. Grant and D. C. Clarke, *IBM J. Res. Dev.*, **15**, 307 (1971).

PROBLEMS

9.1 †Define dissocative ionization, multiple ionization, and isotopic ionization.

9.2 Examine the cracking pattern of CO_2 given in Appendix E.2. List the mass-to-charge ratios resulting from (a) dissociative ionization, (b) multiple ionization, and (c) isotopes.

9.3 Give an example of a mass peak which might be caused by (a) atomic attachment and (b) dissociation-recombination.

9.4 The charge to mass ratio of a singly ionized benzene molecule is 78. A small peak at $M/z = 79$ is observed. What is the origin of this peak and what is the ratio of the two peak heights?

9.5 What is the effect of operating an ion gauge near to a residual gas analyzer in the presence of (a) hydrogen, (b) argon, and (c) nitrogen?

9.6 †Sketch individually the dominant mass-to-charge peaks of (a) HCl, (b) HF, (c) Ar, (d) Ne and (e) air.

9.7 In an ion beam system which uses large amounts of argon, what will interfere with analyzing for trace amounts of water vapor?

9.8 Mass scans shown in Fig. 9.10*a* and 9.10*b* were taken on the same system on adjacent days after, each time, reaching a base pressure. Describe the difference in the condition of the two systems.

9.9 Identify the compound in Fig. 9.11.

9.10 Identify the compound in Fig. 9.12.

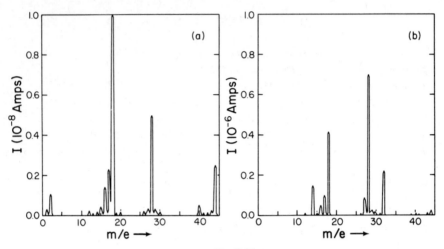

Fig. 9.10

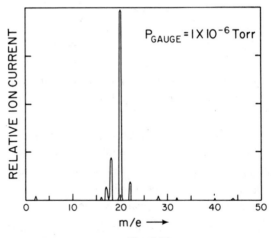

Fig. 9.11

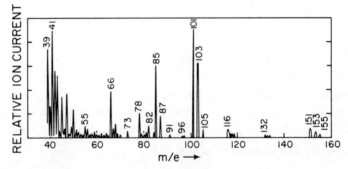

Fig. 9.12

Production

Vacuum pumps are often classified according to the physical or chemical phenomena responsible for their operation. In practice this is a bit awkward because some pumps combine two or more principles to pump a wide range of gases or to pump over a wide pressure range. In this section the discussion of pump operation is divided into six chapters. Chapter 10 discusses mechanical vacuum pumps. Rotary vane, rotary piston, and Roots pumps operate by displacing gas from the work chamber to the pump exhaust. Rotary vane and piston pumps operate in the low vacuum region, whereas the Roots pump operates in the medium vacuum region. The turbomolecular pump is a mechanical high vacuum pump that is the subject of Chapter 11. It transports gas from regions of low pressure to high pressure by momentum transfer from high speed blades.

Chapter 12 is devoted to the diffusion pump, which, like the turbomolecular pump, is a momentum transfer pump. The diffusion pump has been the mainstay of the vacuum industry for many applications and is now being supplanted by cryogenic and turbomolecular pumps for many applications.

Vacuum pump fluids are common to all of the above pumps. In Chapter 13, we describe the properties of the fluids needed for correct operation of each pump.

Capture pumps, or entrainment pumps, bind particles to a surface instead of expelling them to the atmosphere. Chapter 14 describes getter and ion pumps. Getter pumps, like the titanium sublimator, remove gases by chemical reactions that form solid compounds; ion pumps ionize gas molecules and imbed them in a wall. The sputter-ion pump combines gettering and ion pumping. Other entrainment pumps are based on condensation and sorption. Sorption pumps physisorb gas molecules in materials of high surface area. These surfaces are usually cooled to enhance their pumping ability. Another capture pump, the cryogenic pump, is the subject of Chapter 15. A cryogenic pump uses at least two stages of cooling. The warmer stage pumps by condensation or adsorption on a cooled metallic surface; the colder stage uses in addition an adsorbate such as charcoal or zeolite. Shown below are the operating pressure ranges of many pumps and pump combinations.

These three chapters do not cover all of the many techniques by which high vacuum may be achieved. Their purpose is to review the operation of commonly used pumps in a concise manner and to amplify the treatment of turbomolecular and cryogenic pumps. The latter warrant an expanded treatment in view of the rapidity with which they are replacing older pumps in most applications.

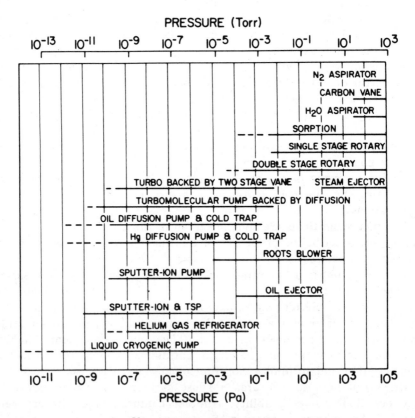

Vacuum pump operating ranges

CHAPTER 10

Mechanical Pumps

In this chapter we review the operation of two low vacuum pumps, the rotary vane and piston, and one medium vacuum pump, the lobe blower or Roots pump. Vane and piston pumps are widely used for backing high vacuum pumps and initial chamber evacuation. Lobe blowers are used with piston or vane pumps to rough large systems, back large pumps, and to pump large quantities of gas in plasma processing systems.

10.1 ROTARY VANE PUMPS

The rotary piston pump and the rotary vane pump are two oil-sealed pumps commercially available for pumping gas in the pressure range of 1 to 10^5 Pa. Of the two, the rotary vane is the most commonly used in small to medium sized vacuum systems. Rotary vane pumps of 10 to 200 m^3/h displacement are used for rough pumping and for backing diffusion or turbomolecular pumps.

In a rotary vane pump (Fig. 10.1), gas enters the suction chamber (A) and is compressed by the rotor (3) and vane (5) in region B and expelled to the atmosphere through the discharge valve (8) and the fluid above the valve. An airtight seal is made by one or more spring or centrifugally loaded vanes and the closely spaced sealing surfaces (10). The vanes and the surfaces between the rotor and housing are sealed by the low vapor pressure fluid, which also serves to lubricate the pump and fill the volume above the discharge valve. Pumps that use a speed reduction pulley operate in the 400-to-600-rpm range, while direct-drive pumps operate at speeds of 1500 to 1725 rpm. The fluid temperature is considerably higher in the direct-drive pumps than in the low speed pumps, typically

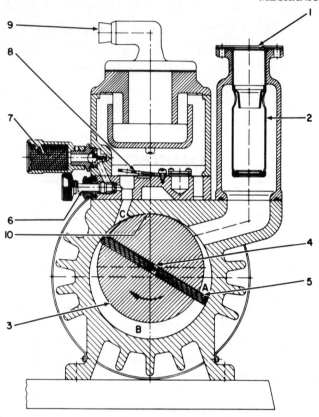

Fig. 10.1 Sectional view of the Pfeiffer DUO-35, 35-m^3/h double-stage, rotary vane pump: (1) intake, (2) filter, (3) rotor, (4) spring, (5) vane, (6) gas ballast valve, (7) filter, (8) discharge valve, (9) exhaust, (10) sealing surface. Reprinted with permission from A. Pfeiffer Vakuumtechnik, G.m.b.H., Wetzlar, West Germany.

80 and 60°C, respectively. These values will vary with the viscosity of the fluid and the quantity of air being pumped.

Single-stage pumps consist of one rotor and stator block (Fig. 10.1). If a second stage is added, as shown schematically in Fig. 10.2, by connecting the exhaust of the first stage to the intake of the second, lower pressures may be reached. The ultimate pressure at the inlet of the second stage is lower than at the inlet to the first because the fluid circulating in the second stage is rather isolated from that circulating in the second stage. The fluid in the second stage contains less gas than the fluid in the reservoir. Physically, the second pumping stage is located adjacent to the first and on the same shaft. The pumping speed characteristics of single-stage and two-stage rotary vane pumps are shown in Fig. 10.3. The free-air displacement and the ultimate pressure are two measures of the performance of roughing pumps. The free-air displace-

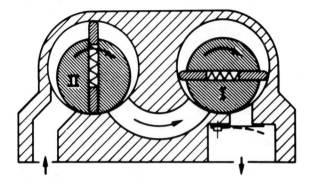

Fig. 10.2   Schematic section through a two-stage rotary pump. Reprinted with permission from *Vacuum Technology*, Leybold-Heraeus, G.m.b.H., Postfach 51 07 60, 5000 Koln, West Germany.

ment is the volume of air displaced per unit time by the pump at atmospheric pressure with no pressure differential. For the two pumps whose pumping speed curves are shown in Fig. 10.3 this has the value of 30 m³/h (17.7 cfm) at a pressure of 10^5 Pa (1 atm). At the ultimate pressure of the blanked-off pump, the speed drops to zero because of dissolved gas in the fluid, leakage around the seals, and trapped gas in the volume below the valve. Rotary vane pumps have ultimate pressures in the 3×10^{-3} to 1 Pa range; the lowest ultimate pressures are achieved with two-stage pumps. The single- and two-stage pumps characterized in Fig. 10.3 have ultimate pressures of 1.4 and 1.5×10^{-2} Pa, respectively. These ultimate pressures are obtained in a new pump with clean, low

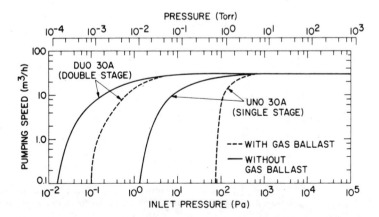

Fig. 10.3   Pumping speed curves for the Pfeiffer UNO 30A and DUO 30A rotary vane pumps. Reprinted with permission from A. Pfeiffer Vakuumtechnik, G.m.b.H., Wetzlar, West Germany.

vapor pressure fluid. As the fluid becomes contaminated and the parts wear, the ultimate pressure will increase.

Kendall [1] demonstrated the effect of dissolved gas in the fluid on the ultimate pressure of a two-stage rotary vane pump. He showed a rotary vane pump could reach an ultimate pressure of 4×10^{-5} Pa when the fluid reservoir was exhausted by another pump. Figure 10.4 shows the effect of prolonged outgassing and the effect of admitting CO_2 to the fluid reservoir after degassing the fluid.

When large amounts of water, acetone, or other condensible vapors are being pumped, condensation occurs during the compression stage after the vapor has been isolated from the intake valve. As the vapor is compressed, it reaches its condensation pressure, condenses, and contaminates the fluid before the exhaust valve opens. Condensation causes a reduction in the number of molecules in the vapor phase and delays or even prevents the opening of the exhaust valve. If condensation is not prevented, the pump will become contaminated, the ultimate pressure will increase, and gum deposits will form on the moving parts. Some compounds will eventually cause the pump to seize. To avoid condensation and its resulting problems, gas is admitted through the ballast valve. The open valve allows ballast, usually room air, to enter the chamber during the compression stage; the trapped volume is isolated from the intake and exhaust valves. This inflow of gas, which can be as much as 10% of

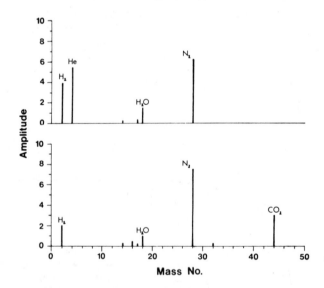

Fig. 10.4 Relative abundances of gases at pump ultimate: (a) After prolonged outgassing of oil and with zero pressure at the exhaust port. Helium pressurization applied intermittently to provide lubricating oil pressure ($P = 7 \times 10^{-7}$ Torr, 9.3×10^{-5} Pa). (b) After oil had been exposed to carbon dioxide and then exhaust pressure reduced to zero ($P = 3 \times 10^{-7}$ Torr, 4×10^{-5} Pa). Reprinted with permission from *J. Vac. Sci Technol.*, **21**, p. 886, B. R. F. Kendall. Copyright 1982, The American Vacuum Society.

the pump displacement, is controlled by valve 6 (Fig. 10.1). The added gas causes the discharge valve to open before it reaches the condensation partial pressure of the vapor. In this manner the vapor is swept out of the pump and no condensation occurs. The ultimate pressure of a gas ballasted pump is not very low with the gas ballast valve open. Figure 10.3 shows the effect of full gas ballast on the performance of a single- and double-stage pump.

Gas ballasting can be used to differentiate contaminated fluid from a leak. If the inlet pressure drops when the ballast valve is opened but drifts upward slowly after the valve is closed, the fluid is contaminated with a high vapor pressure impurity. More details of gas ballasting are covered in the Leybold-Heraeus reference [2], and Van Atta [3] describes alternative methods of pumping large amounts of water.

10.2   ROTARY PISTON PUMPS

Rotary piston pumps are used as roughing pumps on large systems alone or in combination with lobe blowers. They are manufactured in sizes ranging from 30 to 1500 m³/h. A piston pump is a rugged and mechanically simple pump. There are no spring-loaded vanes to stick in a piston pump and all parts are mechanically coupled to a shaft that can be powered by a large motor.

Figure 10.5 shows a sectional view of a rotary piston pump. As the keyed shaft rotates the eccentric (1) and piston (2), gas is drawn into the space A. After one revolution that volume of gas has been isolated from the inlet, while the piston is closest to the hinge box. During the next revolution the isolated volume of gas (B) is compressed and vented to the exhaust through the poppet valve when its pressure exceeds that of the valve spring. Like the vane pump, the piston pump is manufactured in single and compound or multistage types.

The clearance between the piston and housing is typically 0.1 mm, but it is three or four times larger near the hinge box. Because the clearance between moving and fixed parts is greater in a piston than in a vane pump, the piston pump is more tolerant of particulate contamination. A lubricating fluid is used to seal and lubricate the spaces between fixed and moving parts. As in a vane pump, the fluid must have low vapor pressure and good lubricating ability. A rather viscous fluid is used in the piston pump.

The rotational speed of the piston pump is typically 400 to 600 rpm, although some run as slow as 300 rpm and others as fast as 1200 rpm. The maximum rotational speed of a piston pump is limited by vibration from the eccentric. Small piston pumps are air cooled in the same manner as rotary vane pumps. Large pumps are water cooled.

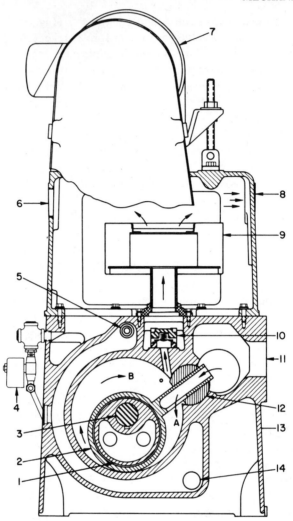

Fig. 10.5 Sectional view of a Stokes 212H, 255-m^3/h rotary piston pump: (1) eccentric, (2) piston, (3) shaft, (4) gas ballast, (5) cooling water inlet, (6) optional exhaust, (7) motor, (8) exhaust, (9) oil mist separator, (10) poppet valve, (11) inlet, (12) hinge gar, (13) casing, (14) cooling water outlet. Reprinted with permission from Stokes Division, Pennwalt Corp., Philadelphia, PA.

The pumping speed curves for a 51-m^3/h single-stage rotary piston pump are shown in Fig. 10.6 with and without gas ballast. The shaft power is also given and it is seen to peak at a pressure of 4 × 10^4 Pa and is independent of ballast. At lower pressures operation of the pump with full gas ballast requires more than twice the shaft power as without gas ballast. The ultimate pressure of the single-stage pump shown in Fig.

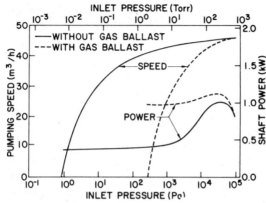

Fig. 10.6 Pumping speed and shaft power for the Stokes 146H, 51-m^3/h rotary piston pump. Reprinted with permission from Stokes Division, Pennwalt Corp., Philadelphia, PA.

10.6 is 1 Pa.   Compound pumps are available with ultimate pressures approaching those of rotary vane pumps.

10.3   LOBE PUMPS

Positive displacement blowers, or lobe blowers, are used in series with rotary fluid-sealed pumps to achieve higher speeds and lower ultimate pressures in the medium vacuum region than can be obtained with a rotary mechanical pump alone.   Lobe blowers, or Roots pumps, consist of two lobed rotors mounted on parallel shafts.   The rotors have substantial clearances between themselves and the housing—typically about 0.2 mm. They rotate in synchronism in opposite directions at speeds of 3000 to 3500 rpm.   These speeds are possible because a fluid is not used to seal

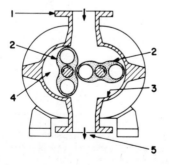

Fig. 10.7   Section through a single-stage lobe blower: (1) inlet, (2) rotors, (3) housing, (4) pump chamber (swept volume), (5) outlet.   Reprinted with permission from *Vacuum Technology,* Leybold-Heraeus G.m.b.H., Köln, West Germany.

the gaps between the rotors and the pump housing. A sectional view of a single-stage lobe blower is given in Fig. 10.7.

The compression ratio, or ratio of outlet pressure to inlet pressure, is pressure dependent and usually has a maximum near 100 Pa. At higher pressures the compression ratio should, theoretically, remain constant. In practice it decreases. Outgassing and the roughness of the rotor surfaces contribute to compression loss at low pressures. Each time the rotor surface faces the high pressure side, it sorbs gas. Some of this gas is released when the rotor faces the low pressure side. The compression ratio K_{omax} for air for a single-stage Lobe blower of 500 m³/h displacement is shown in Fig. 10.8. It has a maximum compression ratio of 44. Large pumps tend to have a larger compression ratio than small pumps, because they have a smaller ratio of gap spacing to pump volume. The compression ratio for a light gas such as helium is about 15 to 20% smaller than the ratio for air. The compression ratio K_{omax} is a static quantity and is measured under conditions of zero flow. The inlet side of the pump is sealed and a pressure gauge is attached. The outlet side is connected to a roughing pump and the system is evacuated. Gas is admitted to the backing line that connects the blower to the roughing pump. The backing pressure P_b is measured at the blower outlet, and the pressure P_i is measured at the inlet. The compression ratio is given by P_b/P_i.

Considerable heat is generated by pumping gas at high pressures with a lobe blower. The heat causes the rotors to expand. If unchecked, rotor expansion could destroy the pump. To avoid overheating, a maximum pressure difference between the inlet and outlet of a lobe blower is specified. This maximum pressure difference is typically 1000 Pa, but that value may be exceeded for a short time without harm to the pump.

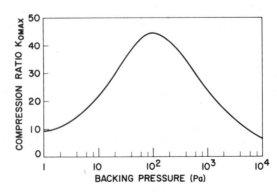

Fig. 10.8 Dependence of the air compression ratio $K_{o\ max}$ of the Leybold WS500 lobe pump on the backing pressure. Values for helium are about 20% smaller. Reprinted with permission from Leybold-Heraeus G.m.b.H., Postfach 51 07 60, 5000 Köln, West Germany.

Lobe blowers are connected as compression or transport pumps to avoid heat generation.

In compression pumping, the common method, a lobe pump is placed in series with a rotary pump whose rated speed is 2 to 10 times smaller than its own speed. When pumping is initiated at atmospheric pressure, a bypass line around the lobe pump is opened, or the pump is allowed to freewheel. All the pumping is done by the rotary pump until the backing pressure is below the manufacturer's recommended pressure difference, at which time the lobe blower is activated and the bypass valve is closed. Some lobe blowers have this bypass feature built into the pump housing. The net speed of a lobe blower of 500-m³/h capacity backed by a 100-m³/h rotary piston pump is shown in Figure 10.9. The speed curve for the mechanical pump alone is shown for comparison. Such lobe blower–rotary pump combinations are often used when speeds of 170 m³/h or greater are required because the combination costs less than a rotary pump of similar capacity.

The second method, transport pumping, used a lobe blower in series with a rotary pump of the same displacement. Figure 10.9 shows the pumping speed of a 60-m³/h lobe blower backed by a 60-m³/h rotary vane pump. The pumping speed of the rotary vane pump is shown for comparison. Both pumps are started simultaneously at atmospheric pressure because the critical pressure drop will never be exceeded.

Detailed calculations of the effective pumping speed of the lobe blower have been carried out by Van Atta [3]. Here we give approximate

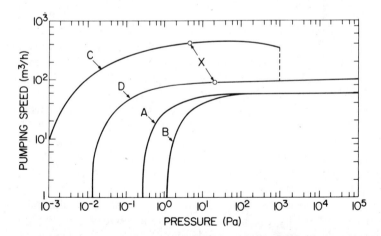

Fig. 10.9 Lobe blower–rotary pump combinations. Transport mode: (a) Leybold RUTA 60 lobe blower and S60 rotary vane; (b) S60 only. Compression mode: (c) Leybold WS500 lobe blower and DK100 rotary piston pump: (d) DK100 only. Reprinted with permission from *Vacuum Technology*, Leybold-Heraeus, G.m.b.H., Köln, West Germany.

formulas for the inlet pressure P_i and the inlet speed S_i [2]:

$$P_i = P_b\left(\frac{1}{K_{omax}} + \frac{S_b}{S_D}\right) \qquad \blacktriangleright (10.1)$$

$$S_i = \frac{S_b S_D K_{omax}}{S_D + S_b K_{omax}} \qquad \blacktriangleright (10.2)$$

where all terms have been defined except S_D. It is the pump displacement, or speed of the lobe blower, at atmospheric pressure. With these approximate equations, the pumping speed curve for the rotary pump, the compression ratio K_{omax}, and the lobe blower displacement, a curve of the speed of the lobe pump versus inlet pressure can be calculated. The line marked X in Fig. 10.9 shows the result of applying (10.1) and (10.2) to calculate one point of this curve. In this example the inlet pressure P_i and the inlet speed S_i were calculated for a backing pressure of 20 Pa. At 20 Pa, $K_{omax} = 30$ (Fig. 10.8) and $S_b = 90$ m³/h (Fig. 10.9d). With (10.1), (10.2), and a displacement of 500 m³/h, we get $P_i = 4.3$ Pa and $S_i = 422$ m³/h.

Lobe blowers are frequently used to back large diffusion or turbomolecular pumps. For example a 35-inch-diameter diffusion pump used to evacuate a 2-m³ chamber is backed by a series combination of a 1300-m³/h lobe blower and a 170-m³/h rotary piston pump.

10.4 MECHANICAL PUMP OPERATION

There are several good rules for operating rotary mechanical pumps. The exhaust should be vented outside the building. Most pumps are supplied with an oil mist separator, but it does not adequately remove all the vapors. In many laboratories and plants safety rules require the use of an outside vent. The vent hose should not run vertically from the exhaust connection because water or other vapors which have condensed on the cooler hose walls will flow into the pump and contaminate the fluid. A satisfactory solution to this problem is the addition of a sump at the exhaust connection to collect the vapors before they can flow into the pump. A vane pump must also be vented at the time it is stopped to prevent fluid from being forced back into the vacuum system by external air pressure. Venting is done automatically in some pumps and can be achieved in others by the addition of a vent valve above the inlet port. The fluid level in mechanical pumps should be checked frequently, especially those that are used on systems regularly cycled to atmosphere. Small pumps of capacity less than 30 m³/h have fluid consumption rates (cm³/h) of about 10^{-6} to $10^{-5}PS$ where P is the inlet pressure and S is the inlet speed in m³/h [4]. Larger pumps will use more fluid. The fluid should be changed when the pump performance deteriorates or when it becomes discolored or contaminated with particulates. Poor fluid maintenance is the major cause of mechanical pump failure. Ninety-five percent of all pump problems can be solved by flushing the pump and changing the fluid.

A discussion of mechanical pump fluids is given in Chapter 13. Vapor pressures and kinematic viscosities of mechanical pump fluids are given in Appendixes F.2 and F.4.

REFERENCES

1. B. R. F. Kendall, *J. Vac. Sci. Technol.*, **21**, 886 (1982).
2. Leybold-Heraeus Publication HU152, Leybold-Heraeus, G.m.b.H., Köln, West Germany.
3. C. M. Van Atta, *Vacuum Science and Engineering*, McGraw-Hill, New York, 1965, Chapter 5.
4. Reference 2, p. H-B 61.

PROBLEMS

10.1 †A simple mechanical piston has a displacement of 1 L. It is connected to a chamber of 10L. If pumping commences at 1 atm, what is the pressure in pascals after four complete strokes?

10.2 Define free-air displacement. Give a formula for the free-air displacement S_D in terms of rotational speed n and swept volume V of (a) a piston pump and (b) a two-stage vane pump.

10.3 Rank the following pumps in order of increasing physical size for equal displacements: (a) rotary vane, two-vane, belt drive; (b) rotary piston pump; (c) rotary vane, two-vane, direct drive.

10.4 †What are three functions of the fluid in a rotary vane or piston pump?

10.5 Plot throughput vs. pressure for the single-stage and two-stage mechanical pumps whose no ballast pumping speed curves are given in Fig. 10.3.

10.6 †What is the hazard in discharging mechanical pump exhaust fumes into the work area? What will happen if the exhaust hose forms a loop like a sink trap before being connected to the exhaust plenum?

10.7 †What will happen if the mechanical pump exhaust hose runs vertically in a direct line from the pump to a connection in the exhaust plenum?

10.8 †What is the most important step in ensuring long, trouble-free operation of a rotary pump?

10.9 The piston pump whose speed characteristic is shown in Fig. 10.9*d* is connected to a chamber by a 400-cm length of 4-cm-diameter pipe. When the pressure at the inlet of the pump is 1 Pa, what is the pressure in the chamber?

10.10 What is the staging ratio of a lobe-piston pump set? What performance differences would you expect between a set with a staging ratio of 10:1 and one of 2:1, provided the two sets use the same mechanical pump?

CHAPTER 11

Turbomolecular Pumps

The axial-flow turbine, or turbomolecular pump as it is known, was introduced in 1958 by Becker [1]. His design originated from a baffling idea with which he had experimented a few years earlier—a disk with rotating blades mounted above a diffusion pump [2]. When it was introduced commercially, the pump had low speed and high cost, as compared to a diffusion pump. It did not, however, backstream hydrocarbons and did not require a trap of any kind. Since its introduction, the turbomolecular pump has undergone rapid development both theoretically and experimentally. The most important theoretical development during this period was the work on blade geometry at MIT in the group headed by Shapiro [3,4]. Many practical advances in lubrication, drive motors, and fabrication techniques have also taken place. Modern turbomolecular pumps have high pumping speeds, large hydrogen compression ratios, and low ultimate pressures. They do not backstream hydrocarbons from the lubricating fluid or mechanical pump and are well suited to pump gas cleanly at high flow rates or low pressures.

This chapter reviews the pumping mechanism in the free molecular pressure range, and the relations between pumping speed, compression ratio, backing pump size and gas flow. The differences between vertical and horizontal rotor designs and the problems concerned with bearings are also discussed. Lubrication techniques are discussed in Chapter 18. The operation and performance of pumps in high vacuum, ultrahigh vacuum and high gas flow are discussed in Chapters 19 to 21.

11.1 PUMPING MECHANISM

The turbomolecular pump is a molecular turbine that compresses gas by momentum transfer from the high speed rotating blades to the gas molecules. It is a high speed molecular bat.   The pumps operate at rotor speeds ranging from 24,000 to 60,000 rpm and are driven by solid state power supplies or motor-generator sets.   The relative velocity between the alternate slotted rotating blades and slotted stator blades makes it probable that a gas molecule will be transported from the pump inlet to the pump outlet.   Each blade is able to support a pressure difference. Because this compression ratio or pressure ratio is small for a single stage, many stages are cascaded.   For a series of stages the compression ratio for zero flow is approximately the product of the compression ratios for each stage.   Figure 11.1 shows a sectional view of a dual-rotor, horizontal-axis, turbomolecular pump.   The blades impart momentum to the gas molecules most efficiently in the molecular flow region; therefore this pump, like a diffusion pump, must be backed by a mechanical pump.

If the foreline pressure is allowed to increase to a point at which the rear blades are in viscous flow, the rotor will be subjected to an additional torque because of viscous drag.   The power required to rotate a shaft in steady state is proportional to the product of the rotor speed and the torque.   The power in some pumps is limited by the supply, so that too large an increase in foreline pressure will cause a sudden reduction in the

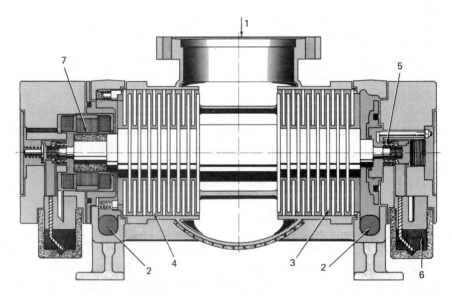

Fig. 11.1   Section view of Pfeiffer TPU-200 turbomolecular pump: (1) inlet, (2) outlet, (3) rotor disk, (4) stator disk, (5) bearing, (6) oil reservoir, (7) motor. Reprinted with permission from A. Pfeiffer Vakuumtechnik, G.m.b.H., Wetzlar, West Germany.

rotor speed and a loss in gas pumping speed. Another design uses a constant–speed motor whose power consumption increases in proportion to the gas load. This design also exhibits a loss of pumping speed as the forechamber enters the viscous flow region. The explanation of pumping speed in the latter case is not clear. It may be related to a conductance limit. Other effects of backing pump size and some rules for selecting backing pumps are discussed later in this section.

11.2 SPEED-COMPRESSION RELATIONSHIPS

Continuum methods that give a reasonable account of pump performance have been developed [2], however they are not discussed. A straightforward way of characterizing the turbomolecular pump has been worked out by Kruger and Shapiro [3-5] and is presented here. This method, which uses probabilistic techniques to calculate the properties of single- and multiple-blade arrays, is valid for all compression ratios. The model used to analyze a single rotor disk is shown in Fig. 11.2. This disk, which rotated with a tip velocity near the thermal velocity of air, imparts a directed momentum to a gas molecule on collision. The blades are slotted at an angle to make the probability of a gas molecule being transmitted from the inlet to the outlet much greater than in the reverse direction. The stator disks are slotted in the opposite direction. Here, Γ_1 and Γ_2 are, respectively, the number of molecules incident on the disk per unit time at the inlet and at the outlet. a_{12} is the fraction of Γ_1 transmitted from the inlet (1) to the outlet (2) and a_{21} is the fraction of Γ_2 transmitted from the outlet to the inlet. Now define the net flux of molecules through the blades to be a function of the Ho coefficient W. Recall from Chapter 7 the Ho coefficient is the ratio of net throughflux to incident flux. In steady state this is

$$\Gamma_1 W = \Gamma_1 a_{12} - \Gamma_2 a_{21} \qquad (11.1)$$

or

$$\frac{\Gamma_2}{\Gamma_1} = \frac{a_{12}}{a_{21}} - \frac{W}{a_{21}} \qquad (11.2)$$

If the gas temperature and the velocity distributions are the same everywhere, the ratio Γ_1/Γ_2 will be equal to the pressure ratio P_1/P_2. The ratio of outlet to inlet pressure is called the compression ratio K:

$$\frac{P_2}{P_1} = K = \frac{a_{12}}{a_{21}} - \frac{W}{a_{21}} \qquad (11.3)$$

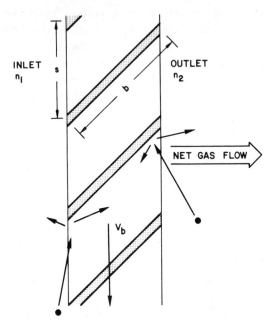

Fig. 11.2 Sectional view of a flat-bladed disk.

If a_{12} and a_{21} are independent of W, the compression ratio will vary in a linear way with the net throughput. Maximum compression occurs at zero flow, while unity compression occurs at maximum speed or mass flow. This is a general property of a fan. Envision the compression and flow in a household vacuum cleaner as you hold your hand over the inlet (zero flow) and release it (maximum flow). The maximum compression, maximum flow, and the general case for the region between these extremes are three regions of (11.3) which we now examine in more detail.

11.2.1 Maximum Compression Ratio

For no gas flow $W = 0$, and (11.3) reduces to

$$K = K_{max} = \frac{a_{12}}{a_{21}} \qquad \blacktriangleright (11.4)$$

Equation (11.4) states the maximum compression ratio is the ratio of forward to reverse transmission probabilities. To maximize the compression ratio the ratio a_{12} to a_{21} is maximized. The important problem solved by Kruger and Shapiro was the calculation of the forward and reverse transmission probabilities by Monte Carlo techniques. They solved for the transmission probabilities as a function of the blade angle

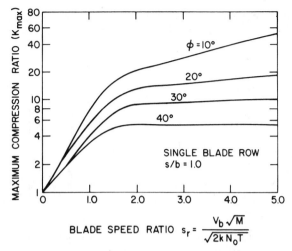

Fig. 11.3 Calculated curve of the compression ratio at zero flow for a single blade row with $s/b = 1$. Reprinted with permission from *Trans. 7th Nat. Vac. Symp.* (1960), p. 6, C. H. Kruger and A. H. Shapiro. Copyright 1961, Pergamon Press.

ϕ, the blade spacing-to-chord ratio s/b, and the blade speed ratio $s_r = V_b (M/2kTN_oT)^{1/2}$. Figure 11.3 sketches the results of a calculation for the single-stage compression ratio at zero flow [4]. From this curve we observe the logarithm of the compression ratio is approximately linear with blade speed ratio for $s_r \le 1.5$, or

$$K_{max} \propto \exp\left[\frac{V_b(M)^{1/2}}{(2kN_oT)^{1/2}}\right] \qquad \blacktriangleright (11.5)$$

The compression ratio is exponentially dependent on rotor speed and $M^{1/2}$. The constant of proportionality is dependent on the blade angle and s/b. In particular hydrogen will have a compression ratio much smaller than for any other gas. For a blade tip velocity of 400 m/s the speed ratio for argon is about unity, and for hydrogen, it is about 0.3. From Fig. 11.3 for $\phi = 30°$ we find this blade velocity corresponds to compression ratios of $K(H_2) = 1.6$ and $K(Ar) = 4$. If two disks (five rotors and five stators) are cascaded, the net compression ratios are approximately 100 and 10^6, respectively. A total of 15 disks would raise $K(H_2)$ to 1000. A stator blade has the same compression ratio and transmission as a rotor. An observer sitting on a stator sees blades moving with the same relative velocity as an observer sitting on a rotor. The linear blade velocity is proportional to the radius and the rotor angular frequency ($V_b = r\omega$). An area closer to the center of the rotor will have a smaller speed ratio and blade spacing-to-chord ratio. The data of Kruger and Shapiro show the net effect of these changes is a low

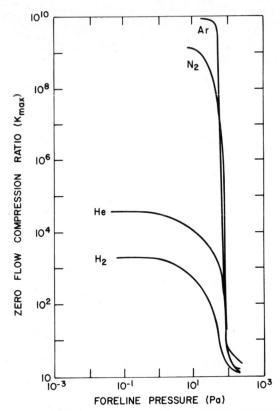

Fig. 11.4 Measured compression ratio for zero flow in a Pfeiffer TPU-400 turbomolecular pump. Reprinted with permission from A. Pfeiffer Vakuumtechnik, G.m.b.H., Wetzlar, West Germany.

compression ratio for the region closest to the rotor axis. Blades designed to have a high $K(H_2)$ should be slotted to a depth of only about 30% of the radius.

Experimental compression ratios are given for a horizontal-axis, dual-rotor pump in Fig. 11.4. These data were taken in a manner identical to lobe blower compression curves, or diffusion pump forepressure tolerance curves. Gas is admitted to the foreline of a blanked-off pump and the measured compression ratio is the ratio of forepressure to inlet pressure. As the foreline or backing line pressure is increased, the rear blades first go into transition flow and then into viscous flow, the rotor speed decreases, and the compression ratio decreases.

11.2.2   Maximum Speed

Maximum speed is achieved when the compression ratio across a blade is unity; that is, when

$$K = 1 = \frac{a_{12}}{a_{21}} - \frac{W}{a_{21}} \qquad (11.6)$$

or

$$W_{max} = a_{12} - a_{21} \qquad \blacktriangleright(11.7)$$

To maximize W the absolute value of $a_{12} - a_{21}$ is maximized. The Ho coefficient W for a single blade is given in Fig. 11.5 as a function of blade-speed ratio for a spacing-to-chord ratio $s/b = 1$ [4]. For $s_r \leq 1.5$, W is approximately linear with s_r,

$$W \propto \left[\frac{V_b(M)^{1/2}}{(2kN_oT)^{1/2}} \right] \qquad \blacktriangleright(11.8)$$

Because the molecular arrival rate is proportional to thermal velocity $(kT/m)^{1/2}$, the net pumping speed of the place is approximately independent of the mass of the impinging molecules, therefore

$$S \propto V_b \qquad (11.9)$$

Close examination of Fig. 11.5 and others of different spacing-to-chord ratios [3] reveals the curves to be sublinear. The pumping speed for light gases will actually be slightly greater than for heavy gases. Approximations methods have been developed for calculating the net Ho coefficient for a series of blades [3]. These calculations show the Ho coefficient to

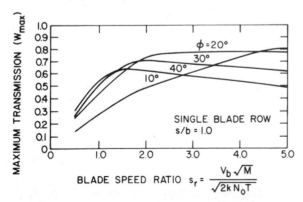

Fig. 11.5 Calculated curve of the Ho coefficient at unity compression ratio for a single blade row with $s/b = 1$. Reprinted with permission from *7th Nat. Vac. Symp. (1960)*, p. 6, C. H. Kruger and A. H. Shapiro. Copyright 1961, Pergamon Press.

increase with the number of stages [6], with a saturation in speed after several blades are operated in series.

11.2.3 General Relation

The maximum compression ratio (11.4) occurs only when the pump is pumping no gas (e.g., at its ultimate pressure), while maximum speed occurs when the pressure drop is zero—a condition that is never reached in actual pump operation. An operating pump works in the region between these two extremes. Equation (11.3) describes the relative speed or Ho coefficient versus pressure for a series of stages and for a single stage provided the proper transfer coefficients are used for the series combination of disks. Equation (11.3) states P_1/P_2 varies linearly between K_{max} and unity, as W is varied linearly between zero and W_{max}. If we plot W versus P_2/K on a semilog scale (Fig. 11.6) the dependence of W on K looks more like the familiar pumping speed curve. The speed goes to zero at the ultimate pressure (P_1/K_{max}) and approaches a constant value at high inlet pressures. In practice W_{max} is not reached, but only asymptotically approached, because the rear blades go into viscous flow at pressures of 1 to 10 Pa and the drag reduces the pump speed. Pumping speed curves for a nominal 400-L/s pump are shown in Fig. 11.7. At a nitrogen inlet pressure of 0.9 Pa the rotor is still running at full speed and the gas throughput is 400 Pa-L/s. This is twice the throughput of a 6-in. diffusion pump operating at a throat pressure of 0.1 Pa. A different view of (11.3) is obtained if we replace the variables a_{12} and a_{21} with K_{max} and W_{max} from (11.4) and (11.7)

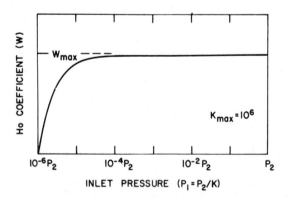

Fig. 11.6 Calculated dependence of Ho coefficient on inlet pressure for $K_{max} = 10^6$.

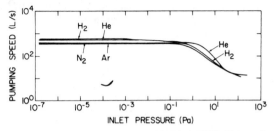

Fig. 11.7 Measured pumping speeds for the Pfeiffer TPU-400 turbomolecular pump. Reprinted with permission from A. Pfeiffer Vakuumtechnik, G.m.b.H., Wetzlar, West Germany

respectively. The result is

$$\frac{W}{W_{max}} = \frac{(1 - K/K_{max})}{(1 - 1/K_{max})} \tag{11.10}$$

For any reasonable K_{max} this reduces to

$$\frac{W}{W_{max}} = \left(1 - \frac{K}{K_{max}}\right) \qquad \blacktriangleright (11.11)$$

Equation (11.11) is plotted in Fig. 11.8 for several values of K in the range $10 \leq K \leq 500$. The compression ratio $K = S_1/S_2 = P_2/P_1$ will be between 50 to 100 for the usual combinations of turbomolecular and forepump. For example, a 500-L/s turbomolecular pump will be backed by a 5- to 10-L/s mechanical pump. In Fig. 11.8 we see the Ho coefficient and the pumping speed will each be at their maximum values if $K_{max} \geq 10^6$. If, however, $K_{max} < 500$, as it is for hydrogen in some pumps, the hydrogen pumping speed will be a function of the forepump speed. A large forepump will be required to make $S/S_{max} = 1$ for hydrogen in a pump with a small K_{max} (H$_2$). A staging ratio (S_2/S_1) of 15:1 to 20:1 is necessary for pumps with K_{max} (H$_2$) ≤ 500 to pump hydrogen at maximum

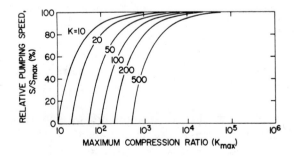

Fig. 11.8   Calculated dependence of relative pumping speed on K and K_{max}.

speed. For large pumps a lobe blower-rotary pump combination is the most economical backing pump system. Pumps with K_{max} (H_2)>1500 will not exhibit a backing pump dependent hydrogen pumping speed for staging ratios of the order 60:1 or less.

Equations (11.10) and (11.11) are not unique to a turbomolecular pump. They are true for any pump that satisfies the conditions of (11.1), that is, in which the reverse leakage is proportional to the pressure difference. These are the equations used to describe the inlet pressure-speed characteristics of a lobe blower-rotary pump combination, (10.1) and (10.2). For a diffusion pump the solutions to these equations are simplified. The maximum compression ratio for hydrogen is high enough (10^5 to 10^6), so that $S = S_{max}$ for all gases regardless of the size of the forepump when it is large enough to keep the foreline below the critical forepressure.

Figure 11.9 shows how the various turbomolecular pump and forepump parameters affect turbomolecular pumping speed. The ultimate pressure, assuming no outgassing, is determined by the forepressure P_2 and the maximum compression ratio K_{max}. The pumping speed in the plateau region is determined by S_{max} for heavy gases, while for light gases it is determined by the maximum compression ratio and the size of the forepump. The pressure at which the constant-throughput region begins is determined by the ratio S_2/S_{max}.

11.3   ULTIMATE PRESSURE

The ultimate pressure of a turbomolecular pump is determined by the compression ratio for light gases and by the amount of outgassing. This is qualitatively similar to a diffusion pump. See discussion Section 12.2. The main difference between the turbomolecular pump and the diffusion pump is the low hydrogen compression ratio in the turbo pump. It is low enough so that the ultimate hydrogen pressure will be determined by K_{max} (H_2) and its partial pressure in the foreline. In some older pumps which have a water-vapor compression ratio of less than 10^4, the water-vapor partial pressure may also be compression ratio limited [7]. The partial pressure of all other gases and vapors will be limited by their respective outgassing rates and pumping speeds. During pumping of an unbaked turbomolecular pump, the slow release of water vapor from the blades closest to the inlet may slightly decrease the rate of water removal. The effective compression ratio for water release from the first few blades is much less than K_{max}.

Henning [8] has shown the partial pressure of hydrogen found in a turbomolecular pump system is dominated by the forepump oil. The ultimate pressure varied from 2×10^{-7} to 5×10^{-7} Pa as a function of the type of oil in the forepump. Because the turbopump oil has a lower

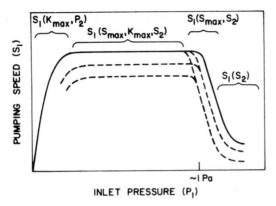

Fig. 11.9 Parameters that control the pumping speed of a turbomolecular pump in its four regions of operation.

vapor pressure than the mechanical pump oil, it will not contribute so much hydrogen to the background as the forepump oil. Henning and Lotz [9] used perfluoropolyether pump fluid for lubricating both the turbomolecular pump and the forepump in the presence of corrosive gases. Using a mass spectrometer they observed distinct fluorine and hydrogen peaks. This decomposition occurred because the local heating of the bearings caused the oil temperature to exceed the range of thermal stability. They concluded from the presence of hydrogen that the ultimate pressure of the pump was not improved with hydrogen-free fluids. They postulated the limiting pressure was caused by the diffusion of hydrogen through the foreline seals.

Ultimate pressures for baked systems between 2×10^{-8} and 5×10^{-9} Pa are possible with high compression turbomolecular pumps without the assistance in pumping from cryobaffles or titanium sublimation pumps. Hydrogen will constitute more than 99% of the residual gas at the ultimate pressure [8].

11.4   DESIGN CONSIDERATIONS

A single blade is inadequate to serve as a high vacuum pump. Multiple-bladed structures with 8 to 20 disks will provide adequate compression and speed to make a functional pump. As in the diffusion pump, the stages nearest the high vacuum inlet serve a purpose different from those nearest the outlet. The flow through each stage is constant or, stated another way, the product of speed and pressure is a constant. The blades nearest the inlet are designed to have a high pumping speed and a low compression ratio. The blades nearest the foreline are designed to have a high compression ratio and a low pumping speed. For economic reasons it would be impractical to make each blade different from its neighbor. A compromise results in groups of two or three blade types. The blades in each group are designed for a specific speed and compression ratio. Each group of blades may be considered analogous to a diffusion pump jet. The pump designer may trade off pumping speed and light gas compression ratio by the proper choice of blade-to-chord ratio and blade angle. Pumps exhibiting an overall large compression ratio for hydrogen use blades that are optically more opaque (s/b, ϕsmall) than those designed to maximize the pumping speed (s/b, ϕlarge).

Figure 11.10 shows a view of the three-stage rotor used in a horizontal-axis, dual-rotor pump. In this design the rotors are individually abrasive-machined and balanced. The rotor disks are positioned on a cooled hub that is allowed to thermally equilibrate with the disks and hold them rigidly in position. Stator disks are formed in a similar manner, cut into half-sections, and mounted stage-by-stage as the rotor is moved into the housing. The top portion of the rotor from the vertical-axis, single-rotor pump shown in Fig. 11.11 was machined from a single block of aluminum. The blade rows were first machined from a single block of aluminum. Next, the individual blades were formed by lengthwise sawing. The desired blade angles were then obtained by twisting. The stators were constructed from stamping which were cut and twisted [6].

Modern turbomolecular pumps are constructed in vertical-axis, single-rotor or horizontal-axis, double-rotor styles. In either configuration the designer is free within limits of material stability to choose the number of

Fig. 11.10 Three-stage rotor from a Pfeiffer TPU-200 turbomolecular pump. Reprinted with permission from A. Pfeiffer Vakuumtechnik, G.m.b.H., Wetzlar, West Germany.

Fig. 11.11   Top stages of a rotor from   Leybold-Heraeus TMP-450 turbomolecular pump. Reprinted with permission from Leybold-Heraeus G.m.b.H., Postfach 51 07 60, 5000 Köln, West Germany.

stages, blade angles, spacings, and blade-to-chord ratios.   The horizontal-axis pump allows for a somewhat more stable bearing design than the vertical pump.   It is possible to optimize both designs for maximum speed or maximum compression.   The single-rotor, vertical-axis pump has little conductance loss between the inlet flange and the rotor, while the horizontal-axis design has a greater conductance loss but pumps from two sides.   Henning [7] estimates the pumping speed of a dual-rotor pump is more than 1.6 times a vertical pump of the same inlet diameter with all other factors constant.

Neither style should be subjected to a steady or transient twisting moment by using the inlet flange to bear the load of a heavy work chamber, especially a cantilever load, or the impulse of a heavy flange closure.   Improper loading can cause premature bearing failure.   All pumps should be suspended from the system by their inlet flanges.   The inlet flange should not be used as a mounting platform for a heavy system. This is usually not a problem for a vertical turbomolecular pump because it looks like a diffusion pump.

The practical upper rotational speed for the rotor is currently about 80,000 rpm. Commercially, the maximum rotational speed is 60,000 rpm, or a blade tip velocity of about 500 m/s. These limits are due to bearing tolerances, thermal coefficients of expansion, and material stress limits. Ball bearings are the component subjected to the greatest wear. Oil, either flowing or in a mist, is used to lubricate and cool the bearings. The oil is in turn either water or refrigeration cooled. Small diameter bearings are desired to increase the bearing lifetime. Some pumps use grease-packed bearings [10]. Air bearings [11] and magnetic bearings [12] with extremely low wear rates have been designed.

REFERENCES

1. W. Becker, *Vac. Tech.*, **7**, 149 (1958).
2. W. Becker, *Vac. Tech.*, **15**, 211 (1966).
3. C. H. Kruger and A. H. Shapiro, *Proc. 2nd. Int. Symp. Rarefied Gas Dynamics*, Berkeley, CA, L. Talbot, Ed., Academic, New York, 1961, pp. 117-140.
4. C. H. Kruger and A. H. Shapiro, *Trans. 7th Nat. Vac. Symp. (1960)*, Pergamon, New York, 1961, pp 6-12.
5. C. H. Kruger, *The Axial Flow Compressor in the Free-Molecular Range*, Ph.D. thesis, Department of Mechanical Engineering, M.I.T., Cambridge, MA, 1960.
6. K. H. Mirgel, *J. Vac. Sci. Technol.*, **9**, 408 (1972).
7. J. Henning, *Proc. 6th Int. Vac. Congr.*, Kyoto, Japan J. *Appl. Phys. Sup. 2, Pt. 1*, 5 (1974).
8. J. Henning, *Vacuum*, **21**, (1971).
9. J. Henning and H. Lotz, *Vacuum*, **27**, 171 (1977).
10. G. Osterstrom and T. Knecht, *J. Vac. Sci. Technol.*, **16** 746 (1979).
11. L. Maurice, *Proc. 6th Int. Vac. Congr.*, Kyoto, Japan J. *Appl. Phys. Sup. 2, Pt. 1*, 21 (1974).
12. Leybold-Heraeus TurboVac 550M turbomolecular pump.   Leybold-Heraeus G.m.b.H., Köln, West Germany.

PROBLEMS

11.1 †Describe the operating mechanism of a turbomolecular pump.

11.2 Why does a turbomolecular pump need a stator?

11.3 A particular turbomolecular pump has a tip velocity of 400 m/s. What is the ratio of average room-temperature molecular velocity to blade tip velocity for: (a) hydrogen, (b) nitrogen, and (c) xenon molecules? What fraction of the molecules have velocities greater than the blade tip velocity?

11.4 †What causes the pumping speed of a turbomolecular pump to fall at its low pressure extreme?

11.5 What happens when the critical discharge pressure of a turbomolecular pump is exceeded during operation?

11.6 †The best place to vent a turbomolecular pump is: (a) in the foreline or (b) in the high vacuum chamber?

11.7 †When a turbomolecular pump is stopped, it is better to: (a) continue pumping on it with the forepump to keep it clean or (b) vent it to air with a dry gas?

11.8 Plot the air throughput of the turbomolecular pump whose speed-vs.-pressure curve is shown in Fig. 11.7.

11.9 Qualitatively, what is the relationship between the thermal velocity of a gas and its maximum compression ratio in a turbomolecular pump?

11.10 Screens are frequently placed in the throat of turbomolecular pumps to prevent objects from falling into the pump and damaging the rotor. One pump contains a double mesh screen in the throat. The first mesh is made from 0.005-in.-diameter wire on 0.025-in. centers, while the second mesh is made from 0.02-in.-diameter wire on 0.150-in. centers. The Ho coefficient or transmission coefficient of the pump is $a = 0.24$ without the screen. Calculate the percent reduction in pumping speed when this screen is used.

CHAPTER 12

Diffusion Pumps

The diffusion pump has been in existence since the early nineteen hundreds. The most important developments, which occurred in the last thirty years, were the discovery of low vapor pressure pumping fluids and the ability to control backstreaming [1]. Today, it is still the most widely used high vacuum pump even though ion, cryogenic, and turbomolecular pumps are desirable for many applications. Because of this long history, this pump has been the subject of more study and literature than any other high vacuum pump. Its problems are thoroughly understood and its performance is, in some cases, understated. Many excellent reviews of the diffusion pump are available. Examples that summarize the pump's properties for practical applications are those of Hablanian [2,3] and Singleton [4]. Review articles by Hablanian and Maliakal [1], Florescu [5,6] and Tóth [7] and books by Dushman [8] and Power [9] cover its theory of operation and design.

This discussion reviews the basic mechanisms of pump operation, pumping speed and throughput, heat effects and backstreaming, baffles and traps. The particular problems associated with the collective operation of a diffusion pump, backing pump, trap or baffle, and work chamber as a complete system are treated in later chapters. The basic high vacuum diffusion-pumped system is discussed in Chapter 19. The special requirements of diffusion pumps for ultrahigh vacuum are treated in Chapter 20 and for high gas flow applications in Chapter 21.

12.1 PUMPING MECHANISM

The name "diffusion pump," first coined by Gaede [10], does not describe the operation of the pump accurately. The diffusion pump is a vapor jet pump which transports gas by momentum transfer on collision with the vapor stream. A motive fluid such as a hydrocarbon oil, an organic liquid or mercury is heated in the boiler until it vaporizes. The vapors flow up the chimney and out through a series of nozzles. Figure 12.1 sketches a sectional view of a metal-bodied diffusion pump. The nozzles, three in this illustration, direct the vapor stream downward and toward the cooled outer wall were it condenses and returns to the boiler. The vapor flow is supersonic. Gases that diffuse into this supersonic vapor stream are, on average, given a downward momentum and ejected into a region of higher pressure. Modern pumps have several stages of compression—usually three to five for small pumps and up to seven for large pumps. Each stage compresses the gas to a successively higher pressure than the preceding stage as it transports it toward the outlet.

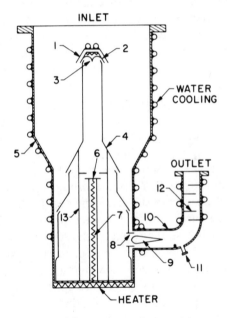

Fig. 12.1 A sectional view of a metal diffusion pump and some of its innovations: (1) cooled hood for prevention of vapor backstreaming [11,12]; (2) heater for the nozzle's cap to compensate for loss of heat [13]; (3) streamlined surface to avoid turbulence [14]; (4) multiple stages to obtain low pressures [15]; (5) enlarged casing to give larger pumping aperture [16]; (6) baffle to impede the access to the jet of liquid splashed up from the boiler [17]; (7) heater for superheating the vapor [18]; (8) lateral ejector stage [19]; (9) conical body allowing operation against higher forepressures [20]; (10) hot maintained diffuser for oil purification [21]; (11) catchment and drain-off of highly volatile oil components [22]; (12) baffle to reduce oil loss [15]; (13) concentric chimneys that allow oil fractionation [23]. Reprinted with permission from *Vacuum*, **13**, p. 569, N. A. Florescu. Copyright 1963, Pergamon Press Ltd.

The boiler pressure in a modern diffusion pump is about 200 Pa. Ideally, the pump cannot sustain a pressure drop any larger than this between its inlet and outlet. The practical maximum value of forepressure tolerated by the pump is less than the boiler pressure. This maximum value called the "critical forepressure," ranges from 25 to 75 Pa and is dependent on pump design and boiler pressure. The latter number is typical of modern pumps. The diffusion pump cannot eject gas into atmospheric pressure, but must be "backed" by another pump in order to keep the forepressure below the critical forepressure. Rotary vane or piston pumps or combinations of rotary and lobe blowers are used as "backing" or "fore" pumps. If the forepressure exceeds the critical value, all pumping action will cease. The pumping action ceases at high pressures because the directed supersonic vapor stream no longer extends from the jet to the wall but is ended in a shock front close to the jet [5]. Those vapor molecules beyond the shock front are randomly directed and cannot stop gas molecules from returning to the inlet. As the critical forepressure is exceeded, the inlet pressure will rise sharply and uncontrollably in response to the cessation of pumping. Needless to say, the critical forepressure should *never* be exceeded. In newer pumps the inlet pressure and the pumping speed will be unaffected by the value of forepressure as long as it is below the critical value and the gas throughput is low. At maximum throughput the critical forepressure will be reduced to about 3/4 of its normal value [2]. The amount of reduction is a function of the pump design, heater power, and pump fluid.

Each stage of the vapor pump has a characteristic speed and pressure drop. Since the jets are in series, the gas flow Q is the same through each stage. The flow, $Q = S\Delta P$, is the product of the speed of the jet times the pressure drop across the jet. The top jet has the largest speed (and the largest aperture) and the lowest pressure drop. The vapor density in the top jet is less than that in the lower jets. Because the gas flow through a series of jets is the same, each successive jet can have a larger pressure drop and a smaller pumping speed. The last jet has the highest pressure drop. Many pumps use a vapor ejector as the last stage because it is efficient at compressing gas in this pressure range. The combination of jets and ejector produces a pump with a higher forepressure tolerance than is possible with vapor jets alone. Fractionating pumps [23] have concentric chimneys and allow pump fluid to be preferentially directed to the lower jets after condensation so that its light fractions can be removed by the forepump. Degassing of the fluid is accomplished by maintaining a section of the ejector walls at an elevated temperature [21]. Pumps incorporating these and other advances which use heavy fluids produced by molecular distillation will reach 5×10^{-5} Pa (untrapped) and 5×10^{-7} Pa when trapped with liquid nitrogen.

12.2   SPEED-THROUGHPUT CHARACTERISTICS

The four operating regions of the diffusion pump are the constant-speed, constant-throughput, mechanical pump, and compression ratio regions. They are graphically illustrated in Fig. 12.2.   In its normal operating range the diffusion pump is a constant-speed device.   Its efficiency of pumping gas molecules (Ho coefficient) is about 0.5 for the pump alone but only about 0.3 when the conductance of traps and valve are included. The usual operating range for constant speed is about 10^{-1} to below 10^{-9} for most gases.   The maximum or limiting inlet pressure is called the critical inlet pressure and it corresponds to the point at which the top jet fails.   In a 6-in. diffusion pump the top jet becomes unstable at pressures of about 0.1 Pa, the middle jet at a pressure of about 3 Pa, and the bottom jet at pressures of about 40 Pa [24].

The gas throughput in the constant-speed range is the product of the inlet pressure and the speed of the pump at the inlet flange.   It rises linearly with pressure until the critical inlet pressure is reached.   Above that pressure the pump throughput is constant until the jets all cease to function.   At higher pressures the throughput again increases in accordance with the speed of the backing pump.   The maximum usable throughput of the diffusion pump corresponds to the product of the inlet speed and the critical inlet pressure.   If that pressure is exceeded, the backstreaming may increase and jet instabilities will appear.   These instabilities make pressure control difficult.   The maximum throughput should not be exceeded in the steady state, although it often happens for short periods of time during crossover from rough to high vacuum pumping.

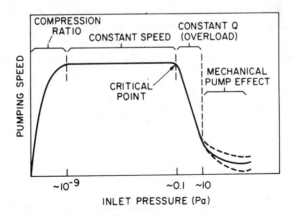

Fig. 12.2 Typical diffusion pump speed curve for a given gas. Four regions are evident: (1) effect of compression ratio limit; (2) normal operation—constant speed; (3) first jet overloaded—nearly constant throughput; (4) effect of mechanical pump. Reprinted with permission from *Japan J. Appl. Phys., Suppl. 2*, Pt. 1, p. 25, M. H. Hablanian. Copyright 1974, Japanese Journal of Applied Physics.

Exceeding the critical forepressure in a well-designed pump is usually the result of equipment malfunction, while the critical inlet pressure is easily exceeded by misoperation. If the pump is equipped with a sufficiently large forepump, the critical forepressure can still be exceeded if a leak occurs in the foreline, the mechanical pump oil level is too low, the mechanical pump belt is loose, or a section of the diffusion pump heater is open. The critical inlet pressure can be easily exceeded by operational error, but otherwise the top jet will continue to pump unless there is a partial heater failure or a large leak.

The speed does not remain constant to absolute zero pressure, but decreases toward zero as shown in the compression ratio region of Fig. 12.2. This curve decreases at low pressures because of the large but finite compression ratio of the diffusion pump jets. The pump whose hypothetical speed curve for one gas is shown in Fig. 12.2 has an ultimate pressure of 10^{-10} Pa. If, at that point, its forepressure were 1 Pa, its compression ratio would be 10^{10}, a value similar to those quoted in the literature. Figure 12.3 shows the air pumping speed for a 6-in. diffusion pump with and without a nitrogen baffle. All diffusion pumps have some small reverse flow of the gas being pumped, and although this reverse flow is exceedingly small for heavy gases it may be feasible for light gases under certain conditions. Because of their high thermal velocity and small collision cross section, the compression ratio of light gases such as hydrogen and helium is lower than that of the heavy gases. Figure 12.4 sketches the relative pumping speeds of several gases and vapors as a function of their inlet pressure and illustrates the effect of a low compression ratio for hydrogen. The compression ratio for heavy gases will be about 10^8 to 10^{10}. For light gases it can be small enough (10^3 to 10^6) in some pumps so that a small foreline concentration can be detected at the inlet [1,25]. It is this phenomenon that explains why hydrogen emanating from an ion gauge in the foreline can be detected at the inlet. The operation of one leak detector [26] is based on this principle. The

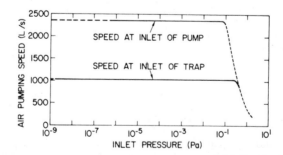

Fig. 12.3 Air pumping speed of the Varian VHS-6 6-in. diffusion pump with and without a liquid nitrogen trap. Reprinted with permission from Varian Associates, 611 Hansen Way, Palo Alto, CA 94303.

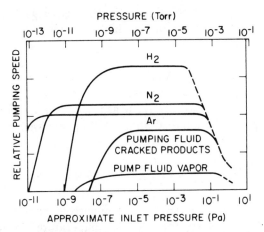

Fig. 12.4 Diffusion pump performance for individual gases. Reprinted with permission from *J. Environ. Sci.*, **5**, p. 7, S. G. Bennet and M. H. Hablanian. Copyright 1964, The Institute of Environmental Sciences.

detector is located at the inlet and the test piece is appended to the foreline. The compression ratio for heavy gases is adequate to produce the required low pressure in the detector while at the same time allowing helium to back diffuse and be counted.

The ultimate or limiting pressure in a diffusion-pumped system can be a result of the compression ratio limit or wall outgassing or both phenomena. For the ideal pump with zero outgassing above the top jet and in the work chamber, and using a perfect baffle to collect all oil vapor fragments, the ultimate pressure would be the sum of each of the partial pressures in the foreline divided by their respective compression ratios:

$$P_u = \frac{P_{f1}}{k_1} + \frac{P_{f2}}{k_2} + \frac{P_{f3}}{k_3} \cdots \tag{12.1}$$

For the case in which the base pressure of the system is achieved in the pump's constant-speed region the ultimate pressure is the sum of each independent gas flow Q_i, divided by the pumping speed for each gas S_i:

$$P_u = \frac{Q_1}{S_1} + \frac{Q_2}{S_2} + \frac{Q_3}{S_3} \cdots \tag{12.2}$$

The individual gas flows can originate from outgassing or leaks. In practice the ultimate pressure is determined by (12.2), but in some situations it can be a combination of the two cases. For example, the partial pressure of hydrogen and perhaps helium may be determined by the compression ratio (12.1), while the partial pressure of heavier gases is determined by wall outgassing (12.2). The pump's ultimate pressure will not be limited by the compression ratio for heavy gases, but rather by outgassing, the vapor pressure of the lightest pump fluid fractions on the baffle, and the release of gases dissolved in the pump fluid. A single pumping speed curve (Fig. 12.2) representative of all gases cannot be drawn because the pumping speed is not the same for all gases (see Fig. 12.4). The pumping speed is greater for light gases, but not in proportion to $m^{1/2}$ as predicted by the ideal gas law. Under normal operating conditions helium will be pumped about 20% faster than nitrogen.

12.3   HEAT EFFECTS

The effect of heat input variation is summarized concisely in Fig. 12.5 [3]. The general trends are that the oil temperature, forepressure tolerance, and throughput increase with boiler power, while pumping speeds decrease at high heat inputs because of the increased density of oil molecules in the vapor stream [7]. It is not possible to optimize the pumping speed for all gases at the same heater power because of the differences in mass and thermal velocity. Each gas reaches maximum speed at a different input power. The pumping speed is a function of the momentum transfer between fluid and gas molecules. Heavy fluid mole-

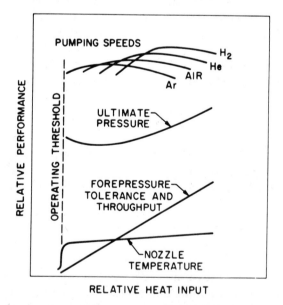

Fig. 12.5   Effect of heat input variations on various diffusion pump parameters. Reprinted with permission from *Japan J. Appl. Phys.*, Suppl. 2, Pt. 1, p. 25, M. H. Hablanian. Copyright 1974, Japanese Journal of Applied Physics.

cules often have lower pumping speeds than light molecules unless the boiler temperature is adjusted. Excessively increasing the boiler temperature also hastens fluid degradation [3]. Modern pumps will accept most fluids equally well but some older pumps, especially those designed to work with older fluids, do not provide enough heat input for pentaphenyl-ether. It should be emphasized that there is a direct relationship between boiler power and throughput. They are dimensionally equivalent; 1000 Pa-L/s = 1 W. The maximum throughput of a pump is proportional to the boiler power. For pumps of efficient design this can be as high as 150 Pa-L/s for each kilowatt of boiler power [2]. A straight-sided pump with a 6-in.-diameter boiler and throat has the same maximum throughput as a pump with a 6-in.-diameter boiler and an expanded top like the one shown in Fig. 12.1. The maximum speed of a pump in the high vacuum region is proportional to its inlet area, but its throughput is proportional to its boiler power.

12.4 BACKSTREAMING, BAFFLES AND TRAPS

For the purpose of this discussion backstreaming is defined as the transport of pumping fluid and its fractions from the pump to the chamber. Hablanian [27] properly points out that the discussion of backstreaming must not be limited to the pump but must include the trap, baffle, and ducts as well, because all affect the transfer of pumping fluid vapors from the pump body to the chamber. First, let us consider the contributions from the pump. Power and Crawley [12] have determined that steady-state backstreaming results from (1) evaporation of fluid condensed on the upper walls of the pump, (2) premature boiling of the condensate before it enters the boiler, (3) the overdivergence of the oil vapor in the top jet, (4) leaks in the jet cap, and (5) evaporation of fluid from the heated lip of the top jet. The backstreaming from (1) can be reduced by the use of low vapor pressure fluids and added trapping over the pump. Modern pump designs eliminate sources (2) and (4). The use of a water cooled cap [1,12] directly over the top jet assembly substantially reduces (3) and (5), which were found to be the major causes of fluid backstreaming. With these precautions the backstreaming can be reduced to approximately 10^{-3} (mg/cm^2)/min a short distance above the pump inlet.

Further reduction of the backstreaming is possible by geometrical considerations and by the use of a baffle or a trap. The words *trap* and *baffle* are often misused. Operationally, a trap is a pump for condensable vapors, and a baffle is a device that condenses pump fluid vapors and returns the liquid to the pump boiler. Today the two words are often used imprecisely, and when the baffle is cryogenically cooled the distinction disappears. Pump-fluid molecules or fragments may find their way through the trap by creeping along the walls, by colliding with gas molecules, and by reevaporation from surfaces. Creep can be prevented by the use of traps with a creep barrier—a thin membrane extending from the warm, outer wall to the cooled surface [28]—or by the use of autophobic fluids such as pentaphenylsilicone or pentaphenylether. Backstreaming due to oil-gas collisions is a linear function of pressure up to the transition region and a function of the trap and pump design. Rettinghaus and Huber [29] have measured this backstreaming. For one 6-in. diffusion pump and trap combination they found the backstreaming to have a peak value of 3×10^{-6} (mg/cm^2)/min at a pressure of 5×10^{-2} Pa. At higher pressures the backstreaming rate was decreased by the flushing action of the gas. In normal operation the diffusion pump will pump through this region quickly. The maximum integrated backstreaming rate from oil-gas collisions is small enough so that contamination from this source is of no concern in an unbaked system.

The problem of reevaporation is subtle. The vapor pressures of diffusion pumps vary widely. See Appendix F.3. The two fluids with the lowest vapor pressures (pentaphenyl silicone and pentaphenylether) have vapor pressures so low that evaporation from a surface at 10°C proceeds at at rate of 5×10^{-10} (mg/cm^2)/min. Some decomposition of the fluid does occur in the boiler and light fractions are generated. Gosselin and Bryant [30] have studied the residual gases in a diffusion-pumped system in which the pump was charged with DC-705. The observed partial pressures of selected light, intermediate, and heavy fragments tabulated by Gosselin and Bryant are plotted in Fig. 12.6. They observed the light

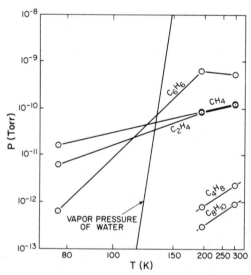

Fig. 12.6 Residual gas analysis of selected mass fragments backstreaming from a diffusion pump filled with DC-705 fluid. Plotted from data reported in *J. Vac. Sci. Technol.*, **2**, p. 293 (1965), C. M. Gosselin and P. J. Bryant.

fractions (methane, ethane and ethylene) were not effectively trapped even on a liquid-nitrogen-cooled surface because of their high vapor pressures. The very heavy fragments (e.g., C_8H_{10}) were quite effectively trapped with only a water-cooled baffle. The partial pressure of an intermediate weight fragment C_6H_6, was reduced by a factor of 1000 when the trap was cooled from 25 to $-196°C$. When using modern low vapor pressure fluids such as pentaphenyl silicone or pentaphenyl ether, the basic operational difference between a liquid nitrogen trap and a cold water baffle is the ability of the liquid nitrogen trap to pump C_6H_6 and to partially trap some of the lightweight fractions. Freon-cooled traps, $T = -35°C$, are of no use when the pump is charged with either of the two aforementioned fluids. They will decrease the partial pressures of the intermediates only slightly, do not pump the lightest fragments and are not needed to pump large nmolecules.

The quantitative effects of various trap, baffle and creep barrier combinations are summarized in Table 12.1 [27]. It was noted that the addition of the chevron baffle water between the liquid nitrogen trap and the pump is not much better than the addition of a piece of straight pipe or elbow of the same length. Rettinghaus [31] has shown for one pump and baffle that the addition of throttling structures below the baffle will further reduce the backstreaming. Figure 12.7 summarizes the results of his measurements on the backstreaming of polyphenylether. The addition of a baffle consisting of three circular half-chevrons was shown to give a

Table 12.1   Diffusion Pump Backstreaming[a]

Conditions	Duration of Test (h)	Backstreaming Rate $(mg/cm^2)/min$
(1) Without baffle	165	1.6×10^{-3}
(2) With liquid nitrogen trap	170	5.3×10^{-6}
(3) Same as (2)	380	6.5×10^{-6}
(4) Item (3) plus water baffle	240	2.8×10^{-7}
(5) Item (4) plus creep barrier	240	8.7×10^{-8}
(6) Same as (5)	337	1.2×10^{-7}

Source. Reprinted with permission from *J. Vac. Sci. Technol.,* **6**, p. 225, M. H. Hablanian. Copyright 1969, The American Vacuum Society.
[a]Measurements made with a 6-in. diffusion pump (NRC HS6-1500), DC-705 pump fluid, and liquid-nitrogen-cooled collectors.

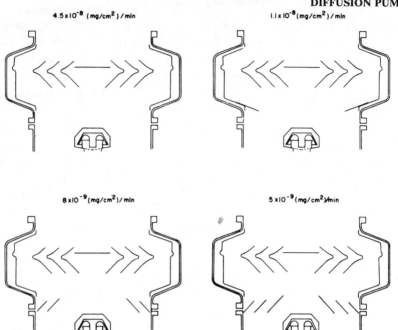

Fig. 12.7 Relationship between backstreaming and added throttling below the baffle for a Balzers 250 diffusion pump stack. Reprinted with permission from G. Rettinghaus, Balzers High Vacuum, Furstentum, Liechtenstein.

net backstreaming rate ten-fold lower than for the baffle alone, but not without further reduction in pumping speed.

The Herrick effect [32] and the fluid burst resulting from the formation and collapse of the top jet are two transient phenomena that also cause backstreaming. The Herrick effect is the ejection of frozen fluid droplets from the surface of a fluid-covered trap during the initial stages of cooling with liquid nitrogen. These fluid droplets ricochet off the walls and land in the chamber or on samples or fixturing. A well-designed cold cap and water-cooled baffle followed by a continuously operating liquid nitrogen trap will operate for more than a year without collecting excessive amounts of fluid on the trap. The transient backstreaming from the top jet during warm-up and cool-down of the pump is well documented [12,19,33]. Figure 12.8 shows an RGA trace of the parent molecule (M/z = 446; Convalex-10 [33]). The backstreaming decreases as the fluid is cooled and reaches a peak of about twice the steady-state rate during heating. Power and Crawley [12] also show a peak as the jet is cooling. The total backstreaming was measured as 6×10^{-4} mg/cm^2 for a complete start-stop cycle [29]. This kind of backstreaming can be avoided by continuous operation of the diffusion pump or by using the gas flushing techniques discussed in Chapter 24.

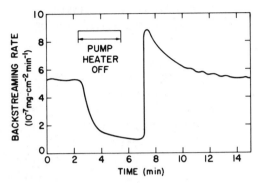

Fig. 12.8 Backstreaming of the parent peak (M/z = 446; Convalex-10) over a liquid nitrogen trap during cool-down and start-up of a diffusion pump. Reprinted with permission from *J. Vac. Sci. Technol.*, **9**, p. 416, G. Rettinghaus and W. K. Huber. Copyright 1972, The American Vacuum Society.

By use of high quality, low vapor pressure, anticreep fluids such as DC-705, Santovac 5, or Convalex 10 and a continuously operating liquid nitrogen trap, the contamination due to pump fluid backstreaming can be made very small. The lowest value of backstreaming shown in Table 12.1 corresponds to a contamination rate of one monolayer per year in a bell jar 500 mm high and 350 mm diameter. This level of organic contamination is below that produced by O-rings and other sources [2]. Fluid backstreaming in a diffusion pump operating at high vacuum is only one source of organic backstreaming. Additional concerns that relate to specific systems are discussed in Sections 19.1, 20.2, and 21.1.

REFERENCES

1. M. H. Hablanian and J. C. Maliakal, *J. Vac. Sci. Technol.*, **10**, 58 (1973).

2. M. H. Hablanian, *Solid State Technol.*, December, 37 (1974).

3. M. H. Hablanian, *Proc. 6th Int. Vac. Congr., Kyoto, Japan J. Appl. Phys.*, Suppl. 2, Pt. 1, 25 (1974).

4. J. H. Singleton, *J. Phys. E.*, **6**, 685 (1973).

5. N. A. Florescu, *Vacuum*, **10**, 250 (1960).

6. N. A. Florescu, *Vacuum*, **13**, 569 (1963).

7. G. Tóth, *Proc. 4th Int. Vac. Congr., (1968)*, Institute of Physics and the Physical Society, London, 300 (1969).

8. S. Dushman, *The Scientific Foundations of Vacuum Technology*, 2nd ed., J. M. Lafferty, Ed., Wiley, New York, 1962, Chapter 3.

9. B. D. Power, *High Vacuum Pumping Equipment*, Reinhold, New York, 1966.

10. W. Gaede, German Pat. 286,404 (filed September 25, 1913).

11. M. Morand, U.S. Pat. 2,508,765 (filed July 27, 1947; priority France, September 25, 1941).

12. B. D. Power and D. J. Crawley, *Vacuum*, **4**, 415, (1954).

13. C. G. Smith, U.S. Pat. 1,674,377 (filed September 4, 1924).

14. W. A. Giepen, U.S. Pat. 2,903,181 (filed June 5, 1956).

15. G. Barrows, Brit. Pat. 475,062 (filed May 12, 1936).

16. J. R. O. Downing, U.S. Pat. 2,386,299 (filed July 3, 1944).

17. B. D. Power Brit. Pat. 700,978 (filed January 25, 1950).

18. J. R. O. Downing and W. B. Humes, U.S. Pat. 2,386,298 (filed January 30, 1943).

19. R. B. Nelson, U.S. Pat. 2,291,054 (filed August 31, 1939).

20. J. J. Madine, U.S. Pat 2,366,277 (filed March 18, 1943).

21. N. G. Nöller, G. Reich, and W. Bächler, *Trans. 4th Nat. Symp. Vac. Technol.*, **6**, (1957).

22. B. B. Dayton, U.S. Pat. 2,639,086 (filed November 30, 1951).

23. C. R. Burch and F. E. Bancroft, Brit. Pat. 407,503 (filed January 19, 1933).

24. L. T. Lamont, Jr., *J. Vac. Sci. Technol.*, **10**, 251 (1973).

25. S. G. Burnett and M. H. Hablanian, *J. Environ. Sci.*, **5**, 7 (1964).

26. Porta Test (R) Varian Associates, 611 Hansen Way, Palo Alto, CA 94303.

27. M. H. Hablanian, *J. Vac. Sci. Technol.*, **6**, 265, (1969).

28. N. Milleron, *Trans. 5th Nat. Vac. Symp. (1958)*, Pergamon, New York, 1959, p. 140.

29. G. Rettinghaus and W. K. Huber, *Vacuum*, **24**, 249 (1974).

30. C. M. Gosselin and P. J. Bryant, *J. Vac. Sci. Technol.*, **2**, 293 (1963).

31. G. Rettinghaus, private communication.

32. M. H. Hablanian and R. F. Herrick, *J. Vac. Technol.*, **8**, 317 (1971).

33. G. Rettinghaus and W. K. Huber, *J. Vac. Sci. Technol.*, **9**, 416 (1972).

PROBLEMS

12.1 †Why do diffusion pumps have several stages with differing jet-to-wall distances?

12.2 †What happens when the critical inlet pressure of a diffusion pump is exceeded during pump operation?

12.3 †What happens when the critical forepressure of a diffusion pump is exceeded during pump operation?

12.4 A diffusion pump has a maximum inlet speed of 1500 L/s and a maximum inlet pressure of 1.3×10^{-1} Pa. Its critical forepressure is 40 Pa. What is the maximum throughput of the pump? What is the minimum pumping speed necessary at the foreline connection to ensure the pump will not exceed the critical forepressure? Would you choose a single-stage or a double-stage vane pump for this application? Why?

12.5 Plot the air throughput-vs.-pressure characteristic for the pump whose speed is depicted in Fig. 12.3 with and without the cold trap.

12.6 What can happen to a diffusion pump if (a) the cooling water flow is too great, say, 2 g/min through a 6-in. diffusion pump instead of 0.4 g/m; the cooling water temperature is (b) too low, say, 7 to 10°C, (c) has too high a soluble salt content from improper water treatment, (d) contains excessive calcium and other insoluble salts (hard water)?

12.7 What must be provided to prevent material collected on traps from migrating to the inlet side of a cooled trap during regeneration,

12.8 In what pressure range will the gas flow into an 20-cm-diameter diffusion pump be in transition between laminar-viscous and molecular flow?

12.9 How does a cold trap reduce backstreaming? What will happen if the trap is warmed for a few hours and then cooled while the high vacuum valve is closed?

12.10 Describe qualitatively how gases are released when a liquid nitrogen trap coated with oxygen, nitrogen, carbon dioxide, carbon monoxide, methane and hydrogen begins to warm.

CHAPTER 13

Vacuum Pump Fluids

Organic fluids are used in rotary vane and piston, lobe blowers, turbomolecular, and diffusion pumps. In a rotary vane or piston pump fluids provide a vacuum seal between the moving surfaces and lubricate and cool the low speed bearings and sliding surfaces. Lobe blowers and turbomolecular pumps use oil or a synthetic fluid to lubricate and cool medium and high speed bearings, but neither pump uses a fluid to make the vacuum seal. Diffusion pumps vaporize liquids to form supersonic jets that transfer momentum to gas molecules. Ideally each fluid should be thermally stable and chemically inert, have a low vapor pressure, and when necessary be a good lubricant. Unfortunately every desirable attribute cannot be realized simultaneously so that some compromise is required in formulating or choosing a fluid for any application.

The fluids currently used in vacuum pumps are highly refined mineral oils and synthetic esters, silicones, ethers and fluorocarbons. In this chapter we review fluid properties, pump fluid types, selection, and reclamation.

13.1 FLUID PROPERTIES

Vapor pressure and lubricating ability are the two most important properties of a vacuum pump fluid. Low vapor pressure is necessary to avoid oil vapor transport to the vacuum chamber, and mechanical pump fluids need to be good lubricants. In Chapter 18 we discuss rheological properties—absolute and kinematic viscosity and viscosity index. Here we review vapor pressure measurements and other physical and chemical properties of vacuum pump fluids.

13.1.1   Vapor Pressure

Regardless of its other qualities, a diffusion or mechanical pump fluid, vacuum lubricant, or additive is of no use if its vapor pressure is so high that it contaminates the working region of the vacuum chamber. A minimum vapor pressure is necessary, and any further reduce will improve performance, simplify trapping and reduce contamination. For example, a mechanical pump fluid should have a vapor pressure less than 0.1 Pa at its operating pressure, while a diffusion pump requires a fluid whose room-temperature vapor pressure is in the range 10^{-3} to 10^{-7} Pa depending on the application. Most of us will never measure the vapor pressure of a pump fluid; however, we do need to know how, when, and *if* the available data were measured.

Many techniques have been devised for the measurement of oil vapor pressure [1-3]. The Knudsen effusion technique [1,2] is considered to be reliable and accurate but there is no standard procedure. Liquid is heated to a constant temperature in a partly filled cell with a small orifice. At equilibrium the rate of vaporization from the fluid surface is equal to the rate of arrival. This equilibrium pressure is the vapor pressure. A tiny hole in the top of the cell allows a small fraction of the vapor to effuse from the cell into a vacuum. The surface area of the liquid needs to be at least ten times the area of the hole in order to maintain the liquid at its equilibrium vapor pressure. Also, the diameter of the opening must be less than the mean free path of the heated molecules. A typical cell size is 1.2×10^{-2} m diameter by 4×10^{-3} m high with a 3×10^{-3}-m-diameter orifice. Outside the cell the pressure must be below about 10^{-3} Pa to prevent molecules from returning to the opening.

If these criteria are met, the vapor pressure can be calculated from (2.9) and (2.13)

$$P = \left(\frac{dm}{dt} \right) \frac{1}{aA} \left(\frac{2\pi kT}{m} \right)^{1/2} \tag{13.1}$$

or

$$P(\text{Pa}) = 2.278 \times 10^4 \left(\frac{dm}{dt} \right) \frac{1}{aA} \left(\frac{T}{M} \right)^{1/2} \tag{13.2}$$

where a is the transmission probability of the orifice. When the orifice thickness is much less than the diameter, $a = 1$, otherwise the appropriate transmission probability must be taken from Table 3.2. The vapor pressure is calculated from the known cell temperature and weight loss (dm/dt). The fractional weight loss is small. For example a 10^{-4}-kg sample of pentaphenyl ether with a molecular weight of 447 and a vapor pressure of 1 Pa at 200°C will lose weight at the rate of 3×10^{-10} kg/s from an effusion orifice of area 7×10^{-6} m^2. Vapor pressure measurements are made over a range of temperatures corresponding to 0.2 to

10% weight loss in 10 to 15 minutes. The temperature range will be different for each fluid, but in all cases 80 to 100°C is the minimum temperature that gives adequate sensitivity. All data reported below that temperature are extrapolated.

The Clapyeron equation

$$\frac{dP}{dT} = \frac{\Delta H}{T\Delta V} \tag{13.3}$$

gives the vapor pressure-temperature relationship for many substances. ΔH is the heat of vaporization of 1 mole of substance, ΔV is the volume change per mole during vaporization. When the specific volume of the gas phase is much greater than that of the liquid phase, and when ΔH is independent of T, the solution is

$$\log P_v = A - \frac{B}{kT} \tag{13.4}$$

This solution is known as the Clausius-Clapyeron equation. If the heat of vaporization is a constant, B is a constant, and P_v versus $1/kT$ will plot as a straight line for temperatures below about half the critical temperature. Data taken by Hickman [4] on one pentaphenyl ether and plotted according to (13.4) are shown in Fig. 13.1. The data were taken over the range 235 to 370°C and extrapolated to lower temperatures.

Equation (13.4) is obeyed by fluids which are single chemical compounds. The pure 5-ring phenyl ether described in Fig. 13.1 is an example of such a fluid. All fluids are not single chemical compounds. Polymers and other mixtures contain molecules of various weights. Polysiloxanes and perfluoropolyether fluids are examples of fluids which contain molecules of similar structure and variable weight. A hydrocarbon mineral oil is an example of a fluid which has an added complication; it contains three distinct structures. It contains paraffinic, naphthenic, and aromatic structures and each has its own distribution of molecular weights. Each fluid contains a distribution of molecular weights; the published molecular weight is an average.

The concept of vapor pressure has little meaning when applied to polymers with a broad molecular weight distribution such as polysiloxane or polyether. A single chemical compound has a distinct boiling point at a particular pressure. Polymers of broadly varying molecular weights with a large deviation about the mean molecular weight have no unique vapor pressure. The measured vapor pressure is dominated by the vapor pressure of the lower molecular weight fractions. When the pump fluid is heated, light fractions preferentially evaporate, and the composition of the residual fluid changes slowly and systematically.

Evaporation of light fractions is known to affect vapor pressure measurements [4]. If a mixture contains a large fraction of light-weight molecules, as do some polymers, a linear Clausius-Clapyeron equation is

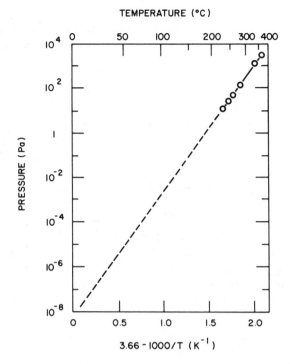

Fig. 13.1 Measured and extrapolated vapor pressure of pentaphenyl ether.

not obtained except at high temperatures. Extrapolating these data to low temperatures by means of a straight-line approximation is invalid for certain applications. It is invalid because it gives a vapor pressure characteristic of the fluid minus the lightweight impurities. Fractionating diffusion pumps systematically eject the lighter weight fractions to the forepump. Extrapolating high temperature effusion data from diffusion pump fluids to lower temperatures yields a vapor pressure that is characteristic of the fluid after it has been operated for some time in a fractionating pump or carefully distilled. Extrapolating the vapor pressures of some mechanical pump fluids to low temperatures is not valid, because the light fractions remain and their vapor pressure is higher than the value predicted by the straight-line approximation.

The Knudsen technique has been used in research laboratories to characterize the vapor pressures of various pump fluids. Each laboratory has its own version and procedure. Currently there is no standard American Vacuum Society or International Standards Organization procedure for measuring pump fluid vapor pressure. Vapor pressure measurements of vacuum pump fluids have been performed historically by a very few investigators. The number of measurements bears little relation to the frequency with which the original data are republished.

As an alternative to the measurement of vapor pressure, commercial pump fluids are characterized by their ultimate or blank-off pressure in a specific pump. Procedures for the measurement of the permanent gas blank-off pressure of a mechanical pump are given in AVS and ISO standards [5,6]. These procedures can be used to characterize fluid performance in a particular pump. Trapped McLeod gauges are used in these standard methods to eliminate the effects of pump oil and condensable vapors. As a result of these procedures, the ultimate pressure measured by the standard depends more upon the pump, and gas solubility of the fluid than on the partial pressure of organic vapors.

Procedures for measurement of the ultimate pressure in a diffusion pump are given in AVS standard 4.4 [7]. The ultimate pressure is dependent on several factors including pump fluid, boiler power, pump design and cooling rate. Relative comparisons of different fluids in the same pump may be made using procedures described in the standard. The vapor pressure and ultimate pressure are not the same, but they do seem to be correlated. The ultimate pressure will usually be within a half an order of magnitude higher than the true vapor pressure [3,4,8-11].

13.1.2 Other Characteristics

Color, pour point, flash and fire points, and gas solubility are properties used to further characterize lubricants and diffusion pump fluids. Color is not directly related to lubricant properties, but it does aid in identification of fluid categories. The color of hydrocarbon pump oils will vary from clear to medium yellow and is characterized by a standard such as the Saybolt color index [12]. Naphthenes are darker than paraffins. Pure straight chain paraffins and most synthetics are transparent, but additives or dyes will cloud or color any fluid.

The pour point is the lowest temperature at which a fluid will flow. Vacuum pump fluids that have been dewaxed behave like Newtonian fluids to very low temperatures. The viscosity at the pour point is of order 10^5 to 10^6 mm^2/s [13]. Flash and fire points are, respectively, the temperatures at which a fluid will burn momentarily and continuously in the presence of a flame [14]. The autoignition temperature is the temperature at which the fluid will ignite spontaneously. The ignition properties of organic fluids are a function of the vapor quantity, the surface-to-volume ratio of the test cell, and the spark energy [15]. They are reproducible, but the conditions in the test may not be identical to those in the pump.

The solubility of gas in a liquid obeys Henry's law and is directly proportional to absolute pressure. Gas solubility increases rapidly with fluid viscosity in the 5 to 50-mm^2/s range. The increase with viscosity becomes less rapid for 100-mm^2/s oils and then levels off around 2000 to 3000 mm^2/s [16]. Typical air solubilities at 1 atm (percent by volume) are: hydrocarbon oil–7 to 10%, phosphate ester–4 to 8%, and silicone–20%. Kendall [17] quotes individual gas solubilities in mineral oil at 20°C as CO_2–80%, N_2–6% and He–2%. Stauffer Fyrquel 220 has a nitrogen solubility of 6.43 % at 26°C, and a hydrogen solubility of 3.23% at 21°C [18]. Fomblin Y-25 has the following solubilities at 20°C: Air–28%, H_2–8%, He–4%, CO_2–327%, ethelyne–122%, while it has a solubility of 350% at 0°C for SF_6 [19]. Chastagner [20] found the solubility of tritium in polyphenyl ether and perfluoropolyether to be 8 and 7% in mineral oil. The solubility of gases in chlorotrifluoroethylene fluid is similar to that for perfluoropolyether. Dissolved gas is undesirable for several reasons. It increases the time required by a mechanical pump to reach its base pressure and increases its base pressure [17]. High gas solubility can also cause fluid foaming and allow a gas to react with one previously pumped. Dissolved gas reduces oil viscosity, while bubbles in oil increase oil viscosity [21]. High fluid viscosity slows the removal of bubbles, whose size is a function of the interface tension.

13.2   PUMP FLUID TYPES

In this section we review the characteristics of currently used mechanical and diffusion pump fluids. Mechanical pumps first used mercury and soon after mineral oils. Mercury was the first working fluid for diffusion pumps. In 1929 Burch [22] distilled low vapor pressure mineral oils and proposed their use as an alternative to mercury. Later, Hickmann [4] distilled synthetic esters and ethers for vacuum use. In the last four decades a host of refined hydrocarbon oils and synthetic fluids have been tried. Not all of them are in use today. Some have been replaced by superior products, eliminated for reasons of safety, or were never accepted because they possessed no advantage over existing fluids. Any unused toxic fluids such as chlorinated byphenyls which are still in stock should be disposed of in a safe manner.

13.2.1   Mineral Oils

The mineral oils used in vacuum pumps are mixtures of paraffinic, naphthenic and aromatic hydrocarbons. The paraffins C_nH_{2n+2} are straight- or branched-chain hydrocarbon structures containing only single bonds. The high boiling point paraffins make excellent lubricants. They are stable at high temperatures, fluid at low temperatures, have reasonably constant viscosity over a wide temperature range (high viscosity index), and are adhesive enough to remain in place under high shear loads. They are not stable in oxygen at high temperatures. Paraffins have many possible isomers, for example $C_{10}H_{22}$ has 75, which have differing properties. Aromatics contain phenyl groups with straight or branched chain structures. They form sludge at high temperatures and have an undesirably low viscosity index. Naphthenes contain rings and chains with no double bonds. Naphthenes have properties between those of paraffins and aromatics. Carbon analysis shows the typical "paraffin mineral oil" to be composed of approximately 65% paraffinic, 30% naphthenic, and 5% aromatic hydrocarbons. Carbon analysis gives the weight percent carbon in singly bonded chains. It's not accurate to say all this carbon is from paraffinic structures, because other structures contain some single bonds.

Preparation of a vacuum fluid begins with a base oil that was vacuum distilled from crude and further purified by solvent extraction and dewaxing [23-25]. The oils supplied to the vacuum fluid distiller are either single cuts with one peak in the molecular weight distribution or blends made from two of the relatively few refined single cuts. For example, a light (30 to 40 mm²/s), medium (60 to 70 mm²/s), or a heavy (100 to 120 mm²/s) oil.

The oil is further purified by additional distillations. The distillation conditions are chosen to produce fluids of the desired viscosity and vapor pressure. The distilled fluid will have a distribution of molecular weights. Stripping both tars and light molecular weight ends produces a single-cut fluid with a distribution of molecular weights centered about the mean molecular weight. Blending produces a fluid with more than one peak in the molecular weight distribution. Viscosity is inversely proportional to vapor pressure, so the vapor pressure of these broad cuts will differ widely. It is not possible to produce a fluid of extremely low vapor pressure from a blend of two oils with widely differing vapor pressures.

The base oil has characteristics unique to its origin, and as a result, all mineral oils are not the same. Not only does the origin determine the paraffinic-naphthenic-aromatic ratio and the impurities but it also determines the type of paraffinic isomers. For example the amount of sulfur and other impurities in mineral oil varies with the geographical origin of the crude. For many applications the residual sulfur is detrimental, but its presence does inhibit oxidation [26]. Refiners of the highest quality, single-cut mineral oils select their base stock from a single source and test each lot for uniformity.

We arbitrarily divide mineral oils for vacuum pumps into four grades: mechanical pump, diffusion pump, fully saturated paraffins, and inhibited fluids. The "mechanical pump" grade is, loosely, composed of blended fluids or single cut fluids that have not been highly refined to remove light ends and tars. Vapor pressure requirements in a mechanical pump are not as severe as in a diffusion pump. This grade is typically used in mechanical pumps which rough pump chambers and back diffusion pumps. "Diffusion pump" grade fluids may be characterized as having a single peak in the molecular weight with narrow mass dispersion. The vapor pressure will be the lowest for those cuts with the highest average molecular weight. Hydrocarbon diffusion pump fluids are used in a variety of high vacuum pumping applications but are not suitable for many ultrahigh vacuum applications.

Fully saturated paraffin oil, or white oil, is made by catalytically hydrogenating a paraffin oil. This fluid will be somewhat more stable in the presence of mild corrosive gases than ordinary mineral oil because its reduced number of dangling bonds reduces its reactivity. White oil is used in mechanical pumps for halogen gases, for example plasma etching systems.

Inhibited fluids contain additives to improve the characteristics of the mineral oil. Mineral oils are not stable in oxygen, have some tendency to sludge and foam, and do not offer adequate protection in boundary layer lubrication. Many additives have been developed to improve mineral oil. Additives will inhibit oxidation, reduce foaming, disperse contamination, reduce wear, depress the pour point, and increase the viscosity index.

Antioxidants are used to to decrease the oil's reactivity to oxygen and extend its useful life. Hindered phenols such as 2,6-di-*tert*-butyl-4-methylphenol, also known as BHT, are often used in concentrations of <1% to reduce oil oxidation. Detergents are used to keep sludge and varnish precursors in suspension. Carboxylates or alkali metal salts of carboxylic acid are one class of detergents used in 1 to 10% concentration to adsorb insoluble particles, hold them in suspension, and prevent them from depositing on metal surfaces. Detergents can introduce foaming, and are usually used with an antifoam agent. Methyl silicone polymers (e.g., dimethyl silicone) are common antifoam agents. They prevent foam buildup by forming tiny droplets that break gas bubbles. Foam inhibitors are added in very low (ppm) concentrations. Pour point depressants and antiwear agents are infrequently used in vacuum fluids. Dyes are sometimes added to mechanical pump fluids to give them a distinctive color. They serve no functional purpose and may raise the vapor pressure. Most additives have high vapor pressures and are useful for a limited number of applications. Examples are diffusion pumps used for decorative coating, mechanical pumps used for pumping corrosives, and turbomolecular pumps.

13.2.2 Synthetic Fluids

Mineral oils lack many properties of the ideal fluid. Their ultimate pressure in an untrapped diffusion pump is unacceptably high for many applications. They are not stable in oxygen and have some tendency to sludge and foam and do not offer adequate protection in boundary layer lubrication. Synthetics were developed to overcome the shortcomings of hydrocarbon oils. Fluids have been synthesized which have low vapor

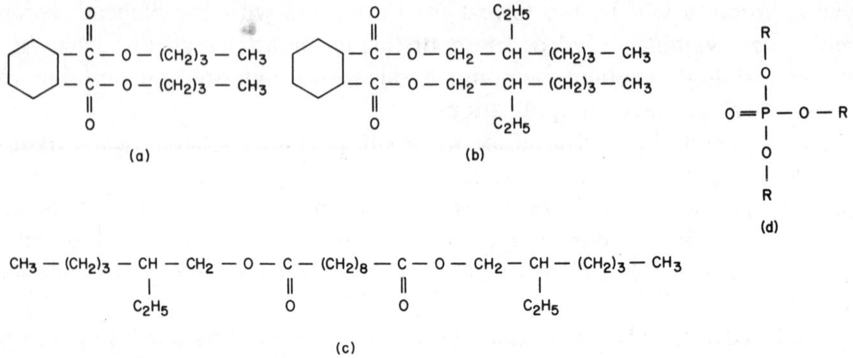

Fig. 13.2 Structures of organic ester pump fluids: (a) dibutyl phthalate, (b) bis(2-ethylhexyl)phthalate, (c) bis(2-ethylhexyl)sebacate, (d) phosphate ester, where R is an aryl or alkyl group.

pressure, high viscosity index, a high degree of oiliness, and chemical inertness.

Esters

Esters are chemicals formed by the reaction of an organic acid and an alcohol. The esters used in vacuum pump fluids all contain the same ester chemical bond, but have differing structures and rather widely varying properties. Dibutyl phthalate (Fig. 13.2a) is a low cost ester which is still used extensively in the vacuum metallurgy industry. Phthalate ester and sebacate ester (Fig. 13.2b,c) are organic esters originally developed as jet engine and aircraft instrument lubricants and at the same time used in diffusion pumps. They are synthesized from benzene or toluene (phthalate), sebacic acid (sebacate) and alcohols. Phthalate ester has only a moderate viscosity index, while sebacate ester has excellent low temperature properties.

Triaryl and trialkyl phosphate esters have been used primarily as mechanical pump fluids. They contain aryl (aromatic radical) or alkyl (hydrocarbon chain radical) groups attached to a phosphorous atom as sketched in Fig. 13.2d. The aryl esters have a lower viscosity index than alkyl esters because of the aromatic components. Both structures have high fire points. When these fluids react with steel at hot spots in boundary flow they form compounds, (e.g. iron phosphide) that plastically flow and redistribute the load. The most serious deficiencies of phosphate esters are their ability to react with the water vapor in air and their incompatibility with many elastomers. See Appendix F.1. Phosphorous containing esters present a disposal problem.

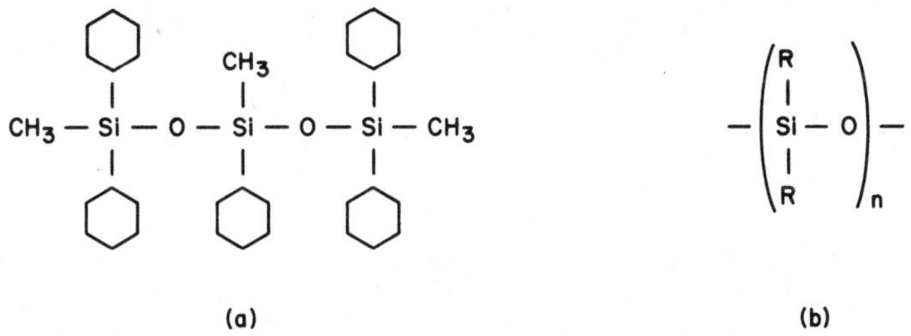

Fig. 13.3 Siloxane structures: (a) one isomer of pentaphenyltrimethylsiloxane, (b) chlorophenylmethylpolysiloxane; R is a methyl or phenyl or chlorine radical.

Silicate esters were tried as pump fluids and found undesirable. Linear silicate esters hydrolize quickly. Polysilicate structures do not have this problem, but they react with halogens.

Silicones

The unique character of the silicon-oxygen bond gives similarly unique properties to silicone-based fluids. Silicones, or siloxane polymers, are made up of repeated silicon-oxygen groups with silicon bonds to side groups. The type of side groups (methyl, phenyl, alkyl, chloro, etc.) and the number of silicon atoms determine the properties and applications of the fluid. The large size of the silicon atom allows the phenyl and methyl side groups great mobility. The high flexibility of the siloxane chain accounts for the high viscosity index of silicones. As a class silicones have the highest viscosity index of any fluid. Trisiloxanes and polysiloxanes are two fluids used in vacuum pumps.

Trisiloxanes are among the most widely used diffusion pump fluids. They consist of three repeating Si-O units with Si-C side bonds to either methyl or phenyl groups. Their stability, resistance to oxidation and low vapor pressure make them outstanding diffusion pump fluids. However, their lack of adhesion to and inability to form a film on steel seriously limits their usefulness in other types of pumps [27].

Trisiloxane fluids are manufactured by controlled hydrolysis of silanes and addition of phenyl groups followed by distillation. The first silicone diffusion pump fluids, DC-702 and DC-703 [28], were mixtures of closely related molecular species with similar boiling points. Further separation leads to the isolation of two specific chemical compounds, tetraphenyl tetramethyl trisiloxane, DC-704 [11], and pentaphenyl trimethyl trisiloxane, DC-705 [8]. One possible isomer of pentaphenyl silicone is sketched in Fig. 13.3a. DC-705 has one of the lowest vapor pressures of any diffusion pump fluid.

(a)   (b)

Fig. 13.4   Ether structures: (a) pentaphenyl ether, (b) alkyl diphenyl ether.

Several polysiloxanes have been developed for use as aircraft hydraulic fluids. A limited number of these have been used in mechanical vacuum pumps. They have chlorine or alkyl radicals attached to the siloxane structure to provide adhesion. Chlorophenylmethyl polysiloxanes Fig. 13.3b, have a high viscosity index. Alkyl siloxane is a good lubricant, but it has a lower viscosity index than chlorinated siloxane. Polysiloxanes are infrequently used and then only in mechanical pumps operating at low temperatures. Siloxanes have several drawbacks which hinder their use in pumps. They are not reactive with fluoroelastomers but cause most others to shrink or swell. The variable length nature of the polymer structure results in a fluid of high vapor pressure. Siloxanes are immiscible with other fluids and they cannot be improved easily by additives.

Ethers

An ether may be regarded as a derivative of a water molecule in which the hydrogens have been replaced by alkyl or aryl groups. Polyphenyl ethers were synthesized in an attempt to develop high temperature jet engine lubricants. Hickman [4] was the first to use them as diffusion pump fluids. He found the five-ring phenyl ether, Fig. 13.4a, to be stable and have extremely low vapor pressure. Commercially available fluids are mixed meta and para isomers of the pentaphenyl ether which contain trace impurities of the four-ring compound. The four-ring compound has a high vapor pressure, while the six-ring compounds are either solids or glasses. Pentaphenyl ether is very viscous at low temperatures but is stable and has excellent high temperature lubricating properties. Its wear, friction and load capacity are similar to or better than ethylhexyl sebacate and in some cases equal to mineral oil [29]. Its chemical stability and low vapor pressure make it an outstanding fluid for critical diffusion pump applications.

Alkyl diphenyl ethers, Fig. 13.4b, have been refined for diffusion pump use. They have properties similar to other moderate quality diffusion pump fluids and are relatively inexpensive.

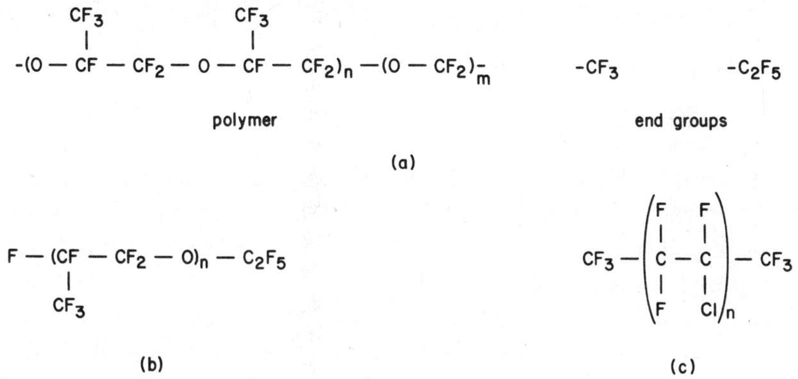

$$CF_3 \qquad\qquad CF_3$$
$$-(O-CF-CF_2-O-CF-CF_2)_n-(O-CF_2)_{\overline{m}} \qquad -CF_3 \qquad -C_2F_5$$

polymer

end groups

(a)

$$F-(CF-CF_2-O)_n-C_2F_5$$
$$CF_3$$

(b)

$$CF_3-\left(\begin{array}{cc} F & F \\ | & | \\ C-C \\ | & | \\ F & Cl \end{array}\right)_n -CF_3$$

(c)

Fig. 13.5 Fluorochemical structures: (a) Fomblin perfluoroalkylpolyether, (b) Krytox perfluoroalkylpolyether, (c) Halovac chlorofluorocarbon.

Fluorochemicals

Fluorochemical fluids are characterized by their inertness to a wide range of chemical compounds. Fluorine oil chemistry studies were initiated on a large scale during the Manhattan project in a search for uranium hexafluoride dilutants and by the Navy in a search for fire resistant fluids. Since then partially and fully fluorinated fluids have found use as lubricants for the space program and in lubricating oxygen compressors and liquid oxygen systems. Fluorocarbon and chlorofluorocarbon fluids are now distilled for use in vacuum pumps.

The completely fluorinated fluids used in pumps are perfluoroalkylpolyethers, (perfluoropolyethers, or PFPE for short) which are currently manufactured by two techniques. Fomblin [19,30] fluids are prepared by the UV-stimulated photooxidation of hexafluoropropylene and oxygen. It is a random copolymer of C_3F_6O and COF_2 with the structure shown in Fig. 13.5a, where n ranges from 10 to 40 and is 10 to 50 times greater than m [31]. Krytox [32,33] fluids are prepared by the polymerization of hexafluoropropylene epoxide. Krytox has the structure shown in Fig. 13.5b. It consists of 20 to 30 repeating C_3F_6O groups.

Raw perfluoropolyethers have a distribution of molecular weights extending as high as 10,000 AMU. They are distilled to yield cuts of average molecular weights from 1800 to 3700 that are suitable for use in rotating mechanical, turbomolecular and diffusion pumps. The resulting cuts are good lubricants with viscosity indexes of 50 to 120. Distillation yields cuts with vapor pressures suitable for use in mechanical, turbomolecular and diffusion pumps.

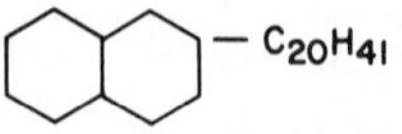

Fig. 13.6   Alkyl naphthalene structure.

Perfluoropolyethers are stable Lewis bases which react with few chemicals. They should not be placed in contact with ammonia, amines, liquid fluorine, liquid boron trifluoride, or sodium or potassium metal. Laboratory experiments have shown PFPE fluids to decompose when heated sufficiently ($>100°C$) in the presence of Lewis acids. The trichlorides and trifluorides of aluminum and boron are examples of Lewis acids that may be generated or are used as etchants in plasma and reactive ion etching. If heated to a high temperature Lewis acids act as depolymerization catalysts and cause the release of toxic fragments.

Chlorofluorocarbon fluids for vacuum pump use are synthesized by polymerization of chlorotrifluoroethylene [34]. Their structure consists of repeating links of $CClF_3$ terminated by CF_3 groups. See Fig. 13.5c. Chlorofluorocarbons currently used for vacuum applications have an average molecular weight of <1000. They have been stripped by distillation of the fraction whose molecular weight is less than 602 ($n < 4$ in Fig. 13.5c). These fluids have a high density, a low viscosity index and a vapor pressure considerably higher than that of PFPE. They are good lubricants for both the full film and boundary lubrication region. Chlorofluorocarbon fluids are nonflammable and inert to a large number of commonly used chemicals. They are reactive to the same general classes of chemicals as PFPE. The extremely high localized temperatures of minute seizures of aluminum which expose unoxidized surfaces have been reported to cause a chemical reaction with chlorofluorocarbon fluids which has resulted in detonation. Chlorofluorocarbon fluids are used only in mechanical pumps because their vapor pressures are too high for use in diffusion pumps. Their vapor pressure-viscosity characteristics make them well suited for piston pumps; however, diffusion pumps operate near the decomposition temperature of these fluids.

Other Synthetics

One other fluid currently produced for vacuum applications which has not been previously discussed is an alkyl naphthalene. See Fig. 13.6. This fluid is chemically resistant, moderately inexpensive and has a low vapor pressure which makes it suitable for use in diffusion pumps [35,36].

13.3   FLUID SELECTION

In this section we describe typical fluids that are commercially available for use in rotary, lobe, and diffusion pumps. We describe how the pump requirements and possible gas reactions limit the choice of pump fluids. Properties of representative fluids are included.

13.3.1   Rotary Vane, Piston, and Lobe Pump Fluids

Fluids used in vane and piston pumps must provide a vacuum seal between the moving surfaces and lubricate the bearings and sliding surfaces. The fluid in a lobe blower is used to lubricate the gear drive. Fluid assists in transferring heat from the bearings to the pump surface or cooling jacket in all three pumps. Fluids should not react with process gases or evolve fragments which could contaminate a process.

The fluid must have a low vapor pressure and be viscous enough to form a film that will fill the gap between the moving surfaces. The viscosity required in a particular pump depends upon the clearances between moving parts, the rotational speed, and the pump operating temperature. Table 13.1 lists typical properties of representative vane, piston, lobe blower and turbomolecular pump fluids. The viscosity indexes tabulated here were obtained from the manufacturer or calculated from the ASTM standard [37]. Table 13.2 gives relative costs for these fluids. Some piston and lobe pumps are machined to fine tolerances, while others are not so closely machined and require a viscous fluid. Viscosity specifications are available from pump manufacturers. Appendixes F.2 to F.4 give vapor pressures and kinematic viscosities.

Interpreting the vapor pressure data is not so easy. Some were taken on a Knudsen cell, some on an isoteniscope, and others by unstated procedures. Manufacturers vapor pressure data may be taken from the "typical" product, the "worst case" product, or from a sample lot. There can be an order of magnitude difference between worst case and typical data. Each manufacturer needs to be consulted to interpret their published data.

"Mechanical pump" grade mineral oils are satisfactory for most routine mechanical pumping applications such as backing a diffusion pump or roughing an air-filled chamber. It is common in research and development labs to find the lighter "diffusion pump" grade mineral oils used in mechanical pumps. Their use results in a somewhat lower base pressure, and less backstreaming provided the pump was flushed several times to eliminate all traces of the previous fluid. However, they ultimately degrade thermally [33]. Inhibited and fully saturated mineral oils (white oils) have been refined for improved resistance to oxidation and mild corrosive gases. Both fluids are flammable in oxygen.

Table 13.1   Typical Properties of Mechanical and Turbomolecular Pump Fluids

Chemical Type	Representative Trade Name	MW (ave)	Sp. Gr. at 25°C	P_v at 20°C (Pa)	Pour Point (°C)	Viscosity Index	Fire Point (°C)
Mineral Oil							
"mech. pump"	Balzers P-3[45]	190	0.88	1×10^{-2}	-16	95	295
	CVC-70/19[46]	325	0.91	1×10^{-3}	-15	130	250
	Duo-Seal 1407[47a]	450	0.88	—	-6.7	>95	240.6
	Inland 19[48]	440	0.88	5×10^{-4}	-15	130	244.5
inhibited	Syn-Lube[49b]	—	1.04	1.2(70°C)	-17.8	145	240.2
	Inland 07[48]	—	0.88	8×10^{-4}	-12	120	272.2
white oil	CVC-TW7[46]	—	0.86	2×20^{-4}	-9	130	270
	Inland-TW[49]	—	0.86	2×10^{-4}	-9	130	270
cleaning fluid	Solvex-10[46]	250	0.91	8×10^{-2}	-20	-50	220
	Inland FF-10[48]	—	—	—	—	25	221
Ester							
phosphate	Fyrquel 220[18c,d]	410	1.15	—	-17.8	18	365.6
sebacate	Balzers T-11[45e]	416	0.92	5×10^{-4}	-60	130	251
	Leybold HE-500[50e]	430	0.92	2×10^{-3}	-62	130	—
Siloxane chloromethyl polysiloxane	Versilube F-50[51c,d,f]	3000	1.05	—	-73	335	337.8
Fluorocabon polychlorotrifluoroethylene	HaloVac 125[52b,g]	<1000	1.95	1.5(69°C)	-17.8	-160	None
PFPE	Fomblin 25/5[19h]	3250	1.9	3×10^{-4}	-35	120	None
	Fomblin 06/6[19h]	1800	1.88	3×10^{-4}	-50	50	None
	Krytox 1514[32h]	3000	1.88	3×10^{-4}	-35	100	None
	Krytox 1506[32h]	2150	1.86	3×10^{-4}	-45	50	None

Source: Reprinted with permission from *J. Vac. Sci. Technol. A,* **2**(2), p. 174, J. F. O'Hanlon. Copyright 1984, The American Vacuum Society.

[a] Vapor pressure data were taken on a trapped McLeod gauge and the data are not representative of the oil;

[b] blank-off pressure;

[c] vapor pressure is a function of the fluid temperature-pressure history; blank-off pressures of order 1 Pa are obtainable;

[d] fire and oxidation resistant but not fireproof;

[e] contains additives;

[f] for low temperature applications;

[g] recommended by the manufacturer and the NAEC [53] for oxygen pumping.

[h] recommended by the manufacturer for oxygen pumping.

Table 13.2   Relative Cost of
Mechanical Pump Fluids[a]

Mechanical Pump Fluid	Cost
Perfluoroether	80
Chlorofluorocarbon	60
Chlorosiloxane	15
Phosphate ester	2
Low P_v hydrocarbon	2-5
Hydrocarbon	1

[a] Normalized to mineral oil.

Light hydrocarbon fluids are used for cleaning contaminants from mechanical pumps. These fluids, called flushing fluids, are simply a light end by-product from the distillation of ordinary mechanical pump oil. They have a high vapor pressure and are thin (about half the viscosity of normal oils). These light by-products are highly aromatic (high viscosity index) and are good solvents. They clean by causing the pump to run hot. The increased temperature aids in the dissolution of sludge from the pump's interior walls. These fluids simplify periodic maintenance of large mechanical pumps.

For some applications a hydrocarbon oil is totally unsuitable because it reacts with the process gas. Many corrosive gases will quickly polymerize mineral oil while others will cause its slow decomposition. It is often necessary to use an inert fluid for pumping corrosives. In some applications inert fluids are required for safety, and for others they last longer than mineral oil, thereby reducing maintenance and operating costs.

Fluorocarbon fluids can be used in vane, piston and lobe pumps. They have vapor pressures equal to mineral oil and are an excellent choice for use in a vane pump for which chlorofluorocarbon fluids have barely acceptable vapor pressures. PFPE does not oxidize and is completely safe for use when pumping oxygen. It is inert to most corrosive gases. However corrosive gases can slowly etch the interior of a pump if the acids they generate are not removed continually. Acid neutralizing, recirculating oil filters are especially important when using PFPE because the fluid is only infrequently changed. Lewis acids are one class of chemicals which can be reactive with PFPE at temperatures over 100°C. Oil temperatures in ballasted or viscous flushed pumps can reach 120°C and vane tip temperatures can be 100°C higher than the bulk oil temperature under any operating condition. Slow decomposition of PFPE results, and a small amount of fluid must be added periodically. Trichlo-rotrifluoroethane is an effective solvent for any of the halogenated mechanical pump fluids.

Chlorofluorocarbon fluids are more suitable for use in piston and lobe blowers than in vane pumps. Pumps for corrosive gas use are typically run with a large nitrogen gas ballast or nitrogen viscous purge. The heat of compression increases the operating temperature of a vane pump to the point where the vapor pressure of the fluid is equal to the viscous sweeping gas. The high viscosity cuts have a low vapor pressure and are suitable for use in piston and lobe blowers.

Until the availability of fluorocarbon fluids, phosphate esters were used to pump oxygen. Phosphate esters are used for pumping chemicals for which they are solvents. For example the vapors released from pumping epoxies would gum a pump lubricated with mineral oil. Phosphate esters react slowly with water during pump operation. This reaction causes the base pressure of the fluid to increase with time, resulting in frequent oil changes. Phosphate ester and polysiloxane fluids have higher fire points than hydrocarbons and are considered fire resistant and not fireproof.

Chloropolysiloxane fluids have been used in mechanical pumps. Their high viscosity index makes them suitable for use at low temperatures. They are not widely used in mechanical pumps and have no unique characteristic other than high viscosity index that would warrant their selection over another fluid.

13.3.2 Turbomolecular Pump Fluids

The requirements for a turbomolecular pump oil are somewhat different from those of mechanical pump oil. Because the bearing loading is not severe, a high shear strength, high viscosity oil is not required. In full-film lubrication the coefficient of friction is proportional to $\eta U/L$. See Fig. 18.1. High speed bearings require a low viscosity oil. Although the average fluid temperature is $70°C$, spot heating on the bearings can cause decomposition of fluids with poor vapor pressure. It is important that the fluid be vacuum degassed before use to prevent foaming.

A single-cut, light viscosity mineral oil which has been highly refined to remove both light ends and tars will work in a turbomolecular pump. The small amount of residual tar is of less concern than hydrogen from light ends which has a low compression ratio and contributes to the background gas spectrum.

Sebacate ester, modified by the addition of an antioxidant, a rust inhibitor, a viscosity index improver and extreme pressure additive, is useful for lubricating high speed turbomolecular pump bearings. MIL-L-6085A aircraft instrument lubricant is one such formulation of sebacate ester that is similar to those used to lubricate turbomolecular pumps. PFPE fluids have been used in many turbomolecular pumps. The manufacturer should be consulted to determine if PFPE is compatible with a specific pump and, if so, what viscosity is required. The properties of several turbomolecular pump fluids are given in Table 13.2.

13.3.3   Diffusion Pump Fluids

The ideal diffusion pump fluid should be stable, have a low vapor pressure, low specific heat, a low heat of vaporization, and be safe to handle, dispose of, and use. It should not thermally decompose, entrap gas, or react with its surroundings. Unfortunately no such fluid exists. Mercury was the first and only elemental fluid used in diffusion pumps. Today it has been supplanted by distilled hydrocarbons and synthetic fluids for almost all applications. The properties of the distilled hydrocarbons and synthetic fluids currently used in diffusion pumps are given in Table 13.3. Relative costs are given in Table 13.4. The vapor pressures of diffusion pump fluids are given in Appendix F.3.

Mercury has a high vapor pressure and must be trapped during operation. It is also toxic. Even so it is still used in some specialized applications like mass spectrometry because it does not decompose or dissolve gas and because it has an easily identifiable mass spectrum. Mercury cannot be used interchangeably with oil in a conventional oil pump. Mercury requires different operating conditions and jets, and it will react with some materials used in the construction of pumps designed for use with organic fluids. The sustained operation of a liquid nitrogen trap over a mercury pump will deplete the boiler of its charge after a few days unless an additional baffle, which is warmer than −40°C, is placed between the liquid nitrogen trap and the pump.

The ultimate pressure obtainable with a mineral oil is limited by its decomposition on heating. Several "diffusion pump" grades are manufactured for different applications. Lighter "diffusion pump" cuts (ave. molecular weight of 300 to 450) are used in applications where high pumping speed, moderate pressure, and low cost are most important. These fluids are sometimes used in mechanical pumps. Heavier diffusion pump grades (ave. molecular weight > 550) are used to reach low ultimate pressures with reduced backstreaming. Inhibited diffusion pump fluids with improved oxidation resistance are used for operations such as vacuum metallizing. All mineral oils oxidize when exposed to air while they are hot.

Tetraphenyl silicone is extensively used in quick-cycled, unbaked systems because of its moderate cost, low backstreaming, thermal stability and oxidation resistance. It has a freezing point of 20.5°C. Occasionally it will be found frozen in an unheated storage area. Pentaphenyl silicone has improved stability and reduced vapor pressure and is used widely in systems which are baked to achieve the lowest ultimate pressures. Petraitis [38] has shown that the ultimate pressure in a diffusion

Table 13.3   Typical Properties of Diffusion Pump Fluids

Chemical Type	Representative Trade Name	MW (ave)	Sp. Gr. at (25°C)	P_v at 25°C (Pa)	Fire Point (°C)	Latent Heat (J/g)	C_g (J/g·°C)	T_{boiler} 100 Pa (°C)
Mineral Oil								
"diff. pump"	Apiezon C[54]	574	0.87	1×10^{-6}	293	217.6	1.92	269
	Balzers-71[45]	280	0.88	3×10^{-6}	325	—	1.88	180
	Convoil 20[46]	400	0.86	5×10^{-5}	258.9	—	—	210
	Invoil 20[48]	450	0.88	5×10^{-5}	259	170	1.88	210
inhibited	Convoil 30[46]	400	0.86	8×10^{-4}	272	—	—	126
	Invoil 30[48]	—	0.88	8×10^{-4}	271	—	—	127
Ester								
butyl phthalate	butyl phthalate	278	1.04	3×10^{-3}	190.5	266	2.1	135
phthalate[a]	Invoil[48]	390	0.98	3×10^{-5}	229	217	2.1	200
sebacate[a]	Octoil-S[46]	427	0.91	3×10^{-6}	247	—	—	220
Silicone								
trisiloxane[b]	DC-704[55]	484	1.07	3×10^{-6}	275	220.5	1.72	220
	DC-705[55]	546	1.09	4×10^{-8}	275	215.9	1.76	250
Ether								
PPE[a,b]	Convalex 10[46]	447	1.2	6×10^{-8}	350	205.8	—	275
	Santovac 5[56]	447	1.2	6×10^{-8}	350	205.8	1.84[d]	275
diphenyl	Neovac Sy[57]	405	0.94	3×10^{-6}	230[e]	209	2.3	220
Fluorocarbon								
PFPE[a,c]	Fomblin 25/9[19]	3400	1.90	3×10^{-7}	None	29.3	1	230
	Krytox 1625[32]	3700	1.88	3×10^{-7}	None	41.8	1	230

Source: Reprinted with permission from *J. Vac. Sci. Technol. A,* **2**(2), p. 174, J. F. O'Hanlon. Copyright 1984, The American Vacuum Society.
[a] Suitable for use where ion or electron beams could cause polymerization;
[b] excellent oxidation resistance;
[c] does not react with oxygen, do not use when pumping Lewis acids;
[d] 4-ring ether;
[e] flash point.

pump charged with tetraphenyl silicone is dependent on purity, in particular on the quantity of low molecular weight impurity. Some suppliers re-distill high quality diffusion pump fluids to remove traces of light impurities. The added expense of such a distillation is not warranted because the extra distillation accomplishes immediately what a fractionating pump will do with time. Pentaphenyl silicone is rapidly degraded by BCl_3 [39] and slowly degraded by CF_4 and CCl_4.

Table 13.4   Relative Cost of
Diffusion Pump Fluids[a]

Diffusion Pump Fluid	Cost
Perfluoroether	50
Pentaphenylether	40
Pentaphenyl silicone	13
Tetramethyl silicone	8
Sebacate ester	6
Phthalate ester	3
Butyl phthalate	1.5
Light diffusion pump grade	1
Heavy diffusion pump grade	10

[a] Normalized to light mineral oil.

Pentaphenyl ether (PPE) was first suggested by Hickmann [4] for vacuum pump use because of its exceptional stability and low vapor pressure. Pentaphenyl ether and pentaphenyl silicone have the lowest vapor pressures of any fluid currently available. Solbrig and Jamison [40] were not able to induce explosions when either DC-705 or pentaphenyl ether was used in a system pressurized to 1/2 atm with pure oxygen. When using pentaphenyl ether it is absolutely necessary to restrict the cooling water flow so that the ejector stage operates at a wall temperature of 45 to 50°C. The ejector stage should be warm in all pumps to achieve adequate fluid degassing. However, if the ejector is too cold in a PPE-charged pump, a large fraction of the very viscous fluid will be hung up on the ejector and inlet walls. Some workers use a slightly larger fluid charge than recommended by the manufacturer to compensate for the fluid that resides on cool walls. PPE is suitable for use in mass spectrometers, leak detectors, residual gas analyzers, electron microscopes, and electron beam mask generation systems because it does not form an insulating film. It is also the most stable diffusion pump fluid available for pumping corrosive gases. Lewis acids will degrade PPE in diffusion pumps [40].

Esters are used for specific applications. Butyl phthalate and ethylhexyl phthalate are used where a rapid pumping, low cost fluid is desired. Ethylhexyl sebacate is an inexpensive fluid and does not polymerize to form an insulator. It is suitable but not the best material for use in leak detectors.

PFPE fluid is suitable for some diffusion pump applications. From activation energy measurements, decomposition in a diffusion pump at 250°C was estimated to be 0.009% (Krytox) and 0.14% (Fomblin) in ten years [40]. Fomblin thermally decomposed at the C-C bond to yield equal molar amounts of C_3F_6O, COF_2 and CF_3COF [31,41]. The decomposition of Krytox was reported to be essentially the same, but with less COF_2 [40]. PFPE decomposes on electron or ion impact into low molecular weight radicals and therefore does not form a film; this makes it useful for diffusion pumped electron beam systems. Conru and Laberge [42] devised a method of quantitatively determining specimen contamination in a scanning electron microscope and used it to show contamination from a sebacate ester was about 20 times that from PFPE. Residual gas analysis has shown the presence of high molecular weight fractions up to $M/z = 240$ advising against their use in heavy ion acceleration systems where exchange processes are dangerous [43]. PFPE is extremely stable for pumping all the usual reactive gases such as oxygen except for Lewis acids and fluorinated solvents. Since the boiler operates at temperatures over 200°C PFPE, a Lewis base, reacts and decomposes into toxic vapors. Under no circumstances should diffusion pumps charged with PFPE be used to pump on BCl_3, AlF_3, etc. Pearson et al. [41] report on diffusion pumps charged with Fomblin Y-H VAC 18/8 used for pumping HF and UF_6. After prolonged use at pressures of 3 Pa, solids of UO_2F_2 and UF_4 built up, and a dark sulfur containing colloidal suspension was observed.

Chastanger [20] studied the effects of tritium radiolysis on the long term performance of tritium pumping systems. His study showed mineral oils and polyphenyl ethers to be the only fluids in which no corrosive products were formed. The degradation products could be removed with a filter. Polyphenyl ethers were found to be the preferred diffusion pump fluid, while mineral oil was the best choice for mechanical and turbomolecular pump systems. Perfluoropolyether was not satisfactory because of the large quantity of corrosive radiolysis products (HF, F, COF_2) formed while pumping tritium.

The interior of a pump must be thoroughly cleaned before changing fluids. Hydrocarbon oils are easier to remove than silicones, but severely contaminated pumps that use either fluid may be cleaned successively in decahydronapthalene, acetone, and ethanol. If the pump is relatively clean, acetone and alcohol are usually adequate. Polyphenyl ether is soluble in trichloroethylene and in 1,1,1-trichloroethane, but the latter is less toxic than the former. Pumps charged with PFPE are cleaned with a fluorinated solvent such as trichlorotrifluoroethane or perfluorooctane [44]. If the pump fluid level is low, it is good practice to drain completely and refill rather than add fluid. During operation the light fractions of broad molecular weight fluid are selectively removed and the fluid's viscosity increases slowly with time. Gas bursting may be observed for several days following cleaning and changing fluid. Have patience while waiting for a newly charged pump to reach its ultimate pressure.

Pump boiler power may have to be changed if the fluid has thermal properties which differ significantly from those of the fluid for which the pump was designed. Most pumps in this country are designed to operate with a tetraphenyl siloxane. The heat required to maintain the boiler at its proper temperature depends upon the heat capacity C, the latent heat of vaporization h_v, of the new fluid, and pump losses. An estimate of boiler power differences due to fluid differences may be made by calculating their respective heat requirements. The heat required to vaporize 1 cc of a fluid is $\rho[C(T_{boiler} - T_{wall}) + h_v]$. Latent heats and heat capacities of representative fluids are given in Table 13.3.

13.4   RECLAMATION

Reclamation uses procedures such as settling, filtering, adsorption and distillation to remove contamination from pump fluids. See Table 13.5.

Table 13.5   Recommended Reclamation Procedures

Composition	Applications	Recommended Procedure
Distilled hydrocarbon	General purpose mech. pump fluids and flushing fluids	Chemical treatment and disposal at EPA approved facility for incineration or landfill
Highly distilled hydrocarbons	High performance mech. pump fluids and diffusion pump fluids	Chemical treatment, settling, filtration, distillation
Chlorofluorocarbons	Synthetic, non-reactive mech. pump fluid	Settling or filtration, vacuum distillation, adsorbent treatment
Perfluorinated polyether fluids	Synthetic, non-reactive mech. & diffusion pump fluid for use with O_2	Settling or filtration, chemical treatment, vacuum distillation, adsorbent treatment.
Polyphenyl ether	Synthetic, very stable low vapor pressure diffusion pump fluid	Settling, filtration, vacuum distillation, adsorbent treatment
Silicones	Very stable synthetic diffusion pump fluids	Settling filtration, vacuum distillation

Source: Reprinted with permission from *J. Vac. Sci. Technol. B,* 5(2), 255, C. B. Whitman. Copyright 1987, The American Vacuum Society.

Expensive fluids can be economically purified, while the cost of reclaiming inexpensive mineral oil is about the same as that of new fluid. Before attemtimg to institute reclaiming, one is advised to consult a reclaimant in order to determine how fluids should be segregated for shipmentuand how to specify the quality of the purified fluid. The cleaning procedures vary with the fluid and the type of contamination. The final cost will be dependent on the degree of contamination, the fluid type and the desired quality of the purified fluid. For example, very low vapor pressure color centers are costly to remove from a silicone fluid, but do not affect pump operation. Therefore, addition of a color specification, to a vapor pressure specification will increase the cost of the reclaimed fluid. The technology and economics of pump fluid reclamation has been reviewed by Whitman [58].

REFERENCES

1. G. W. Thomson and D. R. Douslin in *Techniques of Chemistry*, A. Weissberger and B. W. Rossiter, Eds., Vol. 1, *Physical Methods of Chemistry, Part 5*, Wiley, New York, 1971, p. 74.

2. M. Knudsen, *Ann. Physik*, **28**, 75, 999 (1909); **29**, 179 (1909).

3. J. P. Deville, L. Holland and L. Laurenson, *Proc. 3rd Int'l. Congr. Vac. Sci. Technol,* Pergamon, New York, 1965, p. 153.

4. K. C. D. Hickman, *Trans. 8th. Nat. Symp. on Vac. Technol.,* L. E. Preuss, ed., Macmillan, New York, 1961, p. 307.

5. AVS Standard 5.1, *J. Vac. Sci. Technol.*, **2**, 312 (1965).

6. ISO Standard 1607/2, 1978, American National Standards Institute, Inc., 1403 Broadway, New York, NY 10018.

7. AVS Standard 4.4, *J. Vac. Sci. Technol.*, **8**, 664 (1971).

8. D. J. Crawley, E. D. Tolmie and A. R. Huntress, *Trans 9th Nat'l. Vac. Symp. AVS,* Macmillian, New York, 1962, p. 399.

9. G. Rettinghaus and W. K. Huber, *J. Vac. Sci. Technol.,* **9**, 416 (1962).

10. N. T. M. Dennis, B. H. Colwell, L. Laurenson and J. R. H. Newton, *Vacuum*, **28** 551 (1978).

11. A. R. Huntress, A. L. Smith, B. D. Power, and N. T. M. Dennis, *Trans. 4th. Nat. Symp. on Vac. Technol.,* W. G. Matheson, Ed., Pergamon, New York, 1957, p. 104.

12. ASTM D-156, *1981 Annual Book of ASTM Standards*, Part 23, American Society for Testing Materials, Philadelphia, 1981, p. 111.

13. H. H. Zuidema, *The Performance of Lubricating Oils*, Reinhold, New York, 1959, p. 30.

14. ASTM D-92, *1981 Annual Book of ASTM Standards,* Part 23, American Society for Testing Materials, Philadelphia, 1981, p. 33.

15. J. M. Kuchta, *Summary of Ignition Properties of Jet Fuels and Other Aircraft Combustible Fluids*, U. S. Bureau of Mines, Pittsburgh Mining and Safety Research Center, AFAPL-TR-75-70, Air Force Aero Propulsion Laboratory, Wright-Patterson Air Force Base, Ohio, Sept., 1975, p. 16.

16. L. Laurenson, Private communication.

17. B. R. F. Kendall, *J. Vac. Sci. Technol.*, **21**, 886 (1982).

18. Stauffer Chemical Co., Specialty Chemical Division, Westport, CN 06880.

19. Montedison USA Inc., 1114 Avenue of the Americas, New York, NY 10036.

20. P. Chastagner, *Selection of Fluids for Tritium Pumping Systems*, 13th Annual Symposium on Applied Vacuum Science and Technology, Clearwater Beach, Florida, Feb. 6, 1984.

21. A. Cameron and C. M. McEttles, *Basic Lubrication Theory, 3rd ed.*, Horwood, Chichester, 1980, p. 33.

22. C. R. Burch, *Proc. Roy. Soc.* **A 123**, 271 (1929).

23. J. R. Davy, *Industrial High Vacuum,* Pitman, London, 1951, p. 109.

24. E. R. Booser, in *Kirk-Othmer Encyclopedia of Chemical Technology, 3rd ed.,* **14**, M. Grayson and D. Eckroth, Eds., Wiley, New York, 1980, p. 484.

25. H. H. Zuidema, *The Performance of Lubricating Oils*, Reinhold, New York, 1959, p. 177.

26. K. L. Kreuz, *Lubrication*, **56**, 77 (1970).

27. M. J. Fulker, M. A. Baker and L. Laurenson, *Vacuum*, **19**, 555 (1969).

28. A. L. Smith and J. C. Saylor, *Vacuum Symposium Trans.*, L. E. Preus, Ed., Committee on Vacuum Techniques, Boston, 1955. p.31. 1954.

29. C. L. Mahoney and E. R. Barnum, in *Synthetic Lubricants*, R. C. Gunderson and A. W. Hart eds., Reinhold, New York, 1962, p. 402.

30. L. Holland, L. Laurenson, and P. N. Baker, *Vacuum*, **22**, 315 (1972).

31. D. Sianisi, V. Zambeni, R. Fontanelli and M. Binaghi, *Wear*, **18**, 85 (1971).

32. Du Pont and Co., Chemicals and Pigments Department, Wilmington, DE 19898.

33. N. D. Lawson, *Perfluoroalkylpolyethers*, Report No. A-70020, du Pont, Wilmington, DE, Feb. 1970.

34. W. E. Ashton and C. A. Strack, in *Synthetic Lubricants*, R. C. Gunderson and A. W. Hart eds., Reinhold, New York, 1962, p. 246.

35. L. Laurenson, *J. Vac. Sci. Technol.*, **20**, 989 (1982).

36. L. Laurenson, *Vacuum*, **30**, 275 (1980).

37. ASTM D-2270, *1981 Annual Book of ASTM Standards*, Part 24, American Society for Testing Materials, Philadelphia, 1981, p. 277.

38. D. Petraitis, *Society of Vacuum Coaters, 24th Annual Technical Conf.,* Dearborn, MI, 1981, p. 73.

39. H. W. Lehmann, E. Heeb and K. Frick, *Proc. 3rd. Symp. on Plasma Processing*, **82-6**, J. Dieleman, R. G. Frieser and G. S. Mathad, Eds., 364 (1982).

40. C. W. Solbrig and W. E. Jamison, *J. Vac. Sci. Technol.,* **2**, 228 (1965).

41. R. K. Pearson, J. A. Happe, G. W. Barton, Jr., LLL Report UCID-19571, Sept. 27, 1982, Lawrence Livermore Laboratory, Livermore, CA.

42. H. W. Conru and P. C. Laberge, *Journal of Physics E*, **8**, 136 (1975).

43. A. Luches and M. R. Perrone, *J. Vac. Sci. Technol.,* **13**, 1097 (1976).

44. L. Laurenson, *Ind. Res. Dev.*, November, 61 (1971).

45. Balzers High Vacuum, Furstentum, Liechtenstein.

46. CVC Products, Inc., 525 Lee Rd, Rochester, NY 14603.

47. Sargent-Welch Co., Vacuum Products Division, 7300 N. Linder Ave., Skokie, IL 60077.

48. IVACO, Inc., 35 Howard Ave., Churchville, NY 14428.

49. Synthatron Corp., 50 Intervale Rd., Parsippany, NJ 07054.

50. Leybold-Heraeus G.m.b.H., Koln, West Germany.

51. General Electric Co., Silicone Products Dept., Waterford NY 12188.

52. Fluoro-Chem Corp., 82 Burlews Court, Hackensack, NJ 07601.

53. T. D. Weikel and H. H. Yuen, *Vacuum Pump Explosion Study*, NAEC-GSED-60, Naval Air Engineering Center, Philadelphia, PA., August 1972.

54. Edwards High Vacuum, Inc. 3279 Grand Island Blvd. Grand Island, NY 14072.

55. Dow Corning Co., Inc., 2030 Dow Center, Midland, MI 48640.

56. Monsanto, Co., 800 N. Lindbergh Blvd., St. Louis, MO. 63166.

57. Varian Associates, Lexington Vacuum Division, 121 Hartwell Ave., Lexington, MA 02173.

58. C. B. Whitman, *J. Vac. Sci. Technol. B*, **5**(2), 255 (1987).

PROBLEMS

13.1 Laurenson [*J. Vac. Sci. Technol.*, **20**, 989 (1982)] measures the following pressures in a small mechanical pump with the following gauges and pump fluids. Explain the relative differences in readings.

	Ultimate Pressure (Pa)	
Fluid	Trapped McLeod	Capacitance Manometer
Chlorotrifluoroethylene	4×10^{-1}	2.6
Perfluoropolyether	3.2×10^{-1}	5×10^{-1}
Mineral oil	4×10^{-2}	5×10^{-2}

13.2 For the fluid and Knudsen cell example given in Section 13.1.1, (PPE, 1 Pa, 200°C), calculate: (a) the equilibrium number per unit time of fluid molecules arriving at and departing from the surface within the Knudsen cell, (b) the ratio per unit time of molecules leaving the orifice to those leaving the oil surface, and (c) the ratio of the mean free path of the PPE vapor to the diameter of the Knudsen cell orifice.

13.3 Four pumps are operating in a plant. One uses a saturated white paraffin oil, one a chlorofluorocarbon fluid, one a silicone, and the last a perfluoropolyether. The four fluids are all transparent. How would you identify the fluid in each pump without the use of any sophisticated equipment, e.g., with what is available in a typical pump maintenance area? Keep it simple.

13.4 What are the effects of operating (a) a rotary vane pump, (b) a Roots pump, and (c) a diffusion pump with too much pump fluid and too little pump fluid?

13.5 (a) What happens to the vapor pressure of a diffusion pump fluid after it has been operated for some time in a fractionating diffusion pump? (b) What happens to mechanical pump fluid after it has been operated in a pump for an extended period of time?

13.6 The fluid drained from one rotary vane pump backing a diffusion pump was found to be six parts mineral oil and one part water. Describe the effects of this "fluid" on the operation of the diffusion pump and the vane pump. How do you think the water got into the pump? How would you prevent it from recurring?

13.7 What are the effects of the light ends and the tars in a turbomolecular pump fluid, and in a diffusion pump fluid?

13.8 How can water vapor be removed from a rotary vane pump oil while the pump is operating?

13.9 Compare a high quality mineral oil and a tetraphenyl silicone fluid for use in a diffusion pump used for exhausting an unbaked production evaporating system used to deposit aluminum on plastic at 10^{-3} Pa. Give an evaluation of the advantages and disadvantages of each fluid.

13.10 A 4-in. diffusion pump charged with DC-704 is used for evacuating CRT's before aluminizing the bulb interior. The bulbs to be aluminized have their faceplates coated with phosphor and lacquer and the base of their funnels coated with a carbon containing dag. During routine maintenance, the pump was removed from the system. There was little fluid in the boiler, but the interior walls of the pump casing were covered with a thick colorless deposit that could be removed by a knife or softened with heat. What happened?

CHAPTER 14

Getter and Ion Pumps

Getter and ion pumps are entrainment pumps. They operate by capturing gas molecules and binding them to a surface. The physical or chemical forces that bind molecules to surfaces are sensitive to the gas species, and all gases are not pumped equally well. As a result two or more entrainment processes are usually combined to pump effectively a wide range of active and noble gases.

Getter and ion pumps are often referred to as clean pumps. They are clean in the sense that they do not backstream heavy organic molecules as do diffusion or oil-sealed mechanical pumps. For many applications cleanliness is the freedom not only from large organic molecules, but from hydrogen, methane, carbon oxides, and inert gases as well. In that sense entrainment pumps do contaminate. Certain gases will displace other adsorbed gases, and carbon in some metals can react with water vapor when heated to produce methane or carbon oxides. Hydrogen and other gases may be poorly pumped or thermally released from surfaces on which they were sorbed. Entrainment pumps may produce, or not pump, one or more gas species. The labeling of these gases as contaminants depends on the application.

In this chapter we review the titanium sublimation pump (TSP), the non-evaporable getter pump (NEG), and the ion pump.

14.1 GETTER PUMPS

Many reactive metals rapidly pump large quantities of active gases because they getter (react with) the gases. The gas either reacts to form a surface compound, for example TiO, or diffuses into the bulk of the getter as does hydrogen. The pumping speed of a surface getter is determined by the sticking coefficient of the gas on the surface. In a surface getter pump there is little diffusion into the bulk. It is often cooled to enhance the sticking coefficient, and at reduced temperatures indiffusion is slow. The pumping speed of bulk getter material is limited by the diffusion of gas through the surface compounds, so bulk getters are usually operated at elevated temperatures to enhance the dissolution and diffusion of the surface compounds. The titanium sublimation pump is a surface getter pump, and the non-evaporable getter pump is a bulk getter pump.

14.1.1   Titanium Sublimation Pumps

Many metals including molybdenum, niobium, tantalum, zirconium, aluminum and titanium are surface getters for active gases. They become the active surface of a vacuum pump when they are deposited on a surface in a thin-film layer. Titanium is the choice for commercial pumps because it can be sublimed (changed from the solid to vapor state without being a liquid) at much lower temperatures than most other metals, is inexpensive, and pumps a large number of gases. Figure 14.1 sketches one form of titanium sublimation pump. An alternating current heats the filament which sublimes the titanium and deposits it on adjacent walls. Pump elements are fabricated with three or four separately heated filaments to extend the time between filament replacements. Active gases are captured on the fresh titanium surface which is cooled with water or liquid nitrogen. Because the pumped gases cannot be desorbed by heating, a fresh titanium layer must be deposited periodically to ensure continuous pumping. The pumping characteristics of titanium differ for the active gases, the intermediate gases, and the chemically inactive gases. The active gases (carbon oxides, oxygen, water vapor, and acetyl-

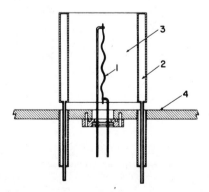

Fig. 14.1   Schematic of a basic titanium sublimation pump. (1) Titanium alloy filament, (2) coolant reservoir, (3) titanium deposit, (4) vacuum wall.

ene) are pumped with high sticking coefficients. Water dissociates into oxygen and hydrogen which are then pumped separately. The temperature of the film has no major effect on the pumping speed of these gases because the sticking coefficients are generally near unity in the 77 to 300 K range. The sticking coefficients of the intermediate gases (hydrogen and nitrogen) are low at room-temperature but increase at 77 K. After adsorbing, hydrogen will diffuse into the underlying film. The chemically inactive gases such as helium and argon are not pumped at all. Methane has the characteristics of an inactive gas and is only slightly sorbed on titanium at 77 K. Figure 14.2 gives the room temperature sorption characteristics of several active and intermediately active gases [1]. The sticking coefficient is the highest for all gases on a clean film, and for the very active gases remains so until near saturation.

The replacement of one previously sorbed gas by another gas is important. It does create a memory effect, and also results in actual sticking coefficients that depend on the nature of the underlying adsorbed gas. Gupta and Leck [1] observed a definite order of preference in gas replacement. Table 14.1 illustrates the order in which active gases replace less active gases. Oxygen, which is the most active gas, can replace all other gases, while methane, which is bound only by van der Waals forces, is displaced by all other active gases.

Gas replacement is a major cause of the large differences in measured sticking coefficients, especially when the films were not deposited under clean conditions. Harra [2] has reviewed the sticking coefficients and sorption of gases on titanium films measured in several independent studies and tabulated their average values in Table 14.2. These coeffi-

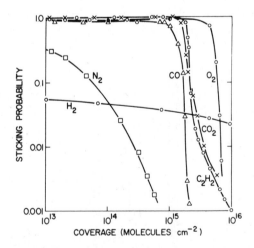

Fig. 14.2 Room-temperature sorption characteristics for pure gases on batch evaporated, clean titanium films. Reprinted with permission from *Vacuum*, **25**, p. 362, A. K. Gupta and J. H. Leck. Copyright 1975, Pergamon Press, Ltd.

Table 14.1 Order of Preference of Gas Displacement on Titanium Films[a]

	Displaced Gas				
Gas Being Pumped	CH$_4$	N$_2$	H$_2$	CO	O$_2$
CH$_4$		N	N	N	N
N$_2$	Y		N	N	N
H$_2$	Y	Y		N	N
CO	Y	Y	Y		N
O$_2$	Y	Y	Y	Y	

Source. Reprinted with permission from *Vacuum,* **25,** p.362, A. K. Gupta and J. H. Leck. Copyright 1975, Pergamon Press, Ltd.
[a] Y = Yes, N = No.

cients represent the average of the values obtained in different laboratories and under different conditions. They are probably more representative of those in a typical operating sublimation pump whose history is not known than are those measured under clean conditions.

The TSP operates at pressures below 10^{-1} Pa. Above that pressure surface compound formation inhibits sublimation. A typical pumping speed curve is sketched in Fig. 14.3. At low pressures there are few collisions between titanium atoms and gas molecules until the titanium atoms reach the surface. At pressures below 10^{-4} Pa, more titanium is sublimed than is needed when the filament is operated continuously. This results in a constant pumping speed that is determined by the surface area of the film and the conductance of any interconnecting tubing. At pressures greater than 10^{-4} Pa titanium-gas collisions occur before the titanium strikes the surface and the pumping speed is determined by the rate of titanium sublimation. Gas throughput is proportional to the titanium sublimation rate. Therefore the pumping speed will decrease as $1/P$, as sketched in Fig. 14.3.

The calculation of the pumping speed in the low pressure region is not easy to do precisely because of the sticking coefficient uncertainty and

**Table 14.2   Initial Sticking Coefficient and Quantity
Sorbed for Various Gases on Titanium**

Gas	Initial Sticking Coefficient (300 K)	(78 K)	Quantity Sorbed[a] ($\times 10^{15}$ molecules/cm^2) (300 K)	(78K)
H_2	0.06	0.4	8-230[b]	7-70
D_2	0.1	0.2	6-11[b]	-
H_2O	0.5	-	30	-
CO	0.7	0.95	5-23	50-160
N_2	0.3	0.7	0.3-12	3-60
O_2	0.8	1.0	24	-
CO_2	0.5	-	4-24	-
He	0	0		
Ar	0	0		
CH_4	0	0.05		

Source. Reprinted with permission from *J. Vac. Sci. Technol.*, **13**, p. 471, D. J. Harra. Copyright 1976, The American Vacuum Society.

[a] For fresh film thicknesses of 10^{15} Ti atoms/cm^2.

[b] The quantity of hydrogen or deuterium sorbed at saturation may exceed the number of Ti atoms/cm^2 in the fresh film through diffusion into the underlying films at 300 K.

the geometry. In molecular flow the pumping speed S of the geometry shown in Fig. 14.1 is given approximately by

$$\frac{1}{S} = \frac{1}{S_i} + \frac{1}{C_a} \qquad (14.1)$$

where S_i is the intrinsic speed of the surface and C_a is the conductance of the aperture at the end of the cylindrical surface on which the titanium is deposited. This conductance can be ignored if the film is deposited on the walls of the chamber. If a valve or connecting pipe is used, the appropriate series conductance should be added. The intrinsic speed is approximately

$$S_i(\text{L/s}) = 1000A\frac{v}{4}s' \qquad \blacktriangleright (14.2)$$

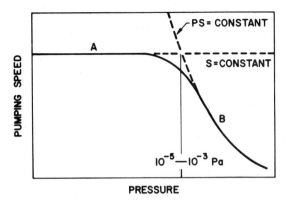

Fig. 14.3 Characteristic pumping speed versus pressure for a TSP: (a) speed determined by the getter area, sticking coefficient, gas species, and inlet conductance; (b) speed determined by pressure and rate of sublimation.

where A is the area of the film in units of m^2, v is the gas velocity in m/s, and s' is the sticking coefficient of the gas. Cooling to 77 K provides little additional pumping speed in pumps whose speed is conductance limited by the geometry.

At high pressures the pumping speed is determined by the rate of titanium sublimation. This theoretical maximum throughput is related to the titanium sublimation rate (TSR) by the relation [3]

$$Q(Pa - L/s) = \frac{10^8 V_o TSR(atoms/s)}{n N_o}$$

or

$$Q(Pa - L/s) = \frac{10^{-18}}{n} TSR \ (atoms/s) \qquad \blacktriangleright (14.3)$$

V_o is the normal specific volume of the gas, N_o is Avogadro's number, and n is the number of titanium atoms that react with each molecule of gas; n = 1 for CO and n = 2 for N_2, H_2, O_2, and CO_2 [2]. For titanium (14.3) may be rewritten as

$$Q(Pa - L/s) = \frac{1.25 \times 10^{-2}}{n} TSR \ (\mu g/s) \qquad (14.4)$$

This theoretical throughput can be reached only when the titanium is fully reacted with the gas. The corresponding pumping speed is obtained by dividing the throughput by the pressure in the pump.

The TSP is used in combination with other pumps that will pump inert gases and methane. Small TSPs are used continuously for short periods to aid in crossover between a sorption pump and an ion pump. Large pumps have been developed for use in conjunction with smaller ion

pumps for long term, high throughput pumping. The TSP is used intermittently for long periods at low pressures to provide high speed pumping
of reactive gases. At low pressures the film need be replaced only periodically to retain the pumping speed. Titanium is sublimed until a fresh
film is deposited. The pump is then turned off until the film saturates.
Figure 14.4 [3] sketches the pressure rise with decrease in pumping speed
as the titanium film saturates in a typical ion pumped system. Here P_i is
the initial pressure with the ion and sublimator operating and P_f is pressure with only the ion pump operating. Not shown on these curves is the
pressure burst due to gas release from the titanium during sublimation.
Hydrogen, methane, and ethane are released from titanium during heating. The methane and ethane result from a reaction between hydrogen
and the carbon impurity in the hot filament [1,4]. After sublimation
begins, hydrogen is pumped. Methane and ethane are marginally pumped
on surfaces held at 77 K.

Commercially available TSPs use directly heated filaments, radiantly
heated sources, or electron-beam heated-sources. The most commonly
used source is a directly heated filament with a low voltage ac power
supply. Filaments were first made from titanium twisted with tantalum or
tungsten and later from titanium wound over niobium and tantalum wire

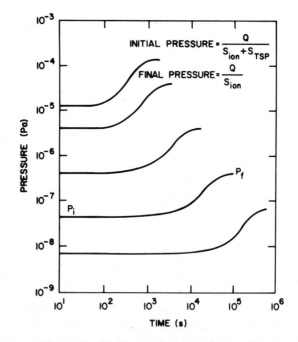

Fig. 14.4 Typical pressure rise due to decrease in pumping speed as a titanium film
saturates. Reprinted with permission from *General Characteristics of Titanium Sublimation
Pumps*, B. E. Keitzmann, 1965, Varian Associates, 611 Hansen Way, Palo Alto, CA 94303.

[5]. Because of thermal contact problems, the sublimation rate proved to be unpredictable. Modern pumps use filaments fabricated from an alloy of 85% Ti and 15% Mo [6-8]. This filament has an even sublimation rate and a long life. A typical filament 15-cm long can be operated at sublimation rates of 30 to 90 $\mu g/s$. Large TSPs have been constructed with radiantly heated titanium at sublimation rates as high as 150 $\mu g/s$ [9,10] and with electron-beam-heated, rod-fed sources at sublimation rates ranging from 300 $\mu g/s$ to 0.15 g/s [11]. Electron-beam heated sources do not operate well at pressures higher than 10^{-3} Pa and for most applications are too expensive to operate at pressures below 10^{-5} Pa. They serve best as a high speed pump in the intermediate region. Radiantly heated sources are best for high speed pumping in the very high vacuum region.

14.1.2 Non-evaporable Getter Pumps

A non-evaporable getter (NEG) pumps by surface adsorption followed by bulk diffusion. Its speed for pumping active gases is determined by the gas diffusion rate into the bulk. For this reason NEG pumps are operated at high temperatures. They do not pump inert gases or methane because these gases do not adsorb on the surface. One effective getter for vacuum use is an alloy of 84% Zr and 16% Al [12-16]. This alloy, when heated to 400°C, has a pumping speed of ~ 0.3 L-s^{-1}-cm^{-2} (N_2), ~ 1 L-s^{-1}-cm^2 (CO_2, CO, O_2), and 1.5 L-s^{-1}-cm^{-2} (H_2) [12]. At room temperature H_2 is pumped at about half the speed that it is pumped at 400°C, provided that no oxide or nitride diffusion barriers exist. Other gases are not pumped at room temperature because the surface compounds quickly form diffusion barriers. All gases except hydrogen are pumped as stable compounds and are entrapped permanently [13]. Hydrogen is pumped as a solid solution and may be released by heating above 400°C. The diffusion of carbon monoxide, carbon dioxide, nitrogen and oxygen has been shown to obey a simple parabolic rate law [17]. Cecchi and Knize [18] have shown the speed of a NEG pump to be constant below 10^{-3} Pa, and decrease as $P^{1/2}$ above 5 × 10^{-1} Pa.

To operate the Zr-Al getter pump, the chamber is first evacuated to a pressure below 1 Pa, after which the pump is activated by heating to 800°C to indiffuse the surface layers. The temperature is then reduced to 400°C. The activation step is repeated each time the pump is cooled and released to atmosphere. Another alloy (Zr-70%, V-24.6%, Fe-5.4%) has demonstrated a high gettering efficiency at room temperature after activation at 500°C [19,20].

NEG pumps have been used as appendage pumps on small systems, as well as primary pumps in large UHV systems, fusion machines and particle accelerators. Appendage pumps equipped with getter cartridges fabricated from a plated steel coated with Zr-Al alloy have pumping speeds as high as 10 to 50 L/s [21]. One getter ion pump package has a combined pumping speed of 1000 L/s [22]. NEG pumps have a large capacity for pumping hydrogen [16]. The NEG pump has been used in fusion machines [19,23,24] because it can operate without a magnetic field, and has a high hydrogen pumping speed at room temperature. With the assistance of an ion pump to handle the methane and argon, the NEG can reach base pressures of 10^{-9} Pa. The CERN large positron collider uses 27 km of linear NEG pump with a speed of 500 L/s per meter of chamber [24]. Getter pumping is also found in Tokamaks to control the density of hydrogen plasmas and remove chemically active impurities. The use of glow discharge cleaning has been shown to have no deleterious effects on the operation of a Zr-Al NEG [25]. Hseuh and Lanni [26] have established a worst case pressure of less than 3×10^{-9} Pa in an accelerator storage ring using a linear Zr-V-Fe alloy and lumped ion pumps 10 m apart. Getter pumps are now finding other applications, such as for purifying gases used in semiconductor device processing equipment.

14.2 ION PUMPS

The development of the ion pump has made it possible to pump to the ultrahigh vacuum region without concern for heavy organic contamination. This pump exploits a phenomenon formerly considered detrimental to vacuum gauge operation—pumping gases by ions in Bayard-Alpert and Penning gauges. Ions are pumped easily because they are more reactive with surfaces than neutral molecules and if sufficiently energetic can physically embed themselves in the pump walls. If the ions were generated in a simple parallel-plate glow discharge, for example, the pumping mechanism would be restricted to a rather narrow pressure range. Above about 1 Pa the electrons cannot gain enough energy to make an ionizing collision and below about 10^{-1} to 10^{-2} Pa the electron mean free path becomes so long that the electrons collide with a wall before they encounter a gas molecule. Ions can be generated at lower pressures if the energetic electrons can be constrained from hitting a wall before they collide with a gas molecule. This confinement can be realized with certain combinations of electric and magnetic fields.

The pumping action of a magnetically confined dc discharge was first observed by Penning [27] in 1937, but it was not until two decades ago that Hall [28] combined several Penning cells and transformed the phenomenon into a functional pump. Some elemental forms of the (diode) sputter-ion pump are shown in Fig. 14.5. [29]. Each Penning cell is approximately 12 mm in diameter × 20 mm long with a 4-mm gap between the anode and the cathode. Modern pumps are constructed of modules of cells arranged around the periphery of the vacuum wall with external permanent magnets of 0.1 to 0.2 T strength and cathode voltages of ~ 5 kV.

The electric fields present in each Penning cell trap the electrons in a potential well between the two cathodes and the axial magnetic field forces the electrons into circular orbits that prevent their reaching the anode.   This combination of electric and magnetic fields causes the electrons to travel long distances in oscillating spiral paths before colliding with the anode and results in a high probability of ionizing collisions with gas molecules.   The time from the random entrance of the first electron into the cell until the electron density reaches its steady-state value of ~ 10^{10} electrons/cm^3 is inversely proportional to pressure.   The starting time of a cell at 10^{-1} Pa is nanoseconds, while at 10^{-9} Pa it is 500 s [30].   The ions produced in these collisions are accelerated toward the cathode, where they collide, sputter away the cathode, and release secondary electrons that in turn are accelerated by the field.   Many other processes occur in addition to the processes necessary to sustain the discharge; for example a large number of low energy neutral atoms are created by molecular dissociation and some high energy neutrals are created from energetic ions by charge neutralization as they approach the cathode, collide, and recoil elastically.

The actual mechanism of pumping in an ion pump is dependent on the nature of the gas being pumped and is based on one or more of the following mechanisms: (1) precipitation or adsorption following molecular dissociation; (2) gettering by freshly sputtered cathode material; (3) surface burial under sputtered cathode material; (4) ion burial following ionization in the discharge; and (5) fast neutral atom burial.   (Ions are neutralized by surface charge transfer and reflected to another surface

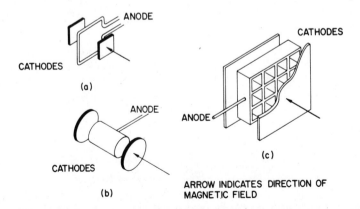

Fig. 14.5 Early forms of the diode sputter-ion pump: (a) ring anode cell [27]; (b) long anode cell [31]; same cell with Ti cathodes [32], (c) multicell anode [28].   Reproduced with permission from *Proc. 4th Int. Vac. Cong. (1968)*, p. 325, D. Andrew. Copyright 1969, The Institute of Physics.

where they are pumped by burial.) The first four of these mechanisms
were elucidated by Rutherford, Mercer and Jepsen [33], and the role of
the elastically scattered neutrals was explained by Jepsen [34]. These
mechanisms are illustrated in Fig. 14.6.

Organic gases, active gases, hydrogen, and inert gases are pumped in
distinctly different ways. There are a few generalities. Initially, gases
tend to be pumped rapidly and their partial pressure decays to a steady
state [34-36]. In steady state reemission rates equal pumping rates. This
is more pronounced with noble than with active gases. Pumping speeds
cannot be uniquely defined for a gas independent of the composition of
other gases being pumped simultaneously. The sputter ion pump is
capable of reemitting any pumped gas. This reemission or memory effect
complicates the interpretation of some experiments.

Organic gases are easily pumped by adsorption and precipitation after
being dissociated by electron bombardment [33].

Active gases such as oxygen, carbon monoxide, and nitrogen are
pumped by reaction with titanium, which is sputtered on the anode
surfaces, and by ion burial in the cathode. These gases are easily
pumped because they form stable titanium compounds [33].

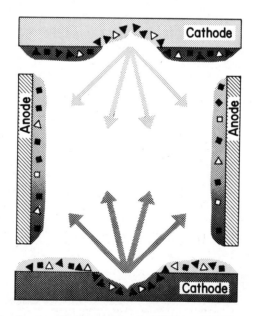

Fig. 14.6 Schematic diagram showing sputter deposition and pumping in a Penning cell:
■ chemically active gases buried as neutral particles; ▶ Chemically active gases ionized
before burial; □ inert gases buried as neutral particles; △ inert gases ionized before burial.
Reprinted with permission from *Proc. 4th Int. Vac. Congr. (1968)*, p. 325, D. Andrew.
Copyright 1969, The Institute of Physics.

Hydrogen behaves differently. Its low mass prevents it from sputtering the cathode significantly. It behaves much like it does in a TSP. It is initially pumped by ion burial and neutral adsorption [34,37] and diffuses into the bulk of the titanium and forms a hydride. Sustained pumping of hydrogen at high pressures will cause cathodes to warp [33] and release gas as they heat. The hydrogen pumping speed does not rate limit unless cathode surfaces are covered with compounds that prevent indiffusion. The pumping of a small amount of an inert gas, say argon, cleans the surfaces and allow continued hydrogen pumping [38], whereas a trace amount of nitrogen will reduce the speed by contaminating the surface [37].

Noble gases are not pumped so efficiently as active gases in a diode pump. They are pumped by ion burial in the cathodes and by reflected neutral burial in the anodes and cathodes. The noble gas pumping on the cathodes is mostly in the area near the anodes where the sputter build-up occurs. Because most of the neutrals are reflected with low energies in the diode pump, their pumping speed in the anode or other cathode is low; for example, argon is pumped only at 1 to 2% of the active gas speed.

Argon, in particular, suffers from a pumping instability. Periodically the argon pressure will rise as pumped gas is released from the cathodes. Figure 14.7 [29] illustrates some of the geometries that were devised as a solution to the problem of low argon pumping speed and its periodic reemission. Brubaker [39] devised a triode pump with a collector surface that operated at a potential between the anode and cathode (Fig. 14.7a).

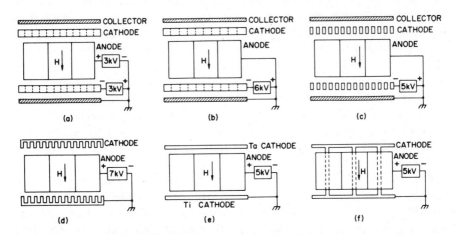

Fig. 14.7 Pump designs for inert gas pumping: (a) The triode pump of Brubaker [39]; (b) triode pump of Hamilton [40]; (c) triode Varian Noble Ion Pump [41]; (d) slotted cathode diode of Jepsen et al. [42]; (e) differential ion pump of Tom and Jones [43]; (f) magnetron pump of Andrew et al. [46]. Reprinted with permission from *Proc. 4th Int. Vac. Congr. (1968)*, p. 325, D. Andrew. Copyright 1969, The Institute of Physics.

Its function was to collect low energy ions that could not sputter. Hamilton [40] showed it worked equally well when the collector surface was held at anode potential (Fig. 14.7b). In the triode pump the argon pumping speeds are as high as 20% of the nitrogen speed. This high speed (high implantation rate) results from the high energy of the neutrals which are scattered at small angles from the cathode walls with little energy loss. Sputtering is much more efficient at these small angles than at normal incidence and sputtering of titanium on the collector is more efficient than in the diode pump. The slotted cathode [42] attempts to accomplish this sputtering with one less electrode than the triode (Fig. 14.7d). This pump has an argon pumping speed of 10% of the speed for air. Tom and Jones [43] devised the differential diode ion pump sketched in Fig. 14.7e. One titanium cathode was replaced by a tantalum cathode. In this manner the recoil energy of the scattered noble gas neutrals, which depends on the relative atomic weight of the cathode material and gas atom, is increased [44,45]. This gives more effective noble gas pumping than in the diode pump. An argon-stable magnetron structure, (Fig. 14.7f), was devised by Andrew et al. [46]. The central cathode rod is bombarded by a high flux of ions at oblique angles of incidence. The sputtering of the rod creates a flux that continually coats the cathode plates and the impinging ions and results in a net argon speed of 12% of the air speed [46]. Among the designs discussed here for increasing the argon pumping speed and reducing or eliminating its instability the triode and differential diode are in most widespread use.

The operating pressure range of the sputter-ion pump extends from 10^{-2} to below 10^{-8} Pa. A characteristic pumping speed curve of a diode and a triode pump are shown in Fig. 14.8. If starting is attempted at high pressures, say 1 Pa, a glow discharge appears and the elements heat and release hydrogen. As the pressure is reduced, the glow discharge extinguishes and the speed rapidly increases. At low pressures the speed decreases because the sputtering and ionization processes decrease. The exact shape of a pumping speed curve is a function of the magnetic field intensity, cathode voltage, and cell diameter-to-length ratio. As the pressure decreases, the ionization current decreases proportionately and the pressure in the pump may be obtained from the ion current without the need for an ionization gauge tube.

The lifetime of a diode pump is a function of the time necessary to sputter through the cathodes. A typical value is 5000 h at 10^{-3} Pa or 50,000 h at 10^{-4} Pa. The triode, which pumps slightly better than the diode at high pressures, is also easier to start at high pressures and has a lifetime of less than half the diode. In both pumps the life may be shorter due to shorting of the electrodes by loose flakes of titanium.

The sputter ion pump has the advantage of freedom from hydrocarbon contamination and ease of fault protecting but does suffer from the

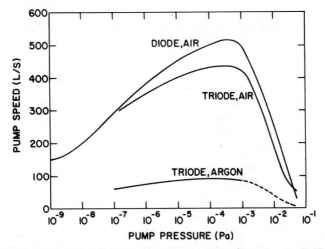

Fig. 14.8 Pumping speeds for air and argon for the 500-L/s Varian diode Vac Ion pump and for the 400-L/s triode Vac Ion pump. Speeds measured at the inlet of the pump. Reprinted with permission from Varian Associates, 611 Hansen Way, Palo Alto, CA 94303.

reemission of previously pumped gases, particularly hydrogen, methane and the noble gases.

REFERENCES

1. A. K. Gupta and J. H. Leck, *Vacuum*, **25**, 362 (1975).

2. D. J. Harra, *J. Vac. Sci. Technol.*, **13**, 471 (1976).

3. B. E. Keitzmann, *General Characteristics of Titanium Sublimation Pumps*, Varian Associates, Palo Alto, CA.

4. L. Holland, L. Laurenson, and P. Allen, *Trans. 8th Nat. Vac. Symp. (1961)*, Pergamon, New York, 1962, p. 208.

5. R. E. Clausing, *Trans. 8th Nat. Vac. Symp. (1961)*, Pergamon, New York, 1962, p. 345.

6. A. A. Kuzmin, *Prib. Tekh. Eksp.* **3**, 497 (1963).

7. G. M. McCracken and N. A. Pashley, *J. Vac. Sci. Technol.*, **3**, 96 (1966).

8. R. W. Lawson and J. W. Woodward, *Vacuum*, **17**, 205 (1967).

9. D. J. Harra and T. W. Snouse, *J. Vac. Sci. Technol.*, **9**, 552 (1972).

10. D. J. Harra, *J. Vac. Sci. Technol.*, **12**, 539 (1975).

11. H. R. Smith, Jr., *J. Vac. Sci. Technol.*, **8**, 286 (1971).

12. P. della Porta, T. Giorgi, S. Origlio, and F. Ricca, *Trans. 8th Nat. Vac. Symp.. (1961)*, Pergamon, New York, 1962, p. 229.

13. T. A. Giorgi and F. Ricca, *Nuovo Cimento Suppl.*, **1**, 612 (1963).

14. B. Kindl, *Nuovo Cimento Suppl.*, **1**, 646 (1963).

15. B. Kindl and E. Rabusin, *Nuovo Cimento Suppl.*, **5**, 36 (1967).

16. W. J. Lange, *J. Vac. Sci. Technol.*, **14**, 582 (1977).

17. S. Parkash and P. Vijendran, *Vacuum*, **33**, 295 (1983).

18. J. L. Cecchi and R. J. Knize, *J. Vac. Sci. Technol.*, **A1**, 1276 (1983).

19. C. Boffito, B. Ferrario, P. della Porta, and L. Rosai, *J. Vac. Sci. Technol.*, **18**, 1117 (1981).

20. C. Boffito, B. Ferrario and D. Martelli, *J. Vac. Sci. Technol.*, **A1**, 1279 (1983).

21. P. della Porta and B. Ferrario, *Proc. 4th Int. Vac. Congr. (1968)*, Institute of Physics and the Physical Society, London, **1**, 369 (1968).

22. S.A.E.S. Getters USA, Buffalo, NY.

23. J. S. Moenich, *J. Vac. Sci. Technol.*, **18**, 1114 (1981).

24. C. Benvenuti, *Nuc. Instrum. & Methods*, **205**, 391 (1983).

25. H. F. Dylla, J. L. Cecchi and M. Ulrickson, *J. Vac. Sci. Technol.*, **18**, 1111, (1981).

26. H. C. Hseuh and C. Lanni, *J. Vac. Sci. Technol.*, **A1**, 1283 (1983).

27. F. M. Penning, *Physica*, **4**, 71 (1937).

28. L. D. Hall, *Rev. Sci Instrum.*, **29**, 367 (1958).

29. D. Andrew, *Proc. 4th Int. Vac. Congr. (1968)*, Institute of Physics and the Physical Society, London. 1969, p. 325.

30. R. D. Craig, *Vacuum*, **19**, 70 (1969).

31. F. M. Penning and K. Nienhuis, *Philips Tech. Rev.*, **11**, 116 (1949).

32. A. M. Guerswitch and W. F. Westendrop, *Rev. Sci. Instrum.*, **25**, 389 (1954).

33. S. L. Rutherford, S. L. Mercer, and R. L. Jepsen, *Trans. 7th Nat. Vac. Symp. (1960)*, Pergamon, New York, 1961, p. 380.

34. R. L. Jepsen, *Proc. 4th Int. Vac. Congr. (1968)*, Institute of Physics and the Physical Society, London, 1969, p. 317.

35. A. Dallos and F. Steinrisser, *J. Vac. Sci. Technol.*, **4**, 6 (1967).

36. A. Dallos, *Vacuum*, **19**, 79 (1969).

37. J. H. Singleton, *J. Vac. Sci. Technol.*, **8**, 275 (1971).

38. J. H. Singleton, *J. Vac. Sci. Technol.*, **6**, 316 (1969).

39. W. M. Brubaker, *6th Nat. Vac. Symp. (1959)*, Pergamon, New York, 1960, p. 302.

40. A. R. Hamilton, *8th Nat. Vac. Sump. (1961)*, Vol. 1, Pergamon, New York, 1962, p. 338.

41. Varian Associates, 611 Hansen Way, Palo Alto, CA 94303.

42. R. L. Jepsen, A. B. Francis, S. L. Rutherford, and B. E. Keitzmann, *7th Nat. Vac. Symp. (1960)*, Pergamon, New York, 1961, p. 45.

43. T. Tom and B. D. Jones, *J. Vac. Sci. Technol.*, **6**, 304 (1969).

44. P. N. Baker and L. Laurenson, *J. Vac. Sci. Technol.*, **9**, 375 (1972).

45. D. R. Denison, *J. Vac. Sci. Technol.*, **14**, 633 (1977). See Ref. 1.

46. D. Andrew, D. R. Sethna, and G. F. Weston, *4th Int. Vac. Cong. (1968)*, Institute of Physics and the Physical Society, 1968, p. 337.

PROBLEMS

14.1 †Why does a non-evaporable getter (NEG) need to be heated to an activation temperature before being operated as a pump?

14.2 †How is a non-evaporable getter restored after saturation?

14.3 A TSP has an oxygen speed of 3000 L/s on a liquid-nitrogen-cooled surface at 1.33×10^{-4} Pa. How many titanium atoms must strike the pump surface per second per unit area?

14.4 Using the average sticking coefficients given in Table 14.2, calculate the room-temperature pumping speeds per unit area for oxygen, hydrogen, carbon dioxide and nitrogen.

14.5 Calculate the nitrogen pumping speed of the pump shown in Fig. 14.1 assuming it has a 30-cm diameter inlet, is 30-cm diameter in length and is cooled with water. Now replace the water with liquid nitrogen and calculate its speed. Why did the speed not increase significantly?

14.6 †How is a TSP restored after it is saturated?

14.7 99.99% argon is admitted into a chamber connected to an ion pump. The RGA shows a methane peak which is 10% of the argon peak. Explain.

14.8 †Why are ion pumps not used for routine pumping in the 10^{-4}-Pa pressure range?

14.9 †Why place a grounded screen over an ion pump inlet?

14.10 †For what reason is dry nitrogen used as a vent or flush gas in ion-pumped systems?

CHAPTER 15

Cryogenic Pumps

Cryogenic pumping is the entrainment of molecules on a cooled surface by weak van der Waals or dispersion forces. In principle, any gas can be pumped, provided the surface temperature is low enough for arriving molecules to remain on the surface after losing kinetic energy. Cryogenic entrainment is a clean form of pumping. The only gas or vapor contaminants are those not pumped or those not released from pumped deposits. Unlike the ion pump, the cryopump does not retain condensed and physically adsorbed gases after the pumping surfaces have been warmed. Proper precautions must be taken to vent the pumped gas load.

Cryopumps are used in a wide range of applications and in many forms. Liquid-nitrogen-cooled molecular sieve pumps are used as roughing pumps. Liquid nitrogen traps are used between diffusion pumps and chambers to pump oil and water vapor. Liquid nitrogen or liquid helium "cold fingers" are used in high vacuum chambers to augment other pumps. High and ultrahigh vacuum cryopumps are cooled by liquid cryogens or closed cycle helium gas refrigerators. Turbomolecular, TSP or NEG pumps are sometimes appended to cryopumps to improve their pumping speed for hydrogen and deuterium. Cryopumping is the only form of pumping by which extremely high speeds (10^7 L/s) can be realized.

Cryogenic pumping is not a new technique. The theory and techniques of pumping on cooled surfaces have been a part of vacuum technology much longer than the helium gas refrigerator. However, knowledge of both is necessary to understand the operation of a He gas cryopump. Several reviews of cryopumping have appeared in recent years. Hands [1] reviewed small refrigerator-cooled cryopumps and very large pumps used for fusion experiments. Bentley [2] explained the operation of the Gifford-McMahon refrigerator, and Haefer [3] has discussed the mechanisms of cryogenic pumping, given system calculations, and examples of pumps and applications.   In this chapter we review the mechanisms of cryocondensation and cryosorption on which all cryogenic pumping is based, and discuss pumping speed, ultimate pressure and saturation effects, refrigeration techniques, and pump characteristics.   We discuss system operation and regeneration techniques in Chapters 19 and 22, the problems of pumping gases at high flow rates in Chapter 21.

15.1   PUMPING MECHANISMS

Low-temperature pumping is based on cryocondensation, cryosorption, and cryotrapping.   In Chapter 4 we define the equilibrium or saturated vapor as the pressure at which the flux of vapor particles to the surface equals the flux of particles leaving the surface and entering the vapor phase, provided all the molecules, solid, liquid, and vapor, are at the same temperature.   The arriving molecules are attracted to condensation sites on the liquid or solid, where they are held for some residence time after which they vibrate free and desorb into the vapor phase.   The vapor pressure and residence time are temperature dependent.   As the temperature is reduced the vapor pressure is reduced and the residence time is increased.   See for example, Table 4.1.   Tables of vapor pressures of the common gases are given in Appendix B.5.   Cryocondensation becomes a useful pumping technique when a surface can be cooled to a temperature

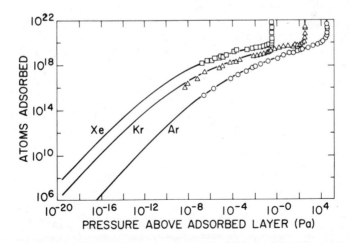

Fig. 15.1 Adsorption isotherms of xenon, krypton, and argon on porous silver adsorbent at 77 K. The lines represent plots of an analytic solution and the points are experimental. Reprinted with permission from *J. Chem. Phys.*, **73**, p. 2720, J. P. Hobson. Copyright 1969, The American Chemical Society.

at which the vapor pressure is so low and the residence time is so long that the vapor is effectively removed from the system. Liquid nitrogen is a good condenser of water vapor because the water vapor pressure at 77 K is 10^{-19} Pa. The probability c that an atom will condense on collision with a cold surface is called the condensation coefficient. Dawson and Haygood [4], Eisenstadt [5], and Brown and Wang [6] have measured the condensation coefficients of many gases at reduced temperatures and found them all to be between 0.5 and 1.0.

Condensation sites on the solid or liquid state are not the only locations at which atoms or molecules can become bound. Any solid surface has a weak attractive force for at least the first few monolayers of gas or vapor. Figure 15.1 describes a typical relationship between the number of molecules adsorbed and the pressure above the adsorbed gas for Xe, Kr, and Ar on porous silver at 77.4 K [7]. These adsorption isotherms tend toward a slope of 1 at very low pressures. This shows the number of adsorbed atoms goes to zero linearly with the pressure. The sorption sites become increasingly populated as the pressure increases. The limiting sorption capacity is reached after a few monolayers have been deposited. A typical monolayer can hold about 10^{15} atoms/cm^2. The actual number depends on the material. The data shown in Fig. 15.1 saturate at 2×10^{19} atoms because the surface area is larger than 1 cm^2. At the vapor pressure, condensation begins and the surface layer can increase in thickness. The thickness of the solid deposit is limited only by thermal gradients in the solid and by thermal contact with nearby surfaces of different temperatures. The density and thermal conductivity of the solid frost are a function of its formation temperature. It will decrease as the condensation temperature decreases.

A curve similar to Fig. 15.1 may be measured for each temperature of the sorbate. The effect of temperature on the adsorption isotherm is illustrated in Fig. 15.2 for hydrogen on a bed of activated charcoal. Gas

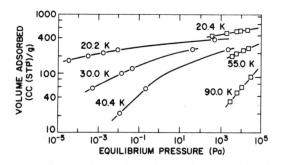

Fig. 15.2 Adsorption of hydrogen on coconut charcoal at low pressures. ○ Gareis and Stern [8], □ Van Dingenan and Van Itterbeek [9]. Reprinted with permission from *J. Vac. Sci. Technol.* 2, p. 165, S. A. Stern et al., Copyright 1965, The American Vacuum Society.

adsorption at a given pressure is increased if the temperature is reduced, because the probability of desorption is less than at higher temperatures.

Adsorption is an important phenomenon because it allows a vapor to be pumped to a pressure far below its saturated vapor pressure. For gases such as helium, hydrogen, and neon this is the only mechanism by which pumping takes place. The data in Fig. 15.1 show ultimate pressures ranging from 10^{-1} to 10^{-12} of the saturated pressure. The ultimate pressure is a function of the surface coverage. The surface coverage can be minimized by pumping a small quantity of gas or by the generation of a large surface area with porous sorbents such as charcoal or a zeolite. Adsorption isotherms have been measured for many materials and several references are given in reference 7. The adsorbent properties of charcoal and molecular sieve, which are most interesting for cryopumping, have been the subject of considerable investigation [8-16]. In some cases the isotherms do not approach zero with a slope of one, which suggests the sorbent was not completely equilibrated. Analytical expressions for adsorption isotherms are discussed elsewhere [7,17,18].

The last mechanism of low temperature pumping has been given the name cryotrapping [19], and has been studied in some detail [3,19-22]. Cryotrapping is simply the dynamic sorption of one gas on and in the porous frozen condensate of another. A gas which would not normally condense will sorb if it arrives simultaneously with another condensable gas. Some of the non-condensable gas molecules are adsorbed on the surface of the condensed gas microcrystallites, while others are incorporated within the crystallites [3]. Cryocondensation takes place only

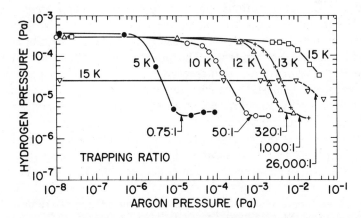

Fig. 15.3 Cryotrapping of hydrogen on solid argon at various temperatures. The drop in hydrogen pressure corresponds to the onset of cryotrapping at a particular argon pressure. Reprinted with permission from *Vacuum*, **17**, p. 495, J. Hengevoss and E. A. Trendelenburg. Copyright 1967, Pergamon Press, Ltd.

between certain pairs of gases. Examples are hydrogen in argon, helium in argon, and hydrogen in carbon monoxide. The cryotrapping of hydrogen in argon is important in sputtering while the pumping of helium in argon is important in fusion work.

Figure 15.3 [20] illustrates the cryotrapping of hydrogen by argon. In this experiment a diffusion pump and a cryosurface pump in parallel on a known hydrogen gas flow. This resulted in a steady-state hydrogen pressure for zero argon flow whose magnitude was determined by the hydrogen flow rate and diffusion pump speed. As the argon flow was increased the cryosurface began to pump the hydrogen and reduce its partial pressure. These data show the efficiency of pumping hydrogen, or the hydrogen-argon trapping ratio, is much higher at 5 than at 15 K. At a temperature of more than 23 K cryotrapping of hydrogen in argon did not occur. Hengevoss [21] showed the density of the solid argon deposit decreased with condensation temperature and that porous argon contained more hydrogen sorption sites than dense argon. He also showed thermal cycling of the argon to a higher temperature irreversibly increased its density, and evolved the previously cryotrapped hydrogen.

15.2 SPEED, PRESSURE, AND SATURATION

In Chapters 2, 3 and 7 we outlined kinetic theory and introduced the concepts of gas flow, conductance, and speed. Whenever the temperature in the system is the same everywhere, these ideas can be used to predict the correct performance of a pump or the state of a system. If the temperature varies throughout the system, as it will in a cryogenic-pumped system, these notions must be applied with care; some are subject to misinterpretation, while others are simply not true. We stated, for example, the mean-free path was pressure-dependent. Strictly speaking, it is particle-density-dependent. The pressure in a closed container will increase if the temperature is increased but the mean-free path will not change because the particle density remains constant. Such a misunderstanding can easily develop because we normally associate pressure change with particle density change.

Lewin [23] points out the definitions of conductance and speed require the throughput Q to be constant in a series circuit. The throughput is constant only in an *isothermal* system. See Section 3.2. Throughput has dimensions of energy. In a non-isothermal system, energy is being added to the gas stream as it flows through a warm pipe, and removed as it flows through a cool region. Particle flow, however, is constant in a non-isothermal system. It is this concern which directs us to formulate the behavior of a cryogenic pump in terms of particle flow rather than throughput.

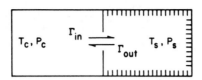

Fig. 15.4 Cryogenic pumping model. The gas in the pump has a temperature T_s equal to the pumping surface and a pressure P_s which is in equilibrium with the gas condensed or adsorbed on the pumping surface.

The sketch in Fig. 15.4 describes a chamber with gas at pressure P_c and temperature T_c, connected by area A to a cryogenic pump whose surfaces are cooled to temperature T_s and in which the gas is in thermal equilibrium with the surface. The temperature of the gas in the chamber is assumed to be greater than the gas in the pump. The net flux of particles into the pump is $\Gamma_{net} = \Gamma_{in} - \Gamma_{out}$. This may be written

$$\Gamma_{net} = \frac{An_c v_c}{4} - \frac{An_s v_s}{4} = \frac{AP_c v_c}{4kT_c} - \frac{AP_s v_s}{4kT_s}$$

$$\Gamma_{net} = \frac{AP_c v_c}{4KT_c}\left[1 - \frac{P_s}{P_c}\left(\frac{T_c}{T_s}\right)^{1/2}\right] \qquad \blacktriangleright (15.1)$$

In this derivation we have assumed the condensation coefficient is unity. Equation (15.1) may be simplified by observing that the term outside the brackets is Γ_{in}. The maximum particle flow into the pump corresponds to $\Gamma_{out} = 0$, or $\Gamma_{in} = \Gamma_{max}$. Equation 15.1 may be written as

$$\frac{\Gamma_{net}}{\Gamma_{max}} = \left[1 - \frac{P_s}{P_c}\left(\frac{T_c}{T_s}\right)^{1/2}\right] \qquad (15.2)$$

Now define

$$P_{ult} = P_s\left(\frac{T_c}{T_s}\right)^{\frac{1}{2}} \qquad (15.3)$$

and express (15.2) as

$$\frac{\Gamma_{net}}{\Gamma_{max}} = c\left(1 - \frac{P_{ult}}{P_c}\right) \qquad \blacktriangleright (15.4)$$

where the condensation coefficient is now included. Equation (15.3) is the thermal transpiration equation (2.37). It relates the ultimate pressure in the chamber of our model P_{ult} to the pressure over the surface. If the pump is a condensation pump, P_s is the saturated vapor pressure. If the pump is a sorption pump, P_s is the pressure obtained from the adsorption isotherm, knowing the fractional surface coverage and temperature of the sorbent.

The ultimate pressure for the cryosorption or cryocondensation pump modeled in Fig. 15.4 can be determined from (15.3) by use of the proper value of P_s. The ultimate pressure for cryocondensation pumping is equal to the saturated vapor pressure multiplied by the thermal transpiration ratio $(T_c/T_s)^{1/2}$. It is a constant during operation of the pump, provided the temperature of the cryosurface does not change. The ultimate pressure for cryosorption pumping will increase with time because the saturation pressure over the sorbent is a function of the quantity of previously pumped gas. In either case the ultimate pressure in the chamber will be greater than the saturated vapor pressure or adsorption pressure by the transpiration ratio. For $T_c = 300\ K$ and $T_s = 15\ K$ the ratio is $P_{ult} = 4.47\ P_s$.

Equation (15.4) may also be used to characterize the speed of the pump because $\Gamma_{net}/\Gamma_{max}$ is proportional to S_{net}/S_{max}. The pumping speed of a cryocondensation pump is constant and near its maximum value when $P_{ult} \ll P_c$, regardless of the quantity of gas pumped. All gases except H_2, He, and Ne have a saturated vapor pressure of less than 10^{-20} Pa at 10 K.

The pumping speed of a cryosorption pumping surface is affected by its prior use because the saturation pressure of the surface increases as the sites become filled. For high vapor pressure gases such as H_2 and He the pumping speed on a molecular sieve at 10 to 20 K can actually diminish from S_{max} to zero as the clean sorbent gradually becomes saturated with

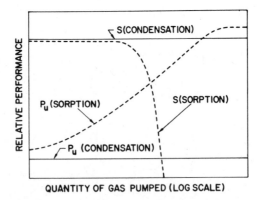

QUANTITY OF GAS PUMPED (LOG SCALE)

Fig. 15.5   Relative variation of pumping speed and ultimate pressure versus quantity of gas pumped for cryosorption pumping and for cryocondensation pumping.

gas during pumping. Figure 15.5 sketches the expected behavior of speed (linear scale) and ultimate pressure (log scale) as a function of the quantity of gas being pumped (log scale) for both cryosorption and cryocondensation pumping. For both cases the net speed goes to zero as the chamber pressure reaches the ultimate pressure.

The simple model presented here is valid for predicting the performance of a cryogenic pump connected to the chamber by an aperture or pumping port when all the gas in the pump is cooled to the temperature of the pumping surface. Unfortunately practical pumps do not meet these criteria. A chamber completely immersed in a liquid cryogen has an ultimate pressure given by P_s, that is, the gas temperature in (15.3) is $T_c = T_s$. Moore [24] has shown the ultimate pressure in a system consisting of a parallel cryopanel and warm wall to be

$$P_u = \frac{P_s}{2}\left[1 + \left(\frac{T_1}{T_2}\right)^{1/2}\right] \qquad (15.5)$$

where T_1 is the temperature of the warm wall and T_2 is the panel temperature. Space chambers are constructed with large cryopanels located inside. The gas density and temperature in those pumping systems are neither uniform nor in equilibrium and the pressure measured depends on the orientation of the gauge [24]. The same is true for a cryogenic pumping unit which contains surfaces cooled to several different temperatures. The model does not account for the heat carried to the cooled surfaces by the gas or by radiation other than to imply an equal amount of energy must be removed from a cooled surface by the cryogen or refrigerant so that its temperature remains constant. The effects of thermal loading on the pumping surfaces are discussed in Section 15.4. This simple model is sufficient to understand conceptually the speed-pressure relationships for the cryocondensation and cryosorption pumping of individual gases.

In most pumping requirements the pump must adsorb or condense a mixed gas load. Cryotrapping is one instance where the pumping of one gas aids the pumping of another. For example, the pumping speed of hydrogen in the presence of an argon flux may be higher than predicted by cryosorption. Its speed may not decrease to zero when the sorbent is completely covered. Cryosorption pumping of mixed gases may also cause desorption of a previously pumped gas, or reduced adsorption of one of the components of the mixed gas. Water vapor will inhibit the pumping of nitrogen [25], and CO has been shown to replace N_2 and Kr on Pyrex glass [26]. This is similar to the gas replacement phenomenon for chemisorption that occurs in TSPs; the gas with a small adsorption energy tends to be replaced or be pumped less efficiently than the gas with a large adsorption energy. Hobson [27] and Haefer [3] have reviewed single gas adsorption processes in cryopumps, while Kidnay and Hiza [28] have summarized the literature on mixture isotherms.

15.3 REFRIGERATION TECHNIQUES

Cryogenic pumping surfaces are cooled by direct contact with liquid cryogens, or gases in an expansion cooler. Liquid helium and liquid nitrogen are used to cool surfaces to 4.2 and 77 K, respectively. Other cryogens such as liquid hydrogen, oxygen, and argon are used to obtain different temperatures for specific laboratory experiments. In a liquid-cooled pump heat is removed from the cryocooled surfaces to an intermittently filled liquid storage reservoir or to a coil through which the liquid cryogen is continuously circulated. In a two-stage closed-cycle gas refrigerator pump gaseous helium is cooled to two temperature ranges, 10 to 20 K, and 40 to 80 K. Cryopumping surfaces are attached to these locally cooled regions. Both methods of removing heat, liquid cooling and gas cooling, require mechanical refrigeration but in different ways. Liquid cryogens are most economically produced in large refrigerators at a central location and distributed in vacuum-insulated dewars, while helium gas refrigerators are economical for locally removing the heat load of a small cryogenic pump. The liquid cryogen requirements of a large cryogenic pumped space chamber warrant the installation of a liquifier at the point of use.

Systems using liquid cryogens are often called open-loop systems because the boiling liquid is usually allowed to escape into the atmosphere. This is not necessary; rare gas collection systems have been in use for decades. Helium gas refrigerators are examples of closed-loop systems. Warm gas is returned to the compressor after absorbing heat at low temperatures.

Many thermodynamic cycles have been developed for the achievement of low temperatures [29-31]. Some produce liquid helium or other liquid cryogens, some cool semiconducting and superconducting devices, and others produce refrigeration of useful capacity at temperatures ranging from 100 K to a few degrees above liquid helium temperature. It is the latter class of refrigerators that is used to cool cryogenic pumps. The refrigerator must be reliable, simple, and easy to manufacture and operate. Cycles embodying these attributes have been developed by Gifford and McMahon [32-35] and by Longsworth [36-38]. These two cycles are variants of a cycle developed by Solvay [29,31] in 1887. Except for some Stirling cycle machines [39], almost every cryogenic pump in use today is operated on one of these two cycles.

Figure 15.6 illustrates a basic one-stage Gifford-McMahon refrigerator. The helium compressor is located remotely from the expander, and is connected to it by two flexible, high-pressure hoses. Within the expander

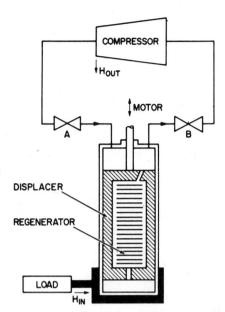

Fig. 15.6 Schematic representation of a single-stage Gifford-McMahon helium gas refrigerator. Adapted with permission from CTI-Cryogenics, Kelvin Park, Waltham, MA 02154.

is a cylindrical piston or displacer made from an insulating material. The piston is called a displacer [33] because the regions at each end are connected to give them little pressure difference.

Inside the displacer is a regenerator—a single-channel heat exchanger through which the gas flows at different times in alternate directions. It is tightly packed with a metal of high heat capacity and large surface-area-to-volume ratio. Alloys of lead, copper (or alloys whose heat capacity is high at cryotemperatures) in the shape of shot or screen are used to pack the regenerator. In the steady state the regenerator will have a temperature gradient. Ambient-temperature helium entering from the warm end will give heat to the metal, and cold gas entering from the cooler end will absorb heat from the metal. Even though the regenerator is tightly packed, there is not much flow resistance. The regenerator can transfer thermal energy from the incoming to the outgoing helium quickly and with great efficiency.

The operation of the Gifford-McMahon refrigerator can be understood by following the helium through a complete steady-state cycle. High pressure gas from the outlet of the compressor is admitted to the regenerator through valve A while the displacer is at the extreme lower end of the cylinder. See Fig. 15.6. During the time valve A is open the displacer is raised. The incoming gas passes through the cold regenerator and is cooled as it gives heat to the regenerator. At this point in the cycle the gas temperature is about the same as the load. Valve A is then closed before the displacer reaches the top of its stroke. Further movement of the displacer forces the remainder of the gas through the displacer. The exhaust valve B is now opened to allow the helium to expand and cool. The expanding helium has performed work. It is this work of expansion which causes the refrigeration effect. No mechanical work is done since expansion did not occur against a piston. Heat flowing from the load, which is intimately coupled to the lower region of the cylinder walls, warms the helium to a temperature somewhat below that at which it entered the lower cylinder area. As the gas flows upward through the regenerator, it removes heat from the metal and cools it to the temperature at which it was found at the beginning of the cycle. The displacer is now pushed downward to force the remaining gas from the end of the cylinder out through the regenerator where it is exhausted back to the compressor at ambient temperature. A single-stage machine of this design can achieve temperatures in the 30 to 60 K range.

Lower temperatures can be achieved with two-stage machines. The first, or warm, stage operates in the range 30 to 100 K, while the second, or cold, stage operates in the range 10 to 20 K. The exact temperatures depend on the heat load and capacity of each stage. A heat-balance analysis of the refrigeration loss has been performed by Ackermann and Gifford [40]. In the Gifford-McMahon refrigerator the gas is cycled with poppet valves; the valves and the displacer are moved by a motor and all are located on the expander. A Scotch yoke displacer drive is used because it applies no horizontal force to the shaft.

Figure 15.7 illustrates the expander developed by Longsworth [37,38]. As on the Gifford-McMahon refrigerator, the remotely located compressor is connected to the expander by hoses of high pressure capacity. The helium is cycled in and out of the expansion head through a motor-driven rotary valve. The expander shown here contains a two-stage displacer and two regenerators. The displacer is gas driven instead of motor driven, as in the original Gifford-McMahon cycle. A slack piston is incorporated to improve timing. Surrounding the valve stem is an annular surge volume. This volume is maintained at a pressure intermediate to the supply and exhaust pressures by a capillary tube connected to the regenerator inlet line; it provides the reference pressure for pneumatic operation of the displacer.

In the steady state the operating cycle proceeds as follows [41]: the valve is timed to admit high pressure helium gas through the stem into the volume below the slack piston and in the regenerators while the displacer is in its lowermost position. Because the pressure over the slack pistons is less than the inlet pressure, the piston compresses this gas as it moves upward. The gas bleeds into the surge volume through the surge orifice at a constant rate. The surge orifice is like a dashpot; it controls the speed of the displacer. As the displacer moves upward the high pressure gas flows through the regenerators and is cooled in the process. The inlet valve stops the flow of high pressure gas just before the displacer reaches the top. This slows the displacer and expands the gas in the displacer. The exhaust valve opens and the gas in the displacer expands as it is exhausted to the low pressure side of the compressor. The slack piston moves downward suddenly until it contacts the displacer, after which it moves at constant velocity as gas flows from the surge volume into the space over the slack piston. The expansion of gas in the displacer causes it to cool below the temperature of the regenerator. This is the refrigeration effect. Like the Gifford-McMahon cycle, this cycle does no mechanical work on the displacer because both ends are at the same pressure. The exiting gas removes heat from the regenerator. Before the end of its stroke the displacer is decelerated by closure of the exhaust valve. This completes one cycle of expander operation. Heat is absorbed at two low temperatures and released at a higher temperature.

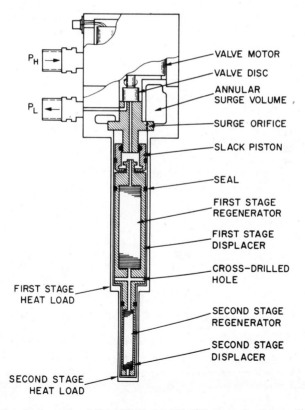

Fig. 15.7 Sectional schematic of the API Model DE-202 expansion head. Reprinted with permission from Air Products and Chemicals, Inc., Allentown PA 18105.

The compressor used in either of the two Solvay cycle variants is sketched in Fig. 15.8. It uses a reliable oil-lubricated, air conditioning type of compressor with an inlet pressure of approximately 7×10^5 Pa (100 psig) and an outlet pressure of about 2×10^6 Pa (300 psig). After the gas is compressed the heat of compression is removed by an air or water aftercooler. Oil lubrication can be used without contaminating the cold stages because it is removed by a two-stage separator and adsorber. Traces of oil which get into the regenerator can cause problems. The oil must be cooled before entering the adsorber, as hot oil vapors are not adsorbed on charcoal. It is imperative that the adsorber cartridge be packed tightly and remain cool or it will not adsorb oil vapor.

The most important attribute of small Solvay-type refrigerators is reliability. The low pressure differential across the seals in the displacers means light pressure loading and long life. Also contributing to long life is the use of room-temperature valves, a reliable compressor, and oil removal techniques. The result is a compact refrigerator that can be isolated from compressor vibration and attached to a vacuum chamber in any attitude.

The coefficient of performance of a refrigerator is defined as the ratio of heat removed to work expended in removing the heat. For an ideal Carnot cycle, the most efficient of all possible cycles, this is [42]:

$$\frac{H_{\text{out}}}{W_{\text{in}}} = \frac{T_1}{T_2 - T_1} \tag{15.6}$$

where heat is being absorbed at T_1 and released at T_2. For $T_2 = 300$ K a Carnot cycle would require a heat input of 2.9 W to remove 1 W at 77 K, and a heat input of 14 W to remove 1 W at 20 K. In practice the efficiency of a refrigerator is defined as the ratio of ideal work to actual work. The refrigerators described here have efficiencies of about 3 to 5% [33,43].

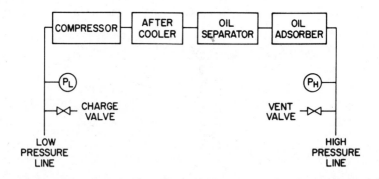

Fig. 15.8 Block diagram of a remotely located helium gas compressor.

There is an interdependence between the refrigeration capacity of the two stages of a cryogenic refrigerator. There is a balance between the heat flow to each stage and the heat removed by each stage. Increasing the load on the warm stage will cause the cold stage to warm slightly also. Each manufacturer will have data on individual compressor performance. The emphasis has been on producing machines with increased capacity in the first stage to isolate radiant heat effectively from the second stage.

15.4   CRYOGENIC PUMP CHARACTERISTICS

In the previous sections we discussed the speed-pressure characteristics of some ideal pumping surfaces. The characteristics of a real cryogenic pump may differ significantly from those ideal cases. A detailed prediction of real pump performance requires a more complete model. The effects of thermal gradients between pumping surfaces and refrigerator, gas and radiant-heat loading, and the geometrical isolation of condensation and sorption stages are three important effects that have not been considered in the ideal model. In the remainder of this section the gas handling characteristics of rough sorption pumps and refrigerator and liquid-cooled high vacuum pumps are related to the materials, geometry, and heat loading of the pumping surfaces.

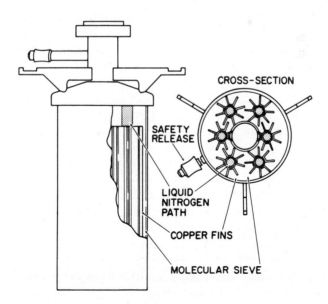

Fig. 15.9 Typical liquid-nitrogen-cooled sorption pump. Reprinted with permission from Ultek Division, Perkin-Elmer Corp., Palo Alto, CA 94303.

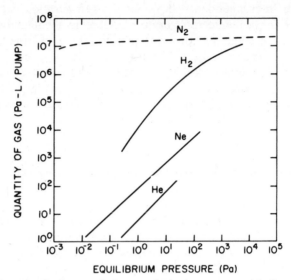

Fig. 15.10 Adsorption isotherms of nitrogen, hydrogen, neon, and helium at 77.3 K in a sorption pump charged with 1.35 kg of molecular sieve. Reprinted with permission from *Cryosorption Pumping*, F. Turner, Varian Report VR-76. Copyright 1973, Varian Associates, 611 Hansen Way, Palo Alto, CA 94303.

15.4.1 Medium Vacuum Sorption Pumps

In the early 1900s Dewar used refrigerated sorption pumping to evacuate an enclosed space. Sorption pumping as we know it today uses high capacity artificial zeolite molecular sieves and liquid nitrogen. A unique feature of cryosorption rough pumping is its ability to pump to 10^{-1} Pa without introducing hydrocarbons into the chamber.

A sorption pump designed for rough pumping is illustrated in Fig. 15.9. It consists of an aluminum body that contains many conducting fins and is filled with an adsorbent. The entire canister is surrounded by a polystyrene foam or metal vacuum dewar filled with liquid nitrogen. Adsorbent pellets are loosely packed in the canister and do not make good thermal contact with the liquid nitrogen. To improve the thermal contact, pumps with internal arrays of metal fins are constructed. Even so, the interior of the pump is not in equilibrium with the liquid nitrogen bath, especially during pumping.

A common adsorbent is Linde 5A molecular sieve. This sieve, with an average pore diameter of 0.5 nm exhibits a high capacity for the constituents of air at low pressure. Figure 15.10 illustrates the adsorption isotherms in a pump containing a charge of 1.35 kg of molecular sieve. The adsorptive capacity for nitrogen is quite high over the pressure range of 10^{-3} to 10^5 Pa, while the capacity for helium and neon is quite low.

Mixed gas isotherms (e.g., neon in air) will show even less pumping capacity for the least active gas (neon) because it will be displaced by active gases. The preadsorption of water vapor will greatly reduce the capacity for all gases; as little as 2 wt% water vapor is detrimental to pump operation [25]. These isotherms are not valid during dynamic pumping because the incoming gas will warm the sieve nonuniformly. They do represent the equilibrium condition of a real pump immersed in a liquid nitrogen bath.

The neon pressure in air is 1.2 Pa and this limits the ultimate pressure. Figure 15.11 sketches the time dependence of the air pressure in a 100 L chamber for a single stage and two sequential stage pumping with pumps each containing 1.35 kg of molecular sieve and prechilled at least 15 min. Turner and Feinleib [45] have shown that more than 50% of the residual gas present after sorption pumping of air was neon. The ultimate pressure attainable by sorption pumping with a single pump may be reduced by using staged roughing. In staged roughing one pump is used until the chamber pressure reaches approximately 1000 Pa. At that time, the pump is *quickly* valved and a second pump is connected to the chamber.

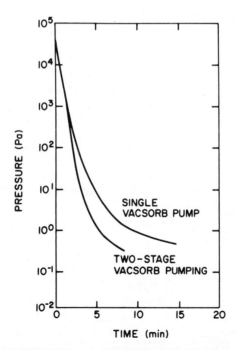

Fig. 15.11 Pumping characteristics of a 100-L, air-filled chamber with (a) one sorption pump, and (b) two sequentially staged sorption pumps. Reprinted with permission from *Cryosorption Pumping*, F. Turner, Varian Report VR-76. Copyright 1973, Varian Associates, 611 Hansen Way, Palo Alto, CA 94303.

Figure 15.11 shows the ultimate pressure to be lower and the pumping speed higher than when only one pump was used. The improved pumping characteristic is a result of adsorbate saturation and neon removal. The first pump removed 10^7 Pa-L (99%) of air, including 99% of the neon, while the second removed only 10^5 Pa-L. The ultimate pressure of the second pump is less than the first stage because it pumped a smaller quantity of gas and because most of the neon was swept into the first pump by the nitrogen stream and trapped there when the valve closed. The valve needs to be closed quickly at the crossover pressure to prevent back diffusion of the neon. Alternatively, a carbon vane or water aspirator pump may be used in place of the first sorption roughing stage.

The ultimate pressure attainable with a sorption pump is a function of its history, in particular the bake treatment and the kinds of gases and vapors. Pressures of an order of 1 Pa are typical. Multistage pumping performance has been described by Turner and Feinleib [44], Cheng and Simpson [45], Vijendran and Nair [46], and Turner [47], while Dobrozemsky and Moraw [48] have measured sorption pumping speeds for several gases in the pressure range of 10^{-4} to 10^{-1} Pa.

Pumps are normally baked to a temperature of 250°C for about 5 h with a heating mantle or reentrant heating element. Miller [49] measured a water vapor desorption maxima (Linde 5A) of 137 to 157°C. All other gases desorb well at room temperature. Baking is required to release water vapor and obtain the full capacity of the sieve.

Each sorption pump has a safety pressure release valve. At no time should the operation of this valve be hindered. Gases will be released when the pump is warmed to atmosphere and when it is baked. A single sorption pump of the size described requires about 5 to 8 L of liquid nitrogen for initial cooling.

15.4.2   High Vacuum Gas Refrigerator Pumps

A typical high vacuum cryogenic pumping array for a two-stage helium gas refrigerator is sketched in Fig. 15.12. The outer surface is attached to the first (warm) or 80 K stage and the inner pumping surface is attached to the second (cold) or 20 K stage. Indium gaskets are used to make joints of high thermal conductance. In practice the temperature of the stages is a function of the actual heat load and thermal path. The warm stage pumps water vapor and sometimes CO_2. First stage pumping of CO_2 is a function of the temperature and the partial pressure. The vapor pressure of CO_2 on a 77 K surface is 10^{-6} Pa; CO_2 will be pumped only if its concentration is large enough or the surface temperature is adequately low. The second stage contains a cryocondensation and a cryosorption pumping surface. The cryosorption surface is necessary to pump helium, hydrogen, and neon. It is shielded from the inlet aperture as much as possible to increase the probability that all other gases will be

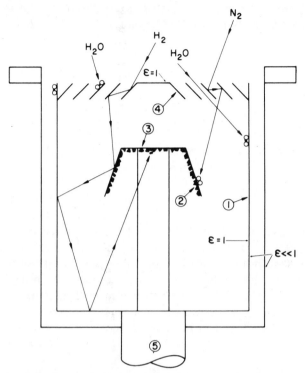

Fig. 15.12 Typical cryogenic pumping array for a two-stage helium gas refrigerator: (1) first stage array, (2) second-stage cryosorption pump, (3) second-stage cryocondensation pump, (4) chevron baffle, (5) refrigerator head.

condensed on the cryocondensation surface and to prevent direct radiation from an external heat source. Charcoal is the most commonly used sorbent because it can be degassed at room temperature. It has a greater capacity and is less affected by impurities than is molecular sieve. Molecular sieve must be degassed at 250°C and this is incompatible with the use of indium gaskets.

The temperature of the two pumping surfaces is determined by the total heat flux to the surfaces. Heat is leaked to these surfaces through the expander housing, radiated from high temperature sources, removed from the incoming gas, and conducted from the warm walls by bouncing gas molecules. Thermal radiation emanates from nearby 300-K surfaces and internal sources such as plasmas, electron-beam guns, and baking mantles. Each condensing molecule releases a quantity of heat equal to its heat of condensation plus its heat capacity. Heat can be conducted from a warm to a cold surface by gas-gas collisions if the pressure is high enough so that Kn < 1. In the high vacuum region the radiant flux is much larger than the gas enthalpy or gas conductance. At a pressure of

10^{-4} Pa a 1000-L/s pump is condensing 0.1 Pa-L/s of nitrogen or a heat load of 0.6 mW. At this pressure gas conductance can also be ignored.

The radiant-heat flow between two concentric spheres or cylinders [50] is given by

$$\frac{H}{A} = \frac{\sigma \varepsilon_1 \varepsilon_2}{\varepsilon_2 + \dfrac{A_1}{A_2}(1-\varepsilon_2)\varepsilon_1}(T_2^4 - T_1^4) \qquad \blacktriangleright (15.7)$$

where the subscripts 1, 2 refer to, respectively, the inner and outer surfaces. If $\varepsilon_1 = \varepsilon_2 = 1$, the heat flow will be maximum and

$$\frac{H}{A} = \sigma(T_2^4 - T_1^4) \qquad (15.8)$$

This yields a heat flow of 457 W/m² from the 300 K surface to the 20 K surface. A typical first-stage area of 0.1 m² for a small cryogenic pump would then absorb a heat load of 45.7 W. Clearly, this would overload the refrigerator. The radiant heat load is reduced by reducing the emissivities of the inside surface of the vacuum wall and the outside surface of the first stage. If the walls are plated with nickel whose emissivity is 0.03 [51], this heat load can be reduced to 0.7 W. Even if these surfaces became contaminated so that $\varepsilon_1 = \varepsilon_2 = 0.1$, the heat flow would still be only 2.4 W.

The first stage isolates the second stage from the radiant heat load and pumps water vapor. Some condensable gas deposits however, have the property of drastically altering the emissivity of the surface on which they condense. Caren et al. [52] have shown as little as 20 μm of water ice on a polished aluminum substrate at 77 K causes the emissivity to increase from 0.03 to 0.8. Their data agree with data taken by Moore [53] at 20 K, and calculations by Tsujimoto et al. [54] over the range 4.2 to 300 K. Carbon dioxide also absorbs thermal energy [53,54]. The two purposes of the first stage, pumping water vapor and thermally shielding the second stage, seem contradictory. If the first stage pumps water vapor, its emissivity will rise and it will adsorb excess heat and overload the refrigerator. This problem is overcome by keeping the outer wall close to the vacuum wall; the water vapor is now pumped along the upper perimeter of the chevron, the chevron itself, and the interior wall of the first stage. It cannot reach the lower portion of the outside wall. Alternatively, for special cases the exterior may be wrapped with multilayer reflective insulation.

The entrance baffle is designed to prevent radiation from illuminating the second stage. It does this by absorbing radiation in the chevron and allowing transmitted radiation to see the blackened inside wall, where it is absorbed. The entrance baffle also impedes the flow of gases to the second stage. An opaque baffle reduces the radiant loading on the

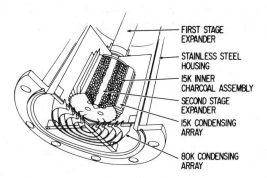

FIRST STAGE
EXPANDER

STAINLESS STEEL
HOUSING

I5K INNER
CHARCOAL ASSEMBLY

SECOND STAGE
EXPANDER

I5K CONDENSING
ARRAY

80K CONDENSING
ARRAY

Fig. 15.13 Cutaway view of the Cryo-Torr 8 cryogenic pump. Reprinted with permission from CTI Cryogenics, Kelvin Park, Waltham, MA 02154.

second stage but also reduces the pumping speed for all gases. Good baffle designs are a compromise between radiant heat absorption and pumping speed reduction. Benvenuti et al. [55] describe an array with an optical transmission of 7×10^{-4} and a molecular transmission of 0.24 which he calculated to have the optimum balance between heat adsorption and speed loss. In some pumps the chevrons are painted black; in others they are highly reflective. It matters little because both will soon be "blackened" with water vapor. Figure 15.13 illustrates a cut-away view of one commercial cryogenic pump. Hands' [56] review of some of the design problems in a cryogenic pump includes the effect of temperature gradients through the deposit and in the arrays on long-term pump performance.

Approximate pumping speed calculations can be made for each species if the temperatures and geometry are known; for example, gases pumped on the second stage must first pass through the chevron baffle, where they are cooled. If the effective inlet area, inlet gas temperature, second-stage area, and species are known, the approximate speed and ultimate pressure can be estimated. The pumping speed of a gas in a cryogenic pump is not only related to the size of the inlet flange and refrigeration capacity. It is also dependent on the pumping array (relative sizes of the warm and cold stages), gas species, and history (ice, hydrogen, and helium load). All pumping speeds will fall off near 10^{-1} Pa as the refrigerator becomes overloaded. Because of these effects, it is not possible to draw an illustration analogous to Fig. 12.4, which represents all gas refrigerator cooled pumps. A "typical" clean pump of 250-mm-diameter inlet flange might have a water vapor pumping speed of 4000 L/s, a nitrogen pumping speed of 1200 L/s, an argon speed of 1000 L/s, a hydrogen speed of 1200 L/s, a carbon dioxide speed of 900 L/s and a helium speed of 700 L/s. The only valid generalization is that helium pumping speeds are usually small and have low saturation values. Hydrogen pumping speeds are larger than those of helium and will be nil if the sorbent has saturated or its temperature is greater than 20 K.

When the pump becomes saturated, it is shut down while the cryosurfaces warm and the pump is regenerated. (Regeneration techniques are discussed in Section 19.4.2.) The vaporizing gases exit through the safety valve or manually operated valve which is connected to a mechanical pump. All cryogenic pumps must have a properly functioning safety release to allow the escape of condensed and adsorbed gases and vapors if the pump fails, is shut-down or looses power. If there were no safety valve, the pump would become a bomb when warmed.

15.4.3 High Vacuum Liquid Pumps

Three-stage cryogenic pumps that use liquid helium (4.2 K), gaseous helium boil-off (20 K), and liquid nitrogen are effective in pumping all gases, especially when the molecular sieve is bonded to the third stage [57]. These pumps have been used for pumping large vessels [58–60] and small vacuum chambers [14,61]. The liquid pump has a higher pumping speed for hydrogen and helium than for a gaseous helium refrigerator because that stage is colder and its temperature is more stable. Ultimate pressures of 10^{-11} Pa [58] have been reported. Liquid pumps are individually designed and constructed for large particle beam projects or small laboratory applications.

REFERENCES

1. B. A. Hands, *Vacuum*, **32**, 603, (1892).

2. P. D. Bentley, *Vacuum*, **30**, 145 (1980).

3. R. Haefer, *J. Phys. E: Sci. Instrum.*, **14**, 159 (1981).

4. J. P. Dawson and J. D. Haygood, *Cryogenics*, **5**, 57 (1965).

5. M. M. Eisenstadt, *J. Vac. Sci. Technol.*, **7**, 479 (1970).

6. R. F. Brown and E. S. Wang, *Adv. Cryog. Eng.* Vol. 10, K. D. Timmerhaus, Ed., Plenum, New York, 1965, p. 283.

7. J. P. Hobson, *J. Phys., Chem.*, **73**, 2720 (1969).

8. P. J. Gareis and S. A. Stern, *Bulletin de l'Institut International du Froid*, Annexe 1966-5, p. 429.

9. W. Van Dingenan and A. Van Itterbeek, *Physica*, **6**, 49 (1939).

10. P. J. Gareis, and S. A. Stern, *Cryogenic Engineering News*, **26** 85 (1967).

11. S. A. Stern and F. S. DiPaolo, *J. Vac. Sci. Technol.*, **4**, 347 (1967).

12. G. E. Grenier and S. A. Stern, *J. Vac. Sci. Technol.*, **3**, 334 (1966).

13. A. J. Kidnay, M. J. Hiza and P. F. Dickenson, *Adv. Cryog. Eng.*, Vol. 13, K. D. Timmerhaus, Ed., Plenum, New York, 1968, p. 397.

14. R. J. Powers, and R. M. Chambers, *J. Vac. Sci. Technol.*, **8**, 319 (1971).

15. C. Johannes, *Adv. Cryog. Eng.*, Vol. 17, K. D. Timmerhaus, Ed., Plenum, New York, 1972, p. 307.

16. H. J. Halama and J. R. Aggus, *J. Vac. Sci. Technol.*, **11**, 333, (1974).

17. J. P. Hobson, *J. Vac. Sci. Technol.*, **3**, 281 (1966).

18. P. A. Redhead, J. P. Hobson, and E. V. Kornelsen, *The Physical Basis of Ultrahigh Vacuum*, Chapman and Hall, London, 1968, p. 37.

19. R. L. Chuan, Univ. South Calif., Eng. Center Rep. 56-101, 1960.

20. J. Hengevoss and E. A. Trendelenburg, *Vacuum*, **17**, 495 (1967).

21. J. Hengevoss, *J. Vac. Sci. Technol.*, **6**, 58 (1969).

22. J. C. Boissin, J. J. Thibault, and A. Richardt, *Le Vide*, Suppl. 157, 103 (1972).

23. G. Lewin, *J. Vac. Sci. Technol.*, **5**, 75 (1968).

24. R. W. Moore, Jr., *8th Nat. Vac. Symp. (1961)*, Vol. 1, Pergamon, New York, 1962, p. 426.

25. S. A. Stern and F. S. DiPaolo, *J. Vac. Sci. Technol.*, **6**, 941 (1969).

26. Y. Tuzi, M. Kobayshi, and K. Asao, *J. Vac. Sci. Technol.*, **9**, 248 (1972).

27. J. P. Hobson, *J. Vac. Sci. Technol.*, **10**, 73 (1973).

28. J. Kidnay and M. J. Hiza, *Cryogenics*, **10**, 271 (1970).

29. S. C. Collins and R. L. Canaday, *Expansion Machines for Low Temperature Processes*, Oxford University Press, Oxford, 1958.

30. R. Barron, *Cryogenic Systems*, McGraw-Hill, New York, 1966.

31. R. Radebaugh, *Applications of Closed-Cycle Cryocoolers to Small Superconducting Devices*, NBS Special Publication 508, U.S. Department of Commerce, National Bureau of Standards, Washington, DC, 1978, p. 7.

32. W. E. Gifford, *Refrigeration Method and Apparatus*, U. S. Pat. 2,966,035 (1960).

33. W. E. Gifford, and H. O. McMahon, *Prog. Refrig. Sci. Technol*, Vol. 1, M. Jul and A. Jul, Eds., Pergamon, Oxford, 1960, p. 105.

34. W. E. Gifford, *Prog. Cryog.*, Vol. 3, K. Mendelssohn, Ed., Academic, New York, 1961, p. 49.

35. W. E. Gifford, *Adv. Cryog. Eng.*, Vol. 11, K. D. Timmerhaus, Ed. Plenum, New York, 1966, p. 152.

36. R. C. Longsworth, *Refrigeration Method and Apparatus*, U.S. Pat. 3,620,029, (1971).

37. R. C. Longsworth, *Adv. Cryog. Eng.*, Vol. 16, K. D. Timmerhaus, Ed., Plenum, New York, 1971, p. 195.

38. R. C. Longsworth, *Adv. Cryog. Eng.*, Vol. 23, K. D. Timmerhaus, Ed., Plenum, New York, 1978, p. 658.

39. For example, Type K-20 Series Cryogenerator, N. V. Philips Gloeilampenfabrieken, Eindhoven, Nétherlands.

40. R. A. Ackermann, and W. E. Gifford, *Adv. Cryog. Eng*, Vol. 16, K. D. Timmerhaus, Ed. Plenum, New York, 1971, p. 221.

41. R. C. Longsworth, *An Introduction to the Elements of Cryopumping*, K. M. Welch, Ed., American Vacuum Society, p. II-1.

42. F. W. Sears, *An Introduction to Thermodynamics, The Kinetic Theory of Gases and Statistical Mechanics*, Addison-Wesley, Reading, MA., 1953, p. 84.

43. T. R. Strobridge, NBS Technical Note 655, U.S. Department of Commerce, National Bureau of Standards, Washington, D.C., 1974.

44. F. T. Turner and M. Feinleib, *8th Nat. Vac. Symp. (1961)*, **1**, Pergamon, New York, 1962, p. 300.

45. D. Cheng and J. P. Simpson, *Adv. Cryog. Eng*, Vol. 10, K. D. Timmerhaus, Ed., Plenum, New York, 1965, p. 292.

46. P. Vijendran and C. V. Nair, *Vacuum*, **21**, 159 (1971).

47. F. Turner, *Cryosorption Pumping*, Varian Report VR-76, Varian Associates, Palo Alto, CA 1973.

48. R. Dobrozemsky and G. Moraw, *Vacuum*, **21**, 587 (1971).

49. H. C. Miller, *J. Vac. Sci. Technol.*, **10**, 859 (1973).

50. R. B. Scott, *Cryogenic Engineering*, Van Nostrand, New York, 1959, p. 147.

51. *Ibid*, p. 348.

52. R. P. Caren, A. S. Gilcrest, and C. A. Zierman, *Adv. Cryog. Eng.*, Vol. 9, K. D. Timmerhaus, Ed., Plenum, New York, 1964, p. 457.

53. B. C. Moore, *9th Nat. Vac. Symp. (1962)*, Macmillan, New York, 1962, p. 212.

54. S. Tsujimoto, A. Konishi, and T Kunitomo, *Cryogenics*, Vol. 22, 603 (1982).

55. C. Benvenuti, D. Blechschmidt and G. Passarde, *J. Vac. Sci. Technol.*, **19**, 100 (1981).

56. B. A. Hands, *Vacuum*, **26**, 11, (1976).

57. H. J. Halama and J. R. Aggus, *J. Vac. Sci. Technol.*, **12**, 532 (1975).

58. C. Benvenuti, *J. Vac. Sci. Technol.*, **11**, 591 (1974).

59. C. Benvenuti and D. Blechschmidt, *Japan. J. Appl. Phys.* Suppl. 2, Pt. 1, 77 (1974).

60. H. J. Halama, C. K. Lam and J. A. Bamberger, *J. Vac. Sci. Technol.*, **14**, 1201 (1977).

61. G. Schafer, *Vacuum*, **28**, 399 (1978).

PROBLEMS

15.1 †Describe three mechanisms by which a cooled surface can pump gases or vapors.

15.2 What surface temperatures are required to condense, respectively, water vapor, carbon dioxide, nitrogen, oxygen and argon to a pressure of 10^{-4} Pa?

15.3 A hydrogen flux is being cryotrapped in an argon flux at a surface temperature of 10 K, A dynamic equilibrium has been reached. The argon "ice" is warmed to 25 K. Describe qualitatively what happened to the pumping speed and background pressure of hydrogen when the equilibrium was disturbed by heating the second stage.

15.4 How does a cryosorption pump saturate? What limits the gas quantity pumped on a cryocondensation surface?

15.5 A vacuum chamber 40 cm in diameter and 178.25 cm high contains nitrogen gas at STP. How much liquid nitrogen does it take to pump this gas load on liquid-nitrogen-cooled molecular sieve whose heat of sorption is 4000 kcal/kg-mole?

15.6 A water aspirator pump is sometimes used as an initial roughing pump before operating a sorption pump. It is constructed from a venturi jet of water into which a vacuum inlet is connected. What is the ultimate pressure of this pump?

15.7 A small sorption pump containing 1.35 kg of molecular sieve is saturated with 1.2×10^7 Pa-L of nitrogen at a pressure of 10 Pa and a temperature of 77 K. The internal volume of the pump is 2 L. What is the pressure inside an isolated pump when the temperature is raised to 300 K? (Assume the safety valve did not open.)

15.8 A spherical dewar of 50 L capacity is vacuum insulated with a second spherical shell whose radius is 5 cm larger than the dewar. The emissivity of the outer side of the 50-L dewar and the inner side of the vacuum jacket are each 0.02. The dewar is filled with liquid nitrogen at 77.35 K and atmospheric pressure. The dewar is sealed with a 20 psi (gauge) relief valve. (a) What is the equilibrium pressure of the liquid nitrogen after the pressure rises to that of the relief valve? (b) How long does it take for the liquid nitrogen to reach equilibrium? Note: The heat capacity of liquid nitrogen is temperature dependent—assume it to be 57.7 kJ/(kmole-K) in the temperature range of interest.

15.9 A cryogenic pump is regenerated by first removing power to the compressor, then pumping with a mechanical pump to its base pressure overnight. Is this a good regeneration procedure? If not, why not?

15.10 †How is the pumping performance of a helium-gas-cooled cryo-pump dependent on pumping history?

Materials

Knowledge of the basic properties of the materials from which vacuum systems are fabricated is essential to operate and maintain vacuum systems properly. One of the most troublesome properties of materials used in vacuum applications is gas release from solids at low pressures. Chapter 16 discusses the origins of gas released from metals, glasses, ceramics and polymers commonly used in systems, and how it can be removed. Chapter 17 discusses techniques for joining and sealing materials, including valves. Chapter 18 describes lubricants used in vacuum systems. There is nothing to vacuum, it's all in the packaging.

CHAPTER 16

Materials in Vacuum

A superficial examination of a high or ultrahigh vacuum pumping system gives an impression of simplicity; clean, polished metal or glass surfaces, view ports, electrical and motion feedthroughs, piping and pumps. A close examination reveals that many requirements are placed on materials in vacuum environments and these requirements sometimes conflict. The chamber walls must support a load of 10,335 kg/m², a load that is present on the surfaces of all vacuum systems, even those merely roughed to 1000 Pa. Metals must be easy to machine and to join by welding, brazing, soldering or demountable seals. Methods are needed for sealing glasses, ceramics, and other insulators to metals, when optical and electrical feedthroughs are required. The outgassing load from the fixturing of high vacuum chambers and tooling must be reduced to obtain low pressures. High vacuum systems contain a large internal surface which cannot be baked. Therefore it is essential they be fabricated from materials that can be processed to yield low outgassing rates. Baking is necessary to reduce the outgassing rate and reach the lowest possible pressure in an ultrahigh vacuum system. Baking reduces the maximum stress limit and increases the strain or deformation of stressed parts, and thermal decomposition limits the temperature at which some materials can be baked. The interdependence of cleaning, joining, construction and application needs to be clearly understood when choosing materials for use in a vacuum environment.

In this chapter we review the outgassing and structural properties of metals, glasses, ceramics and polymeric materials used in the construction of vacuum equipment. Stainless steel is the dominant material of construction and its structural properties are discussed in some detail. Perkins [1] and Elsey [2,3] have reviewed the literature on outgassing of materials used in vacuum. A comprehensive review of materials for ultrahigh vacuum has been given by Weston [4].

Techniques for measuring outgassing rates are have been described by Elsey [3]. It is important to note that most of the outgassing rates reported in the literature are *net* outgassing rates rather than *true* outgassing rates. To make this distinction clear we describe a model presented by Hobson [5]. Hobson considered a model of a hollow sphere which is made of the material under study. q_{net}, the net outgassing rate per unit area of the material, was obtained by properly measuring the flux evolving from the inside of the sphere, Q_T, and dividing it by the area of the sphere A. However the non-zero pressure inside the sphere indicated there was a flux incident on its surface. If the flux had a sticking coefficient s, then it could be shown that the total or true outgassing rate per unit area of the material was $q_{true} = q_{net} + \Gamma s$. It is typically assumed in outgassing measurements that $s = 0$, and that the net outgassing is the same as the true outgassing. Hobson noted that there are conditions where the geometry and application of the material are similar to the measurement conditions and the assumption of zero sticking coefficient is valid. He also cited examples of a proton storage ring and a space shuttle wake shield where a non-zero sticking coefficient caused a very significant difference in the calculated pressures near the outgassing material.

16.1 METALS

Metals are used in the chamber and to form its walls. Metals used for vacuum chamber walls should be joinable and sealable, have high strength, low permeability to atmospheric gases, low outgassing rates and low vapor pressures. Metals used within the chamber should have a low outgassing rate and a low vapor pressure. The particular application in the chamber, such as the filament, radiation shield, and thermal sink, will add other constraints. In this section we discuss the vaporization, permeation, and outgassing properties of several metals and some structural properties of aluminum and austenitic stainless steels.

16.1.1 Vaporization

The vapor pressure of most metals is low enough to allow their use in vacuum systems. The vapor pressures of the elemental metals are found in Appendix C.6. A few metals should not be used in vacuum construction because their vapor pressures are high enough to interfere with normal vacuum baking procedures. Alloys containing zinc, lead, cadmium, selenium, and sulfur, for example, have unsuitably high vapor pressures for vacuum applications. Zinc is a component of brass, cadmium is commonly used to plate screws, and sulfur and selenium are used to make the free machining grades of 303 stainless steel. These materials should not be used in vacuum system construction.

16.1.2 Permeability

The permeability of a gas in a metal is proportional to the product of the gas solubility and the diffusion constant (4.12). Hydrogen is one of the few gases that permeate most metals to a measurable extent. Its permeation rate is proportional to the square root of the pressure difference. Figure 16.1 shows the temperature dependence of the permeation constant of hydrogen through several metals [6]. Hydrogen permeation is the least in Al. Other metals through which hydrogen permeates are, in order of increasing permeability, Mo, Ag, Cu, Pt, Fe, Ni and Pd. The addition of chrome to iron allows the formation of a chrome oxide barrier that reduces the hydrogen permeation rate. The influx will be greater than this value if rusting occurs on the external wall because the reaction of water vapor with iron creates a high partial pressure of hydrogen.

Perkins [1] has described the experimental techniques used to measure the permeation and diffusion constants in solids and has tabulated the permeability and diffusivity of hydrogen in palladium, stainless steels,

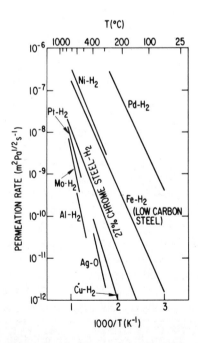

Fig. 16.1 Permeation constant for hydrogen through various metals as a function of temperature. Reprinted with permission from *Trans. 8th Nat. Vac. Symp. (1961)*, p. 8, F. J. Norton. Copyright 1962, Pergamon Press, Ltd.

Fe-Co-Ni alloys, copper, and nickel and the permeability and diffusivity of nitrogen in tungsten and molybdenum. Begeal [7] has measured hydrogen permeation in copper and copper alloys.

16.1.3 Outgassing

The gas load in vacuum fixturing and chamber walls is adsorbed on and dissolved in metals. This gas load will affect the performance of high and ultrahigh vacuum systems if it is not removed. Gas is dissolved in a metal during its initial melting and casting. It consists mainly of hydrogen, oxygen, nitrogen and carbon oxides. Gas is also physi- and chemisorbed on the interior surfaces from exposure to ambient atmosphere. It consists of a large quantity of water vapor, with carbon oxides, oxygen and some nitrogen. The nature and quantity of the adsorbed layer is also a function of the metal, the gas used to release the system to atmosphere and the time and extent to which it was exposed to the surrounding air.

In SI the outgassing rate (quantity of gas evolved per unit time per unit surface area) has units of $(Pa-m^3)/(m^2-s)$ or W/m^2. Factors for converting old units to $W-m^2$ are given in Appendix A.3. The pressure in a chamber with net outgassing rate q, and area A, when pumped at a speed S, is given by

$$P = 1000 \frac{q(W/m^2)A(m^2)}{S(L/s)} \qquad (16.1)$$

The factor of 1000 is included because the pumping speed is expressed in L/s rather than m^3/s. The outgassing rates for various metals are given in Appendix C.1.

Dissolved Gas

Efflux of dissolved gas from within the metal fixturing and walls can be eliminated by rendering it immobile, reducing its initial concentration, or erecting a barrier to its passage. Dissolved gas can be rendered immobile by completely immersing a system in liquid helium. At 4.2 K the diffusion constant for any gas is so small (4.10) that no special precautions need to be taken. The initial concentration can be substantially reduced by vacuum melting, by first degassing parts in a vacuum furnace, or by in-situ bake of the completed system. A barrier to this outgassing flux can be created by incorporating a layer of metal such as copper, which has a low permeability, or by forming an oxide barrier such as chrome oxide on stainless steel. An oxide barrier to hydrogen diffusion can be formed by an air or oxygen bake or by a multistep chemical treatment such as Diversey [8] cleaning. The latter method leaves some water vapor on the surface. Less thorough cleaning methods are needed after the system has been initially treated.   Either glow discharge cleaning ·or a vacuum bake can be used to clean a chamber after each exposure to ambient.   The nature and duration of the cleaning depends on the materials, construction and desired base pressure.

Vacuum melting is an excellent technique for removing dissolved gas under certain conditions.   It is expensive and used for specialized applications that require hydrogen and oxygen-free material in small quantities such as certain internal parts and charges for vacuum evaporation hearths.

Vacuum firing of components and subassemblies will effectively remove the dissolved gas load in cleaned and degreased parts.   Hydrogen firing is traditionally used for this purpose because it reduces surface oxides.   It has the disadvantage of incorporating considerable hydrogen in the metal at the firing temperature which can slowly outdiffuse at lower temperatures.

Vacuum or inert gas firing is preferred for vacuum components, especially those in ultrahigh vacuum systems.   The maximum firing temperatures for several metals are given in Table 16.1.   Iron and steel are usually fired at temperatures of 800 to 1000°C, while copper and its zinc

Table 16.1   Firing Temperatures
for Some Common Metals

Material	Firing Temperature (°C)
Tungsten	1800
Molybdenum[a]	950
Tantalum	1400
Platinum	1000
Copper and alloys[b]	500
Nickel and alloys (Monel, etc.)	750-950
Iron, steel, stainless steel	1000

Source.   Reprinted with permission from *Vacuum Technology*, p. 277, A. Guthrie.   Copyright 1963, John Wiley & Sons.
[a] Embrittlement takes place at higher temperatures. The maximum firing temperature is 1760°C.
[b] Except zinc bearing alloys which cannot be vacuum fired at high temperatures because of excessive zinc evaporation.

free alloys are fired at 500°C. Beavis [9] recommends firing at temperatures as high as possible which will not destroy the material.

It is important to use oxygen-free, high conductivity copper (OFHC)
because any copper oxide will react with hydrogen to form water vapor.
This vapor can cause voids in the material when it is heated. These voids
will create a porous leaky metal. It is not good practice to fire both a
screw and the part containing the tapped hole because the screw will bind
when it is tightened.

Vacuum firing of metal at low temperatures will not reduce the outgassing rate. Gas depletion time is given by $t \propto d^2/(6D)$ [10], a 1-h bake
of stainless steel at 1000°C is equivalent to a 2500-h bake at 300°C.
Calder and Lewin [11] have calculated the time required to reach an
outgassing rate of 10^{-13} W/m^2 for a stainless steel sample 2 mm thick
with an initial concentration of 4×10^4 Pa. Their results, Table 16.2,
show the hydrogen diffusion in stainless steel at intermediate temperatures of 420 to 570°C is rapid enough to make processing at these
temperatures practical.

Adsorbed Gas

Adsorbed gas may be removed from metal surfaces by thermal desorption, chemical cleaning or energetic particle bombardment. The procedures used for cleaning metal vacuum system parts depend on the system
application. The outgassing rate of unbaked, untreated stainless steel is
about 5×10^{-5} W/m^2 after 10 h of pumping. An unbaked system with a
0.5-m^3 work chamber may have as much as 6 m^2 of internal tooling and

Table 16.2 The Theoretical Time to Reach an Outgassing Rate of 10^{-13} W/m^2 in Stainless Steel

t (s)	D (m^2/s)	T (°C)
10^6 (11 days)	3.5×10^{-12}	300
8.6×10^4 (24 h)	3.8×10^{-11}	420
1.1×10^4 (3 h)	3.0×10^{-10}	570
3.6×10^3 (1 h)	9.0×10^{-10}	635

Source. Reprinted with permission from *Brit. J. Appl. Phys.*, **18**, p. 1459, R. Calder and G. Lewin. Copyright 1967, The Institute of Physics.

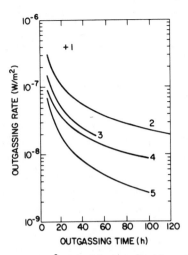

Fig. 16.2 Total outgassing per m^2 of sample as a function of pumping time and treatment. (1) New sample contaminated with lard oil, no cleaning. (2) New sample cleaned by the Diversey process. (3) New sample electropolished only. (4) Vapor degreased following the electropolishing. (5) New sample machined, degreased and contaminated with lard oil, then plasma arc cleaned. Reprinted with permission from *Vacuum*, **20**, p. 1, R. S. Barton and R. P. Govier. Copyright 1970, Pergamon Press, Ltd.

wall area. If the high vacuum pump has a baseplate pumping speed of 2000 L/s, the pressure after 10 h of pumping will be 1.5×10^{-4} Pa, not a good base pressure for such a system. Clearly the outgassing rate of stainless steel must be reduced by a factor of 10 to 100 over its untreated value to be suitable for high vacuum applications and by a factor of 10^4 to 10^6 for ultrahigh vacuum applications. Unbaked systems with base pressures of 5×10^{-6} Pa are generally cleaned by chemical or glow discharge cleaning techniques and perhaps a mild bake at temperatures of 40 to 80°C, while ultrahigh vacuum chambers require some form of high temperature vacuum baking and may also require glow discharge cleaning.

Barton and Govier [12] have measured the outgassing rates of stainless steel when treated by several methods that did not involve baking. Some of their results are shown in Fig. 16.2. The electropolishing treatment gave an outgassing rate as low as 10^{-8} W/m^2, a value similar to those obtained from honed and turned surfaces. Trichloroethylene vapor degreasing was reported to be the best technique for general use. After pumping for 70 h, the vapor-degreased specimen showed a mass spectrum that consisted of 58% H$_2$, 32% carbon oxides, 9% hydrocarbons and 1% trichloroethylene. Today, Freon and ethanol or 1,1,1-trichloroethane are used in place of trichloroethylene. The chrome-oxide-rich surface of a Diversey-cleaned sample yielded an outgassing rate three times greater than that of the vapor-degreased sample. Water vapor was the dominant product of desorption. Plasma arc cleaning gave the lowest observed outgassing rate, but the intense heat makes the process unusable. CO is reported to be the major species desorbed at high temperatures from the surface of Fe-Ni-Co, Mo, Ni and Cu-2%Be and H_2 is the major species desorbed from Cu after air exposure following cleaning and a 500-700°C bake [9].

Govier and McCracken [13] have studied the cleaning of stainless steel (AISI 321) in rare gas glow discharges. The total outgassing rate was about the same after discharge cleaning as after baking, but the composition of the desorbing species was drastically altered. They found the outgassing rate after 20 h of pumping to be 1.7×10^{-7} W/m^2 after the same pumping time. However, the residual gas load from the sample that was not discharge cleaned was 68% H_2O, 7% CO_2, 5% CO and 20% H_2. The residual gas composition from the discharge-cleaned sample was 98% H_2 and 2% Ar. Halama and Herrera [14] reported argon and oxygen glow-discharge cleaning of aluminum 6061 alloy gave an outgassing rate of 5×10^{-10} W/m^2. After a 24 h 200°C vacuum bake, an outgassing rate of $<1.3 \times 10^{-11}$ W/m^2 was measured. The residual gas composition also consisted of more than 99% hydrogen. They observed argon imbedded in the metal that was not released at room temperature. Its release rate reached a maximum between 280 and 300°C. At this temperature it was rapidly released from the metal. Halama and Herrera found oxygen combined with aluminum to form stable oxides and removed carbon from the surface by forming CO and CO_2; the hydrogen remained on the surface. Glow discharge cleaning has been adopted as the final surface treatment for the beam tubes in the Brookhaven colliding beam accelerator [15], and at the CERN intersecting storage ring [16].

The Ar:O_2 plasmas used to clean accelerators also remove material by sputtering. In large Tokamak systems this sputtering is undesirable because it deposits materials on insulators and windows. In these systems H_2 glow discharge cleaning has been used at an energy of 400 eV. The atomic hydrogen ions therefore have energies no greater than 200 eV which is below their sputtering threshold [17]. Knize and Cecchi [18] found the desorption rate of a dilute hydrogen isotope could be enhanced by a high constant pressure of hydrogen. This considerably speeded surface cleaning. See Section 4.2.2.

In Chapter 4 we noted the efficiency of electrons in desorbing molecules from steel surfaces. A simple electron degassing scheme can be arranged by heating a filament and biasing it at −1 kV from the system walls.

Vacuum baking of aluminum and stainless steel has been extensively studied because of the wide use these metals enjoy in vacuum system construction. The data of Calder and Lewin [11], Young [19], Dobro-

Table 16.3 Outgassing Rates of 316L Stainless Steel After Different Processing Conditions[a]

Sample	Surface Treatment	Outgassing Rates (10^{-10} W/m^2)			
		H_2	H_2O	CO	ArCO$_2$
A	Pumped under vacuum for 75 h.	893	573	87	-13.
	50 h vacuum bakeout at 150°C	387	17	6	-0.4
B	40 h vacuum bakeout at 300°C	83	0.7	2.2	-0.01
C	Degassed at 400°C for 20 h in a vacuum furnace (6.5×10^{-7} Pa)	19	0.3	0.44	0.160.11
D	Degassed at 800°C for 2 h in a vacuum furnace (6.5×10^{-7} Pa)	3.6	-	0.07	-0.05
	Exposed to atmosphere for 5 mo, pumped under vacuum for 24 h	-	73	67	-13.
	20-h vacuum bakeout at 150°C	3.3	-	0.08	-0.04
E	2 h in air at atmospheric pressure at 400°C	17	-	1.12	-0.4
	Exposed to atmosphere for 5 mo, pumped under vacuum for 24 h	-	80	69	-33.
	20-h vacuum bakeout at 150°C	17	0.75	0.37	-0.17
F	2 h in oxygen at 27,000 Pa at 400°C	600	253	-	123-
	20-h vacuum bakeout at 150°	5.2	0.09	0.4	0.51-
G	2 h in oxygen at 2700 Pa at 400°C	-	20	13	8.7-
	20-h vacuum bakeout at 150°C	-	0.9	0.64	0.45-
H	2 h in oxygen at 270 Pa at 400°C	-	16	52	19-
	20-h vacuum bakeout at 150°C	5.7	3.2	0.36	2-

Source. Reprinted with permission from J. Vac. Sci. Technol., **14**, 210, R. Nuvolone. Copyright 1977, The American Vacuum Society.
[a] All samples were first degreased in perchloroethylene vapor at 125°C, ultrasonically washed for 1 h in Diversey 708 cleaner at 55°C, rinsed with clean water, and dried.

zemsky and Moraw [20], Samuel [21] and Strausser [22] for stainless steel and aluminum fall in the range 10^{-9} to 10^{-11} W/m^2. The exact values are dependent on the effects of treatment before vacuum bake as well as on its time and temperature. Nuvolone [23] has systematically compared several of these treatments on 316L stainless steel under identical precleaning and measurement conditions. This work, which is consistent with earlier work, is described in Table 16.3. The lowest outgassing rate was obtained with oxidation in pure oxygen at 2700 Pa. The oxide barrier effectively reduced the hydrogen outgassing rate. Most importantly, this shows samples cleaned by an 800°C vacuum bake or a 400°C air bake can be stored for long periods before use, provided they are given a low temperature bake (150°C) after assembly.

16.1.4 Structural Metals

Aluminum and stainless steel are the two metals most commonly used in the fabrication of vacuum chambers. Aluminum is inexpensive and easy to fabricate, but hard to join to other metals. It is often used in the fabrication of vacuum collars for glass bell jar systems which are sealed with elastomer O-rings, and in some internal fixturing. It is difficult to seal via a metal gasket. Aluminum has been largely bypassed in modern vacuum system construction because of these difficulties. The properties of a few common alloys are given in Table 16.4. Cast jig plate is readily available form which cannot be welded and should never be used in a vacuum system because it is porous.

Aluminum has been reexamined for use in the construction of chambers for very large high energy particle accelerators and storage rings [13]. Its high electrical and thermal conductivity and low cost are an asset in the construction of beam tubes. Its residual radioactivity is less than that of stainless steel because its atomic number is less than that of

Table 16.4   Selected Properties of Common Aluminum Alloys

	% Alloying Element[a]							
Alloy	Cu	Si	Mg	Cr	Common Forms	TIG Weld	Bend	Vac.
4043	-	5.0	-	-	weld filler	yes	yes	yes
5052	-	-	2.5	0.25	sheet, angle, tube	yes	yes	yes
6061	0.25	0.6	1.0	0.2	sheet, angle, tube	yes	no	yes
Cast		proprietary			jig plate	no	no	no

[a] Welding Alcoa Aluminum, Aluminum Co. of America, Pittsburgh, PA, 1966.

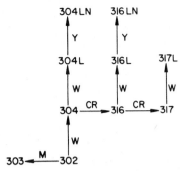

Fig. 16.3 Stainless steels used in vacuum equipment (AISI designation). CR = corrosion resistance, W = ease of welding, Y = yield strength, and M = ease of machining. Reprinted with permission from *Vacuum*, **26**, p. 287, C. Geyari. Copyright 1976, Pergamon Press, Ltd.

iron, chrome or nickel. In this application explosively bonded aluminum-to-stainless steel sections are used to make the transition to stainless steel flanges [13]. Ishimaru [24] has attempted to eliminate stainless steel altogether by designing a system of flanges and bolts of high strength aluminum alloy for use with aluminum O-rings.

For ordinary laboratory high vacuum systems stainless steel is the preferred material. It has a high yield strength, is easy to fabricate, and is stable. The stainless steels used in vacuum systems are part of a family of steels characterized by an iron-carbon alloy that contains greater than 13% chrome. The 300 series of austenitic steels is the most frequently used in vacuum and cryogenic work because it is corrosion resistant, easy to weld and non-magnetic.

The AISI series is an "18-8" steel that contains 18% chrome and 8% nickel. To this basic composition additions and changes are made to improve its properties. The low outgassing rate and oxidation resistance of stainless steel is due to the formation of a Cr_2O_3 layer on the surface. Figure 16.3 outlines some of the 300 series alloys and their characteristics. Appendix C.8 contains additional properties and applications of the series.

Types 304 and 316 are the most commonly used grades. Type 303 is easy to machine, but is not used in vacuum systems because it contains sulfur, phosphorous or selenium. For some applications such as cryogenic vacuum vessels it is desirable to reduce the thickness of the structural steel walls to reduce heat losses. This can be accomplished without loss of any other properties by the use of a nitrogen-bearing alloy such as 304LN or 316LN, or by cold stretching and annealing. The maintenance of corrosion resistance, methods of increasing strength, and reduction of porosity on AISI 300 series stainless steels have been discussed by Geyari [25] and should be thoroughly understood by anyone needing stainless steel for a unique application where it is to be stressed to its limits. Adams [26] has reviewed the characteristic surface of various stainless steels. He describes methods for modifying the protective chrome oxide layer by heating, chemical treatments and ion bombardment.

16.2 GLASSES AND CERAMICS

A glass is an inorganic material that solidifies without crystallizing. The common glasses used in vacuum technology are formulated from a silicon oxide base to which other oxides have been added to produce a product with specific characteristics. Soft glasses are formed by the addition of sodium and calcium oxides (soda-lime glass) or lead oxide (lead glass). Hard glasses are formed by the addition of boric oxide (borosilicate glass). Table 16.5 lists the chemical composition and physical properties of glasses often encountered in vacuum applications.

The physical properties of a glass are best described by the temperature dependence of the viscosity and expansion coefficient. The important properties are described by specific viscosities because it has no definite melting point. At the strain point ($10^{15.5}$ Pa-s), stresses are relieved in hours. The annealing temperature is defined as a viscosity of 10^{14} Pa-s at which stresses are relieved in minutes. At the softening point the viscosity is about $10^{8.6}$ Pa-s, and the working point corresponds to a viscosity of 10^5 Pa-s. Glass is brittle, and because of its high thermal expansion and low tensile strength it can shatter if unequally heated. Its expansion coefficient is important when selecting the components of a glass-to-glass or glass-to-metal seal.

The viscosity-temperature and thermal expansion characteristics determine the suitability of a glass for a specific application. Borosilicate glasses are used whenever the baking temperature exceeds 350°C, while fused silica is required for temperatures higher than 500°C. The thermal expansion coefficient and strength determine the maximum temperature gradient that a glass can withstand and to what it can be sealed. For example a borosilicate glass dish can be heated to 400°C in a few minutes, while a large, 1-in.-thick vessel fabricated from lead glass will require 24 h to reach the same temperature. Glasses are used for bell jars, Pirani gauges, U-tube manometers, McLeod gauges, ion gauge tubes, cathode ray tubes, controlled leaks, diffusion furnace liners, view ports, seals, and electrical and thermal insulation.

A ceramic is a polycrystalline, non-metallic inorganic material formed under heat treatment with or without pressure. Ceramics are mechanically strong, have high dielectric breakdown strength and low vapor pressure. Ceramics include glass-bonded crystalline aggregates, and single-phase compounds such as oxides, sulfides nitrides, borides and carbides. Ceramics contain entrapped gas pores and are not so dense as crystalline

Table 16.5   Properties of Some Glasses Used in Vacuum Applications

Property	Fused Silica	Pyrex 7740	7720[a]	Soda 7052[a]	0080	Lead 0120
Composition						
SiO_2	100	81	73	65	73	56
B_2O_3		13	15	18		
Na_2O		4	4	2	17	4
Al_2O_3		2	2	7	1	2
K_2O				3		9
PbO			6			29
LiO				1		
Other				3	9	
Viscosity characteristics						
Strain Point °C	956	510	484	436	473	395
Annealing Point °C	1084	560	523	480	514	435
Softening Point °C	1580	821	755	712	696	630
Working Point °C	-	1252	1146	1128	1005	985
Expansion coefficient $\times 10^{-7}/°C$	3.5	35	43	53	105	97
Shock temp., 1/4-in. plate °C	1000	130	130	100	50	50
Specific gravity	2.20	2.23	2.35	2.27	2.47	3.05

Source. Reprinted with permission from Corning Glass Works, Corning, NY.
[a] 7720 glass is used for sealing to tungsten and 7052 glass is used for sealing to Kovar.

materials. Their physical properties improve as their density approaches that of the bulk. Alumina is made with densities that range from about 85 to almost 100% of its bulk density. Most ceramics have a density of about 90% of the bulk. The important physical properties of ceramics are their compression and tensile strength and thermal expansion coefficient. High density alumina, for example, has a tensile strength four to five times greater than glass, and a compression strength 10 times greater than its tensile strength. The properties of some ceramics are listed in Table 16.6. Alumina is the most commonly used ceramic in applications such as high vacuum feedthroughs and internal electrical standoffs. Machinable glass ceramic also finds wide application in the vacuum industry for fabricating precise and complicated shapes. It is a recrystallized mica ceramic whose machinability is derived from the easy cleavage of the mica crystallites.

Borides and nitrides have found applications in vacuum technology. Evaporation hearths are made form titanium diboride and titanium nitride, alone or in combination. They are available in machinable or pyrolitically deposited form. Forsterite ceramics ($2MgO:SiO_2$) are used in applications where low dielectric loss is needed, and beryllia (BeO) is

Table 16.6   Physical Properties of Some Ceramics

Ceramic	Main Body Composition	Expansion Coefficient ($\times 10^{-7}$)	Softening Temp. ($^{\circ}$C)	Tensile Strength (10^6 kg/m^2)	Specific Gravity
Steatite	MgOSiO$_2$	70-90	1400	6	2.6
Forsterite	2MgOSiO$_2$	90-120	1400	7	2.9
Zircon porcelain	ZnO$_2$SiO$_2$	30-50	1500	8	3.7
85% alumina	Al$_2$O$_3$	50-70	1400	14	3.4
95% alumina	Al$_2$O$_3$	50-70	1650	18	3.6
98% alumina	Al$_2$O$_3$	50-70	1700	20	3.8
Pyroceram 9696[a]	Corderite ceramic	57	1250	[b]14	2.6
Macor 9658[a]	Fluro-phlogophite	94	800	[b]10	2.52

Source. Reprinted with permission from *Vacuum*, **25**, p. 469, G. F. Weston.   Copyright 1975, Pergamon Press, Ltd.
[a] Reprinted with permission from Corning Glass Works, Corning, NY.
[b] Modulus of rupture.

used when high thermal conductivity is necessary.   Beryllia must be machined while carefully exhausting the dust, because it is extremely hazardous to breathe.   General reviews of ceramics and glasses are attributed to Espe [27] and Kohl [28].

Permeation of gas through glasses and ceramics occurs without molecular dissociation.   The permeation constant, which is given by (4.12) depends on the molecular diameter of the gas and the microstructure or porosity of the glass or ceramic.   Norton [29] has shown that the permeation of gases through a glass is a function of their molecular diameter. Figure 16.4 contains data for the temperature dependence of the permeation rate of He, D$_2$, Ne, Ar and O$_2$ through silicon oxide glasses [1]. With the exception of deuterium and hydrogen, it shows that the permeation rate decreases as the molecular diameter increases.   Norton suggested the measured permeation rate of hydrogen was much larger than predicted because of surface reactions and solubility effects.   The permeation of a gas through a glass depends on the size of the pores in relation to the diameter of the diffusing species.   Permeation is minor through a crystalline material such as quartz but increases with lattice spacing.   The non-network-forming Na$_2$O, which is added to SiO$_2$ to form soda glass, plugs these openings (Fig. 16.5) and causes the permeation rate to decrease.   Permeation rates for some glasses are given in Fig. 16.6.   Shelby [34] has reviewed the diffusion and solubility of gases in glass.   The thermal properties and helium permeation of Corning Macor

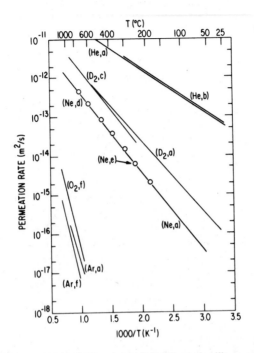

Fig. 16.4 Permeability of He, D_2, Ne, Ar and O_2 through silicon oxide glasses.   Data taken from (a) Perkins and Begeal [30]; (b) Swets, Lee and Frank [31]; (c) Lee [32]; (d) Frank, Swets and Lee [33]; Shelby [34]; and Norton [29].   Reprinted with permission from *J. Vac. Sci. Technol.,* **10**, p. 543, W. G. Perkins.   Copyright 1973, The American Vacuum Society.

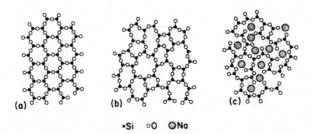

Fig. 16.5 (a) Atomic arrangement in a crystalline material possessing symmetry and periodicity;   (b) the atomic arrangement in a glass;   (c) the atomic arrangement in a soda glass.   Reprinted with permission from the *J. Am. Ceram. Soc.,* **36**, p. 90, F. J. Norton. Copyright 1953, The American Ceramic Society.

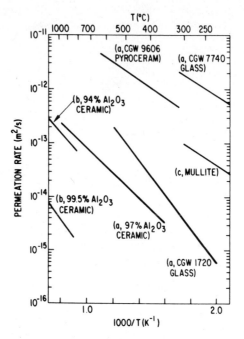

Fig. 16.6 Helium permeability through a number of glasses and ceramics. After Perkins. Data taken from (a) Miller and Shepard [35]; (b) Edwards [36]; (c) Shelby [34]. Reprinted with permission from *J. Vac. Sci. Technol.*, **10**, p. 543, W. G. Perkins. Copyright 1973, The American Vacuum Society.

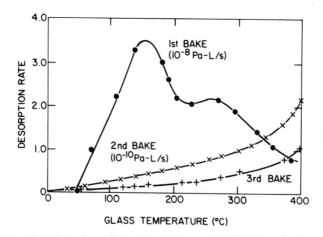

Fig. 16.7 Desorption of water from a Pyrex-glass surface of 180 cm^2 at increasing temperature. Adapted with permission from *Vacuum*, **15**, p. 573, K. Erents and G. Carter. Copyright 1965, Pergamon Press Ltd.

machinable glass ceramic have been reviewed by Altemose and Kacyon [38].

Gases are physically and chemically soluble in molten glass. The gas on the surface of glass is primarily water with some carbon dioxide. Water vapor may exist on glass in layers as thick as 10 to 50 monolayers [39]. Erents and Carter [40] have shown the first bake of a glass releases considerable surface water, while the second and succeeding bakes release structural water (Fig. 16.7). Todd [39] has shown this release of structural water is proportional to $t^{1/2}$ and indicates a diffusion-controlled process. Todd concluded a high temperature bake should completely eliminate outgassing of water from glass because all the surface water is released in a high temperature bake and the diffusion constant of water vapor is negligible at room temperature. At 25°C gas evolution from glass is dominated by the permeation of helium. Outgassing rates of some unbaked ceramics and glasses are given in Appendix C.3. In addition Colwell [41] has tabulated the outgassing rates of more than 80 untreated refractory and electrical insulating materials used in the construction of vacuum furnaces.

16.3 POLYMERS

Polymeric materials find applications in several areas. The generic, trade and chemical names of common materials are given in Table 16.7. Elastomers such as Buna-N, Viton, Silicone and Kalrez are used to form O-ring gaskets for static, sliding, or rotary seals. Other polymers such as Vespel and Kapton are used for high voltage vacuum feed throughs. Mechanical and other properties of elastomers which are important when these materials are used as seals are discussed in Section 17.2.1.

Permeation and outgassing are important physical properties of elastomers. Gases diffuse through voids in the intertwined polymer chains by a thermally activated process. The dimensions of the voids are larger than in a glass or metal, and as a result the permeation constants are much larger than for those materials. Comparison of the permeation rates of gases through polymers with those for glasses shows the diffusion process in a polymer is not as sensitive to molecular diameter as it is in a glass. This implies the diffusion of air and other heavy gases through polymers is a serious problem, while helium is the only gas of any consequence to diffuse through glass. Elastomers will swell when in contact with certain solvents used for cleaning and leak detection. This swelling or increased spacing between molecules results in an increased permeability [2].

Laurenson and Dennis [42] have studied the temperature dependence of gas permeation in three elastomers. Their results, presented in part in Fig. 16.8, have two important conclusions. One is the change of permea-

Table 16.7   Generic Trade and Chemical Names of Polymer Materials
Frequently Used in Vacuum

Generic	Trade	Chemical
Fluoroelastomer	Viton,[a] Fluorel[b]	vinylidene fluoride-hexafluoropropylene copolymer
Buna-N (nitrile)		butadiene-acronitrile
Buna-S		butadiene-styrene copolymer
Neoprene		chloroprene polymer
Butyl		isobutylene-isoprene copolymer
Polyurethane	Adiprene[a]	polyester or polyether di-isocyanate copolymer
Propyl	Nordel[a]	ethylene-propylene copolymer
Silicone	Silastic[d]	dimethyl polysiloxane polymer
Perfluoro-elastomer	Kalrez[a]	tetrafluoroethylene-perfluoromethylvinyl ether copolymer
PTFE	Teflon,[a] Halon[e]	tetrafluoroethylene polymer
PCTFE	Kel-F[b]	chlorotrifluoroethylene copolymer
Polyimide	Vespel,[a] Envex[c]	pyromellitimide polymer

Source. Reprinted with permission from *J. Vac. Sci. Technol.,*
17, p. 330, R. N. Peacock. Copyright 1980, The American
Vacuum Society.
[a] E. I. du Pont de Nemours and Company.

[b] 3-M Company.
[c] Rogers Corporation.
[d] Dow Corning Corporation.
[e] Allied Chemical Company.

tion with temperature is not the same for all gases so that the ratio of
permeation between two gases at room temperature is not the same at
elevated temperature. Second, the change of permeation with temperature for a given gas is not the same in different materials. the material
with the lowest permeability to a given gas at room temperature is not
necessarily the best material to use at 100°C.

Permeation rates for some materials commonly used in vacuum are
tabulated in Appendix C.6. More complete data on the permeation of
gases through AFLAS and Epichlorohydrin are given in the paper by
Laurenson and Dennis [42].

The outgassing of unbaked elastomers has been studied extensively [1,
42-59] and shown to be dominated by the evolution of water vapor.
Figure 16.8 illustrates the time dependence of the outgassing rates of
several elastomers. Baking will reduce this gas load, as revealed in the
mass scans of Figs. 9.7 and 9.8 taken during the heating of Buna-N and
Viton. These mass scans show plasticizers added to the polymer before
vulcanization were released as the temperature was increased. Unreacted
polymer was also evolved. At a higher temperature, the elastomer began
to decompose. de Csernatony [56,57] reviewed the elastomers Kalrez,
and the current Viton E60C. Viton E60C appears to be similar in its
outgassing to Viton A, while Kalrez has a low outgassing rate and can
withstand temperatures up to 275°C.

Outgassing data for elastomers commonly used in vacuum are given in Appendixes C.4 and C.7. Schalla [60] gives data on the outgassing of cellular foams and stranded elastic cords used in the medium vacuum range, while Glassford et al. [61,62] have reported the outgassing rates of aluminized mylar and other multilayer insulation materials. Erickson, et al., [63,64] give outgassing characteristics of several optically black solar absorbing coatings, as well as elastomers and carbon foam [60].

Sigmond [65] noted the pertinent outgassing properties were not dominated by the polymer, but by water, and the plasticizers and stabilizers of proprietary composition and quantity that were altered from lot to lot without the consumers knowledge. This led to his observation that one derives from a polymer whatever the manufacturer chooses to put into it, often for reasons unconnected with the user's expectations [66].

A volatile organic of mass number 149 was found to be present in ethyl alcohol stored in polyethylene bottles and in samples stored in test tubes stoppered with plastic caps [65]. This organic vapor turned out to a fragment of dibutylphthalate, a commonly used plasticizer. Because

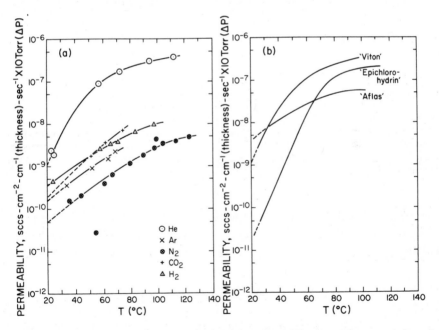

Fig. 16.8 (Left) Relation between permeation and temperature for various gases through Viton. (Right) A comparison of the helium permeability of three elastomers, Viton, Epichlorohydrin, and AFLAS, over the range 20 to 100°C. Reprinted with permission from *J. Vac. Sci. Technol. A*, **3**, p. 1707, L. Laurenson and N. T. M. Dennis. Copyright 1985, The American Vacuum Society.

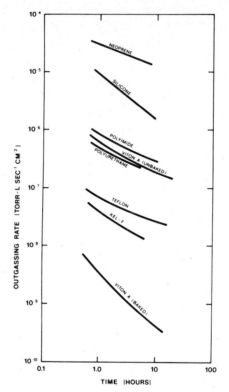

Fig. 16.9 Room-temperature outgassing rates for several polymer materials. The values shown are selected from the literature as typical. The rates decline between one to two decades from one to ten hours of pumping. Moderate baking is even more helpful as a comparison of the two Viton curves shown [46,49,55,56]. Reproduced with permission from *J. Vac. Sci. Technol.*, **17**, p. 330, R. N. Peacock. Copyright 1980. The American Vacuum Society.

this fragment was easily removed by a 100°C bake, it was suggested that it contaminates all materials stored in plastic bags or boxes or touched by plastic gloves.

REFERENCES

1. W. G. Perkins, *J. Vac. Sci. Technol.*, **10**, 543 (1973).
2. R. J. Elsey, *Vacuum*, **25**, 299 (1975).
3. R. J. Elsey, *Vacuum*, **25**, 347 (1975).
4. G. F. Weston, *Vacuum*, **25**, 469 (1975).
5. J. P. Hobson, *J. Vac. Sci. Technol.*, **16**, 84 (1979).
6. F. J. Norton, *Trans. 8th Nat. Vac. Symp. and Proc. 2nd. Int. Congr. Vac. Sci. Technol. (1961),* Vol. 1, Pergamon, New York, 1962, p. 8.

7. D. R. Begeal, *J. Vac. Sci. Technol.*, **15** 1146, (1978).

8. The Diversey-Wyandotte Corp., Wyandotte, MI.

9. L. Beavis, *J. Vac. Sci. Technol.*, **20**, 972 (1982).

10. W. A. Rogers, R. S. Buritz and D. L. Alpert, *J. Appl. Phys.*, **25**, 868 (1954).

11. R. Calder and G. Lewin, *Brit. J. Appl. Phys.*, **18**, 1459 (1967).

12. R. S. Barton and R. P. Govier, *Vacuum*, **20**, 1 (1970).

13. R. P. Govier and G. M. McCracken, *J. Vac. Sci. Technol.*, **7**, 552 (1970).

14. H. J. Halama and J. C. Herrera, *J. Vac. Sci. Technol.*, **13**, 463 (1976).

15. H. C. Hseuh, T. S. Chou and C. A. Christianson, *J. Vac. Sci. Technol. A*, **3**, 518 (1985).

16. R. Calder, A. Grillot, F. LeNormamd and A. Mathewson, *Proc. 7th Int. Vac. Congr.*, R. Dobrozemsky, Ed., Vienna, 1977, p. 231.

17. Ph. Staib, H. F. Dylla and S. M. Rossnagel, *J. Vac. Sci. Technol.*, **17**, 291 (1980).

18. R. J. Knize and J. L. Cecchi, *J. Vac. Sci. Technol. A*, **1**, 1273 (1983).

19. J. R. Young, *J. Vac. Sci. Technol.*, **6**, 398 (1969).

20. R. Dobrozemsky and G. Moraw, *Electron. Fis. Ap.*, **17**, 235 (1974).

21. R. L. Samuel, *Vacuum*, **20**, 195 (1970).

22. Y. E. Strausser, *Proc. 4th Int. Vac. Congr. (1968)*, Inst. of Physics and Physical Soc., London, 1969, p. 469.

23. R. Nuvolone, *J. Vac. Sci. Technol.*, **14**, 1210 (1977).

24. H. Ishimaru, *J. Vac. Sci. Technol.*, **15**, 1853 (1978).

25. C. Geyari, *Vacuum*, **26**, 287 (1976).

26. R. O. Adams, *J. Vac. Sci. Technol. A*, **1**, 12 (1983).

27. W. Espe, *Materials of High Vacuum Technology*, Vol. 2, Pergamon Press, New York, 1966.

28. W. H. Kohl, *Handbook of Materials and Techniques for Vacuum Devices*, Reinhold, New York, 1967.

29. F. J. Norton, *J. Am. Ceram. Soc.*, **36**, 90 (1953).

30. W. G. Perkins and D. R. Begeal, *J. Chem. Phys.*, **54**, 1683 (1971).

31. D. E. Swets, R. W. Lee and R. C. Frank, *J. Chem. Phys.*, **34**, 17 (1961).

32. R. W. Lee, *J.Chem. Phys.*, **38**, 448 (1963).

33. R. C. Frank, D. E. Swets and R. W. Lee, *J. Chem. Phys.*, **35**, 1451 (1961).

34. J. E. Shelby, *Phys. Chem. Glasses*, **13** 167 (1972).

35. C. F. Miller and R. W. Shepard, *Vacuum* **11** 58 (1961).

36. R. H. Edwards, M. S. Thesis, Univ. of Calif., Berkeley, 1966.

37. J. E. Shelby, *Molecular Solubility and Diffusion*, in *Treatise on Materials Science and Technology: Glass II*, Vol. 17, M. Tomozawa and R. H. Doremus, Eds., Academic, New York, 1979.

38. V. O. Altemose and A. R. Kacyon, *J. Vac. Sci. Technol.*, **16**, 951 (1979).

39. B. J. Todd, *J. Appl. Phys.*, **26**, 1238 (1970).

40. K. Erents and G. Carter, *Vacuum*, **15**, 573 (1965).

41. B. H. Colwell, *Vacuum*, **20**, 481 (1970).

42. L. Laurenson and N. T. M. Dennis, *J. Vac. Sci. Technol. A*, **3**, 1707 (1985).

43. R. S. Barton and R. P. Govier, *J. Vac. Sci. Technol.*, **2**, 113 (1965).

44. R. R. Addis, Jr., L. Pensak and N. J. Scott, *Trans. 7th Vac. Symp. (1960)*, Pergamon, New York, 1961, p. 39.

45. R. Geller, *Le Vide*, No. 13, **71**, (1958).

46. M. M. Fluk and K. S. Horr, *Trans 9th Vac. Symp. (1962)*, Macmillan, New York, 1963, p. 224.

47. M. Munchhausen and F. J. Schittko, *Vacuum*, **13**, 548 (1963).

48. P. W. Hait, *Vacuum*, **17**, 547 (1967).

49. J. Blears, E. J. Greer and J. Nightengale, *Adv, Vac. Sci. Technol.*, Vol. 2, E. Thomas, Ed., Pergamon, 1960, p. 473.

50. L. de Csernatony, *Vacuum*, **16**, 13 (1966).

51. L. de Csernatony, *Vacuum*, **16**, 129 (1966).

52. L. de Csernatony, *Vacuum*, **16**, 247 (1966).

53. L. de Csernatony, *Vacuum*, **16**, 427 (1966).

54. L. de Csernatony and D. J. Crawley, *Vacuum*, **17**, 55 (1967).

55. T. L. Edwards, J. R. Budge and W. Hauptli, *J. Vac. Sci. Technol.*, **14**, 740 (1977).

56. L. de Csernatony, *Proc. 7th Int. Vac. Congr. & 3rd Int. Conf. Solid Surf.*, R. Dobrozemsky, Ed., Vienna, 1977, p. 259.

57. L. de Csernatony, *Vacuum*, **27**, 605 (1977).

58. B. B. Dayton, *1959 Sixth Nat. Symp. on Vac. Technol. Trans.* Pergamon, New York, 1960, p. 101.

59. G. Thieme, *Vacuum*, **17** 547 (1967).

60. C. A. Schalla, *J. Vac. Sci. Technol.*, **17**, 705 (1980).

61. A. P. M. Glassford and C-K. Liu, *J. Vac. Sci. Technol.*, **17**, 696 (1980).

62. A. P. M. Glassford R. A. Osiecki and C-K. Liu, *J. Vac. Sci. Technol. A*, **2**(3), 1370 (1984).

63. E. D. Erickson, D. D. Berger and B. A. Frazier, *J. Vac. Sci. Technol. A*, **3**, 1711 (1985).

64. E. D. Erickson, T. G. Beat, D. D. Berger and B. A. Frazier, *J. Vac. Sci. Technol. A*, **2**, 206 (1984).

65. T. Sigmond, *Vacuum*, **25**, 239 (1975).

66. Murphy, as quoted by Sigmond in ref. 65.

PROBLEMS

16.1 What are three ways (other than leaks) by which gas can enter a vacuum system?

16.2 Describe the properties of materials which influence (a) the rate of gas evolution from a surface, (b) their use as vacuum vessel walls, and (c) their use within the vacuum environment.

16.3 The increased gas load from improperly handled or cleaned metal parts is predominantly a result of: (choose two) permeation, desorption, diffusion or evaporation? By what additional processes will gas be evolved from an improperly cleaned or handled elastomer?

16.4 What hydrogen flux can be obtained through a leak constructed from a 10,000-cm^2 palladium film 0.02 mm thick held at a temperature of 400°C?

16.5 When an arc breakdown occurs in a vacuum switch, dissolved gas will be released from the electrodes at a point where the arc makes contact. Assume 0.04 g of copper evaporates from the electrodes during a particular arc. Calculate the number of molecules of gas released from the copper, if the copper were contaminated with dissolved gas at a solubility of 1 ppm. Assuming an initial pressure of 10^{-5} Pa, what instantaneous pressure results from this arc in a switch whose interior volume is 1 L?

16.6 (a) What is the difference between force and pressure? (b) 50-cm-diameter "Magdeburg hemispheres" are sealed together and exhausted to $P = 2000$ Pa. What is the force (newtons) required to separate the hemispheres? (c) Do high vacuum chambers need to be structurally any stronger than those used in the hemispheres?

16.7 A 50-cm-diameter vacuum bell jar is sealed to a baseplate with an elastomer L-gasket. The maximum deflection of the center of such a freely mounted circular plate is given by $x(m) = 0.696Pr^4/(Et^3)$ where P is the pressure in Pa, r is the radius and t is the baseplate thickness in meters. E is the modulus of elasticity. Calculate the baseplate thickness required if it is made from (a) aluminum ($E = 6.9 \times 10^{10}$ Pa) and (b) stainless steel ($E = 1.9 \times 10^{11}$ Pa) such that the maximum deflection of the baseplate in the center is limited to 0.25 mm when the interior of the bell jar is pumped to 10^{-5} Pa.

16.8 The maximum shear stress in a freely mounted circular disk is given by $s_m = 1.24Pr^2/t^2$ where P is the uniformly applied pressure in Pa, and s_m is the maximum shear stress in Pa. The maximum stress of a typical piece of borosilicate glass containing scratches and imperfections is about 6.9×10^7 Pa. However, glass exhibits a time loading effect that reduces its strength when under constant stress for long periods of time. An adequate design load is 6.9×10^6 Pa. Calculate the thickness required for (a) a 2-cm-diameter window and (b) a 20-cm-diameter window on an unbaked high vacuum chamber.

16.9 The major concern for the use of plastic materials in unbaked vacuum systems is not vapor pressure but ... .

16.10 Which has the greatest outgassing flux after a 200°C bake: (a) 10 cm^2 of stainless steel or, (b) 1 cm^2 of Viton A?

CHAPTER 17

Joints, Seals and Components

In years past a discussion of joining and sealing techniques would have been incomplete without a description DeKothinshy cement, Glyptal and black wax. The times when valve stems were sealed with string soaked in grease have also passed. Materials technology has advanced to the place where glass-to-metal seals, demountable elastomer and metal-to-metal gaskets, brazed joints, welded joints, valves and motion feedthroughs are routinely fabricated with excellent reliability. Reliable joints, seals and components are so common we often take them for granted, and we sometimes contribute to their misuse. Occasionally we use a rubber hose as a flex connector, bake a brass valve, grease an elastomer static seal, bake Viton to an excessively high temperature, try to save money by reusing copper gaskets, or apply stop leak to a leaky weld. Singleton [1] remarked that in the early days of vacuum technique some workers preferred the clear variety of Glyptal so as to hide their mistakes. Today's mistakes can be hidden from management with a degree of sophistication. Leaky welds have been known to be coated with a neat bead of low vapor pressure vacuum epoxy and hidden forever beneath a coat of paint.

In this chapter we review welded and brazed metal joints, metal, glass and ceramic joints, elastomer and metal-sealed flanges, valves, and motion feedthroughs. We emphasize proper selection and use of joining and sealing techniques, and not design. The technology of sealing and joining has been extensively discussed in publications by Roth [2-4].

17.1 PERMANENT JOINTS

Welding, brazing and soldering are used to make permanent joints in vacuum chambers, pumping lines and components. The technique depends on the materials to be joined, and the thermal and vacuum environment to which the parts will be exposed. Welded joints of stainless steel or aluminum, flame sealed glass joints, glass-to-metal and ceramic-to-metal joints are commonly used in high and ultrahigh vacuum systems.

17.1.1 Welding

Metals can be joined permanently by welding - locally melting - closely mating pieces. Welding of vacuum components is most commonly done by the tungsten inert gas (TIG) process to avoid oxidation. A detailed review of TIG and other welding techniques is given in *Metals Handbook* [5]. TIG welding is not the only technique, plasma arc and electron beam welding also can be used to make high quality joints. However, these techniques are not as readily available or applicable for all types of joints. TIG welding, also known as heliarc or argon arc welding, is a technique for forming clean, oxide-free, leak-tight joints by flooding the area immediately around the arc with an inert gas, usually argon. TIG welding is used in vacuum fabrication to make joints in materials such as stainless steel, aluminum, nickel, copper and titanium. It is not suitable for joining alloys with high melting point components, for example, brass and certain aluminum or stainless steel alloys. Stainless steel is the most common material used in the construction of high- and ultrahigh-vacuum vessels. The 300 series alloys, except for 303, are easily welded. It is more difficult to weld aluminum than stainless steel, but leak-tight joints can be made by a skilled welder. Dissimilar metals also can be welded. Combinations of stainless steel, copper, nickel and Monel can be joined by TIG welding. Aluminum and stainless can be joined by explosion bonding. The welding of austenitic stainless steels has been reviewed in extensive detail by Rosendahl [6] and Geyari [7].

Figure 17.1 depicts some common joints welded with techniques acceptable for vacuum use. Welding from the vacuum side, or through welding from the atmospheric side, and welding walls of equal thickness are basic in all the illustrations of correct welding practice. It is important to make the weld on the surface which will be exposed to vacuum. If the weld is made on the atmospheric surface, a gap may be created which is impossible to clean. For structural reasons some joints need to be welded on the atmospheric side as well. If this is necessary it should be a discontinuous weld. See for example, the dashed welds in Fig. 17.1. If the outer weld were continuous and the inner joint leaked, the inner joint leak could not be detected, and a trapped gas pocket would make a slow, virtual leak. In some geometries, tubing for example, the weld

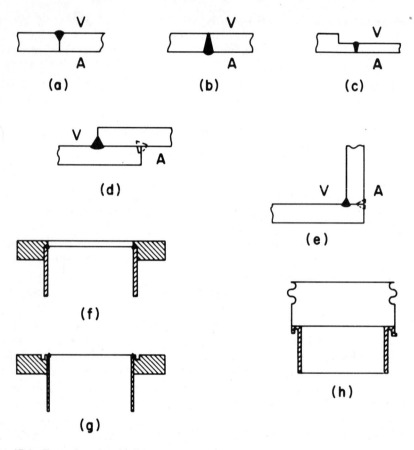

Fig. 17.1 Examples of welded joints: (a) butt welded from the vacuum side; (b) through welded from the atmospheric side; (c) joining parts of unequal thickness; (d) lap weld illustrating proper tack welds on the atmospheric side; (e) corner weld; (f) thick-wall tubing-to-flange weld. The flange is machined so that the bead depth is approximately equal to the wall thickness. (g) Thin-wall tubing-to-flange weld showing relief groove; (h) bellows-to-pipe adapter. The right side is shown before welding. The raised lip (0.025 cm high × 0.035 cm thick) provides filler metal for the weld. A copper heat sink is attached to the bellows end section during welding.

cannot be made on the vacuum side. In these instances a through weld will prevent the formation of regions where cleaning solvents can collect. Continuous butt welding or orbital welding is a superior process for joining long lengths of tubing. With this technique the two ends of the tubing are forced together and purged during welding. Large pipe can be chemically cleaned following welding. Oxidation can be prevented on the opposite surfaces by flushing with argon or an argon-hydrogen mixture during welding. It is also important for the two surfaces to be as equal in thickness as possible. A weld relief groove should be used to match the thickness when welding a flange to a thinner tube. This is particularly important in the welding of thin-wall tubing and bellows. See Fig. 17.1. These two ideas, welding walls of equal thickness, and welding from the vacuum side or through to the vacuum side, are basic to all properly designed and executed weld joints.

Carbide precipitation and inclusions in ASIA 300 series stainless steels are potential problems in fabricating vacuum vessels. Carbon that has precipitated at grain boundaries because of welding, or improper cooling after annealing removes with it a substantial fraction of the chrome from nearby regions. See Fig. 17.2. The nearby regions then contain less than 13% chrome and are no longer stainless steel. They are subject to corrosion if exposed to a corrosive atmosphere. Subsequent baking in the presence of water vapor can increase hydrogen permeation through the affected region. The formation of microscopic cracks is also a concern for stainless steel subjected to low temperatures. A crack in a carbide-rich zone can cause a leak in a cold trap or cold finger.

Carbide precipitation may be prevented with an alloy containing a low carbon content, a stabilized alloy, or a minimum-heat welding technique. A good solution is the use of low-carbon steel alloys such as 304L and 316L, but they are not as strong, require more nickel, and are more

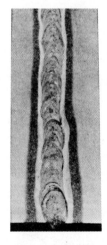
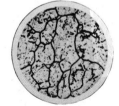

Fig. 17.2 Photograph showing the appearance and location of carbide-rich zones bordering a weld bead in 304 stainless steel. The sample has been etched in a hot acid to make the carbide-rich regions visible. The dark zones on either side of the weld are regions in which the metal cooled from 900°C to 500°C in more than 5 min. The right inset shows a magnified view of the carbide-rich zone. The left inset shows a magnified view of the grain structure of normal annealed 304 stainless steel. Reprinted with permission from Stainless Steel Fabrication, Allegheny Steel, Pittsburgh. Copyright 1959, Allegheny Steel Company.

expensive to manufacture than their higher carbon counterparts. Titanium, niobium and tantalum form carbides more easily than chrome, so an alternative solution is the use of an alloy (321, 347 or 348) stabilized with one of these elements. A minimum heat weld will also reduce the time that the metal weld region spends in the dangerous 500 to 900°C region. Minimum heat welds are simplified by use of weld relief grooves like those shown in Fig. 17.1. An 18-8 alloy of 0.06% carbon will not precipitate carbides at the grain boundaries if it is cooled from 900 to 500°C in less than 5 min. As the carbon content is reduced, the steel can remain in the critical temperature region for additional time without carbide precipitation.

A most important concern of the ultrahigh vacuum user, and occasionally the high vacuum user, is the proper fabrication and joining of stainless steel components to eliminate leaks through inclusions in the metal. These minute inclusions, which occur in the process of making steel, are masked by grease and other impurities until the steel wall is thoroughly baked. At that time tiny leaks will appear. The impurities in a cooling ingot distribute themselves at the top and center. See Fig. 17.3. The portion with most of the oxide and sulfide impurities is removed before rolling and any remaining impurities are stretched into long narrow leak paths. The inclusions are in the direction of rolling. It is important to

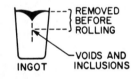

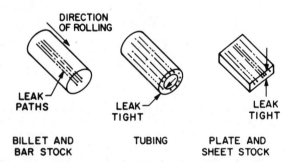

Fig. 17.3 Schematic inclusions in steel during casting and rolling. Reprinted with permission from Varian Report VR-39, *Stainless Steel for Ultra-high Vacuum Applications,* V. A. Wright. Copyright 1966, Varian Associates, 611 Hansen Way, Palo Alto, CA.

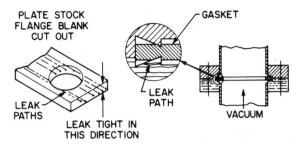

Fig. 17.4 Porosity in high vacuum flanges. Reprinted with permission from Varian Report VR-39, *Stainless Steel for Ultrahigh Vacuum Applications*, V. A. Wright. Copyright 1966, Varian Associates, 611 Hansen Way, Palo Alto, CA.

know the inclusion direction when selecting the raw stock from which components are to be made. In particular inclusions can cause virtual leaks when sheet or tubing is welded on the atmospheric side without full penetration. Figure 17.4 illustrates the origin of leak paths in high vacuum flanges made of plate stock. To avoid such potential leaks modern flanges are made from bar stock. Also the material is usually forged to break up the long filamentary inclusions and reduce the possibility of such leak paths. For critical applications a section from each end of the billet will be individually examined before fabrication.

17.1.2 Soldering and Brazing

Soldering and brazing are techniques for joining metal parts with a filler metal whose melting point is lower than the melting point of the parts to be joined. The definitions of soldering and brazing are not universal. One definition states brazing is done with fillers whose melting point is greater than 450°C, while soldering uses fillers which melt at lower temperatures. The term "hard soldering" is obsolete. Brazing and soldering are used where alloy or metal combinations cannot be welded, or where warping of parts during welding would produce unacceptable distortion. Soldering is little used in vacuum technique because low melting point solders contain high vapor pressure materials which are unsuitable for use in baked systems.

Furnace brazing in a hydrogen or vacuum ambient is a common method for joining large pieces without introducing thermal stress. Bellows were traditionally brazed to heavier tubing before joints of the type illustrated in Fig. 17.1*h* were developed. Brazing can be done by torch, by dipping, by induction heating or by heating in a furnace. An excellent review of practical brazing techniques is contained in the pamphlet by Peacock [8].

The parts to be brazed or soldered should be closely machined and touch over a large surface area. The filler has a lower shear stress than

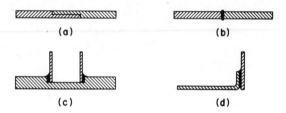

Fig. 17.5 Examples of brazed joints: (a) one form of a strong butt-lap joint; (b) a weak butt joint in which the stress is placed on a small filler area; (c) a poor tube-to-flange joint with excessively large clearances; (d) a strong corner joint.

the metals being joined, and a large surface is necessary to place the stress on the metal rather than on the filler. The metal parts need to fit closely so the filler can flow into the joint by capillary action. Thick gaps between parts will result in voids. Figure 17.5 illustrates proper and improper construction of brazed and soldered joints. The filler should melt at a temperature below the melting point of the pieces to be joined.

Pure metal and alloy braze fillers are available in a wide range of melting points ranging from –40°C (mercury, which has been used in cryogenic joints) to 3180°C (rhenium). Several copper-silver and copper-gold alloys with melting points in the range 800 to 1000°C are available for brazing stainless steel, copper and other alloys used in the construction of valves, diffusion pump casings, feedthroughs, and internal fixturing. Cu-Ag is commonly listed as a braze filler for stainless steel, however Cu-Au fillers flow more readily between closely mating parts. Tables of braze alloys and their applications are given in Kohl [9], the American Welding Society *Brazing Manual* [10], and the Handy and Harmon *Brazing Book* [11].

The vapor pressure of some metals is high enough to preclude their use in baked vacuum systems. Most common solders are lead based, while many brazing alloys with melting points under 700°C contain Pb, Zn, Cd or P. These alloys are limited to torch brazing of assemblies that will not be baked. If components containing these elements are baked, the high vapor pressure components will vaporize and contaminate the vacuum system. In the same manner they will contaminate the brazing furnace. Besides vapor pressure limitations, there are well-known metal incompatibilities. Gold and aluminum bonds form an intermetallic known as the purple plague. Gold and silver cannot be used in contact with mercury, and brazing alloys containing silver cannot be used in contact with iron-nickel-cobalt alloys such as Kovar without causing intergranular corrosion.

17.1.3   Joining Glasses and Ceramics

The techniques for joining glasses and ceramics recognize the different expansion coefficient, tensile strength, and shear strength of each material. Glasses have widely differing expansion coefficients, ranging from $105 \times 10^{-7}/°C$ for soda-lime glass to $3.5 \times 10^{-7}/°C$ for fused quartz. Glasses are not as strong in tension as in compression, and they are not as strong as ceramics or metals. Here we review techniques for glass-to-glass, and glass- and ceramic-to-metal seals.

Glasses can be fused in a flame, or joined by frit or cane solder glass if their expansion coefficients differ by less than 10%. Flame sealing is done by melting the pieces with a torch. Frit seals use ground solder glass mixed in a slurry that evaporates as the frit melts. Pre-shaped cane glass seals are made from frit or cane solder glass. The parts to be sealed with solder glass are aligned, loaded with a suitable weight, and thermally cycled to the melting temperature of the solder glass. The solder glass and the parts need to have the same expansion coefficient, but the solder glass melting point must be less than the sealed parts. A review of solder glasses has been published by Takamori [12].

Glass pairs with widely dissimilar expansion coefficients such as quartz and Pyrex, or lead glass and Pyrex cannot be directly sealed. Such combinations are joined with a graded seal. A graded seal, also known as a step seal, is made by successively joining glasses whose expansion coefficient differs by about 10 to 15%. Adjacent glasses must also have nearly equal solidification temperatures or large thermal expansion differences will be frozen in during cooling.

Glasses can be sealed to metals if the metal has the same expansion coefficient as the glass, if the metal holds the glass in compression, or if the metal is very thin so that it can plastically deform. Glass does not adhere directly to metal (except for platinum), but rather to the metal oxide. The metal oxide must be stable and adhere well to the parent metal. The most common glass-to-metal seals use matching expansion coefficients. For example platinum, or an iron-nickel-chrome alloy such as Sealmet 4, is used to seal to lead glass. Tungsten is used to seal to 7720 and 3320 uranium glass, and Kovar is used to seal to 7052 glass. Ring seals are sometimes used where a metal band or ring of appropriate expansion coefficient contracts on cooling to hold the glass in compression. A seal for joining glass to a feathered, deformable copper edge—the Housekeeper seal has been out of vogue for many years. Today a stainless steel - 7052 glass version designed by Benbenek and Hoenig [13] is commercially available. A cross sectional view is shown in Fig. 17.6a. Construction details are given in the original paper and amplified in Rosebury [14]. The direct sealing of glass to stainless steel avoids the corrosion problems associated with the use of iron-nickel-cobalt alloys. Espe [15] describes the alloys used in glass-to-metal seals,

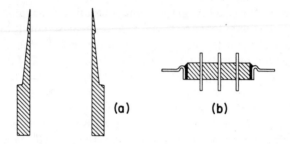

Fig. 17.6 (a) Tapered stainless tubing for fabricating a 7052 or Pyrex glass-to-stainless Housekeeper seal. Reprinted with permission from *Rev. Sci. Instr.*, **31**, p. 460, J. E. Benbenek and R. E. Honig. Copyright 1960, The American Institute of Physics. (b) Ceramic-to-metal seal.

while Kohl [9] and Rosebury [14] discuss many types of glass-to-metal seals.

Ceramic-to-metal seals are more easily made and stronger than glass-to-metal seals. Ceramics have high compression strength so it is not necessary to have as close an expansion coefficient match between ceramic and metal as it is between glass and metal. The high compression strength of a ceramic results in a very rugged seal. Ceramic seals are made by firing a thin layer of refractory metal on the ceramic at high temperatures. The metal is then furnace brazed to the refractory using a filler with a lower melting point than the refractory. For example, alumina is brazed to Kovar via a fired molybdenum paste. See Fig. 17.6b.

17.2 DEMOUNTABLE JOINTS

The most reliable, long-lived connection which can maintain its integrity in vacuum and related environments is a permanent, welded joint. Special purpose welded metal joints have been designed to be cut open, and glass can always be cut and reworked. However, most vacuum systems have a requirement for easy access which cannot be met in this way. Demountable joints are therefore a practical necessity.

A demountable seal must be designed and constructed to form a leak-tight joint and maintain it until the time it is to be opened. Forming a leak-tight joint requires an initial contact force large enough to make the joined material merge and fill the irregularities in the harder of the two materials. Maintaining the joint requires a force sufficient to overcome the effects of differential expansion and long term plastic flow. Depending on the relative expansion coefficients of the two materials, a joint may develop an unwanted leak if it is heated or cooled [4]. The restoring force must be maintained by the stored energy in the joint. Stored energy is a fundamental requirement in a demountable seal. (There are one or two cases where two metals form a diffusion joint or bond, and the force can be removed, e.g., copper-steel shear seals; however they have to be pulled apart.) A joint consists of three components: the seal or gasket material, the flange pair and the clamping means. The required energy may be stored in any one of the components or shared between them [8]. Besides maintaining the seal force the joint must be able to withstand the thermal, chemical and radiation environment of the system. It must be constructed from materials whose vacuum properties (vapor pressure, permeation and outgassing rate) are compatible with the vacuum operating range.

Elastomeric compounds are excellent choices for the seal or gasket material. They store elastic energy, conform to fit surface irregularities of the flange, are resistant to many chemicals and can be repeatedly used. Unfortunately elastomers cannot withstand the high baking temperatures or plasma bombardment schemes used to clean UHV systems without decomposing or drastically altering their properties. For these systems the only acceptable gasket material is a metal. In this section we review the techniques for demountable joints using both elastomers and metal gaskets. Elastomer gaskets have been reviewed by Peacock [16], while metal seals have been reviewed by Roth [4].

17.2.1   Elastomer Seals

Elastomer seals between metal or glass flanges are made by deforming the elastomer freely between two flat surfaces or confining it in a groove of rectangular, triangular or dove-tail crosssection. Elasticity, plasticity, hardness, compression set, seal loading, outgassing and gas permeation are important attributes of an elastomer seal. A summary of various elastomer seal properties is given in Table 17.1. Elastomers are formed by grinding the starting polymer, mixing with plasticizers and stabilizers, and vulcanizing to a state that is largely elastic. The materials are incompressible. That is, any deformation or compression in one direction must be accompanied by motion in another, so that the total volume remains constant. If the compound is completely elastic, it will return to its exact shape after the force has been removed. If it has some degree of plastic behavior it will flow and not return to its original shape. The measure of its shape change is called compression set.

Compression set occurs in an elastomer which has been deformed for a long time. It is more pronounced in elastomers which have been heated because compression set is strongly temperature dependent. Figure 17.7 gives comparative compression set data for five elastomers, while the time effects of compression set in Kalrez 1050 and Vitons A and E-60C

are described in Table 17.2. From these data it might appear that a
Kalrez or Viton seal would leak excessively after baking. Peacock [16]
comments that many of the problems with compression set in perfluoropolyether may be because of improper groove design. As this material is
heated it expands; expansion can induce excessive compression set if
inadequate room for expansion is left in the groove.

Adequate seal loading is rarely a problem with an elastomer. However
too much compression can limit the upper temperature because of thermal expansion. O-rings are typically compressed 15 to 20% of their
diameter. For a nominal 0.318-cm-diameter O-ring of 75 Shore hardness, this translates into a seal force of about 2.7 kg/cm of O-ring length
[16]. The initial pressure will be slightly reduced with time as the elastomer undergoes some plastic deformation, but it will not affect the
integrity of the vacuum seal. Because sealing is determined by contact
pressure, the compressive force, deformation and percentage groove
filling must all decrease as the hardness of the O-ring or the expansion
coefficient is increased. A chord compression of 20% is typical for
Viton O-rings, while Kalrez should not be compressed more than 12%.

Table 17.1   A Summary of Various Mechanical and General Considerations
Regarding the Selection of Polymer Seal Materials

Seal Material	Lin. coeff therm exp ($\times 10^{-5}/°C$)	Max op T (°C)	Cold flow at T_{op}	Gas perm	Wear/ abrasion resistance	Prime seal appl'n
Fluoroelastomer						
Viton E-60C	16	150	good	mod	good	a
Viton A	16	150	fair	mod	good	a
Buna-N	23	85	good	mod	v good	b
Buna-S	22	75	good	high	good	c
Neoprene	24	90	good	mod	v good	d,e
Butyl	19	-	good	mod	good	f
Polyurethane	3-15	90	poor	mod	excel	g,h
Propyl	19	175	good	high	v good	h
Silicone	27	230	poor	v high	poor	i
Perfluoroelastomer	23	275	poor		excel	j
Teflon	5-8	280	v poor	mod	excel	j
Kel-F	4-7	200	good	low	v good	j
Polyimide	5	275	good	mod	v good	j,k

Source. Reprinted with permission from *J. Vac. Sci. Technol.*, **17**, 330, R. N. Peacock.
Copyright 1980, The American Vacuum Society. (*a*) Generally used vacuum seal; (*b*) best
all around low cost; (*c*) little vacuum application; (*d*) oil resistance; (*e*) low cost; (*f*)
specific chemical application; (*g*) radiation resistant; (*h*) mechanical properties; (*i*)
electrical applications; (*j*) chemical resistance; (*k*) high temperatures.

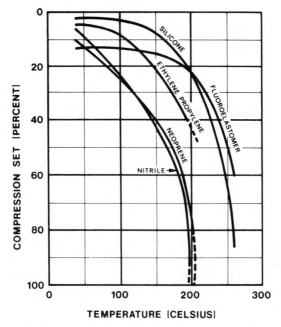

Fig. 17.7 Compression set as a function of temperature measured according to ASTM D-395 (22 h.) Reprinted with permission from *J. Vac. Sci. Technol.*, **17**, 330 , R. N. Peacock. Copyright 1980, The American Vacuum Society.

Many vacuum technology texts and O-ring manufacturers supply tables of groove depths and widths with information for a range of O-ring sizes. Sessink and Verster [17] observed the empirical tabulations from nine sources to vary widely. They found chord compressions up to 38% of the chord diameter, and groove filling ratios (chord cross section area/groove cross section area) of 74% to 102%. In their study they found the general criterion for high vacuum sealing to be a minimum initial contact pressure of 13 kg/cm^2 for gaskets in the hardness range 60 to 75 Shore [17,18]. The tables listed in the various references are not all for the same elastomer. Expansion coefficient, temperature range and hardness affect the groove depth and width. It is best to obtain a table from an O-ring supplier to verify the groove dimensions for a particular elastomer.

Outgassing and permeation data for elastomers are given in Chapter 16 and in Appendix C. Outgassing data are given in Fig. 16.9 and Appendixes C.4 and C.7. The important thing to remember about outgassing from elastomers is the flux is predominantly water vapor [19].

Permeation data are given in Appendix C.6. Permeation of gases through elastomers is significant. For example oxygen and nitrogen from atmospheric air can permeate Viton in almost equal fluxes. Naturally occurring water vapor from air of 50% RH will permeate Viton with a flux over 2 times larger than that from nitrogen. de Csernatony [20] observed a 480-fold reduction in the outgassing of an immersed Viton O-ring after a 16-h, 100°C bake; when the O-ring was used as a seal, the reduction was only 6-fold. His measurements illustrate how permeation, and not outgassing, dominates the ultimate pressure in a vacuum system containing a large number of O-rings after long pumping times. Gas permeation can limit the time available for leak checking and limit the ultimate pressure in a system containing a large number of O-rings. Laurenson and Dennis [21] have studied the permeation of several gases through the elastomers Viton, Aflas and Epichlorohydrin at elevated temperatures. Some of their data are presented in Fig. 16.8. These authors did not see any water permeation. However, it takes a long time for water vapor to reach steady state permeation levels.

Seal materials can also become degraded by radiation. Peacock [16] gives a brief review of gamma radiation damage. He notes the trend in elastomers is to become brittle, take a large compression set and increase in hardness after being subjected to high doses of gamma rays. Teflon appears affected at lower doses than other elastomers, while polyimide (e.g. Vespel) can withstand considerably more flux than other elastomers. Wheeler and Pepper [22] show that a high x-ray flux (8×10^6 rad/s)

Table 17.2 Compression set for Viton and Kalrez for Various Times and Temperatures Measured per ASTM D-395B

Temperature (°C)	Time (h)	Compression Set (percent)		
		Kalrez 1050	Viton A	Viton E-60C
24	70	20	21	8
100	70	32	-	-
204	70	71	63	13
204	360	-	90	30
204	960	-	-	55
204	7200	-	-	100
232	70	71	-	30
260	70	6	-	-
288	70	74	-	100

Source. Reprinted with permission from *J. Vac. Sci. Technol.*, **17**, p. 330, R. N. Peacock. Copyright 1980, The American Vacuum Society.

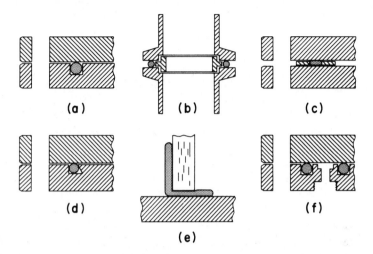

Fig. 17.8 Elastomer seal geometries: (a) rectangular groove, (b) ISO-KF flange with centering ring, (c) confined gasket, (d) dovetail groove, (e) L-gasket, (f) double gasket with differential pumping port.

decomposes Teflon into saturated fluorocarbon gases and polymer fragments short enough to desorb from the surface.

Gaskets are shaped in many ways. Figure 17.8 illustrates several ways which gaskets are used between metal and metal-to-glass joints. The rectangular O-ring groove (17.8a) is a common joint in the United States, while the ISO-KF (kleinflansch) with a centering ring, 17.8b, is the European standard and becoming common in the U.S. The confined gasket (17.8c) is commercially available and useful, especially on non-circular joints. A half- or full-dovetail groove (17.8d) is especially useful for vertical doors. An L-gasket (17.8e) is freely squeezed between a glass pipe or bell jar and a metal surface. Double gaskets (17.8f) are used to reduce permeation between atmosphere and the chamber.

Cleaning an elastomer by a solvent wash is an ineffective way to reduce outgassing. One effective way is a simple vacuum bake. An unbaked Viton O-ring will have an initial outgassing rate of 10^{-3} W/m^2. See Appendix C.4. After a 4-h bake at 150°C and 12 h of pumping, this value is reduced to 4×10^{-7} W/m^2. This latter value corresponds to 2×10^{-8} Pa-L/s per linear centimeter of 0.4-mm-diameter gasket material, assuming about 1/3 of the surface is exposed to the vacuum. If the gasket is not given a vacuum bake, its initial outgassing rate will be about 2500 times higher, or 5×10^{-5} Pa-L/s per linear centimeter. Its ultimate, unbaked rate will be about 2×10^{-6} Pa-L/s per linear centimeter. Re-exposure to atmosphere will result in increased water outgassing. All elastomers are "sponges" for water and other gases; a fact easily visible on a mass spectrometer when compressing a gasket in a valve seat. O-ring grooves need to be carefully cleaned and both groove and ring wiped clean with a lint-free cloth. Grease is not needed to make a static seal between an elastomer and a metal surface. It will cause pressure bursts as trapped gas pockets are released. Occasionally we use grease on a main seal flange that has become scratched with misuse. Remember to apply the *very thin* coating of grease with lint-free cloth. Finger oils have a very high vapor pressure.

The most commonly used O-ring elastomers are Buna-N and Viton E-60C. Most of the published data are for Viton A, but it has not been used for many years. Viton E-60C has a lower room temperature outgassing rate and less compression set than Viton A. Buna-N is used for low cost applications, while Viton is used where a moderate bake and low outgassing are needed. A 200°C bake will release adsorbed gases, unreacted polymer and plasticizers from Viton [16]. It cannot be used at this temperature because of compression set; at higher temperatures it will decompose. Silicone has an unusually high permeation rate and is infrequently used as a gasket material in very high vacuum systems. It is better suited to high temperature vacuum furnaces operating in the high vacuum region. Silicone compounds are formulated for a range of high temperature applications. Polyimide has a low outgassing rate [23,24] but adsorbs large amounts of water when re-exposed. Hait [25] and Edwards et al. [26] describe flange seals made from thin polyimide films. Elastomer gaskets are widely used in systems that pump to the 10^{-6} Pa range. The development and widespread use of metal gaskets has eliminated the use of cooled elastomers [27] where lower ultimate pressures are desired.

17.2.2 Metal Gaskets

The thermal, radiation, outgassing and permeation properties of elastomers make them unsatisfactory for many seal applications. Metal gaskets must be used in high quality UHV systems and are often used in high vacuum systems on parts which do not have to be frequently opened to reduce the total outgassing load.

Many metal gasket-flange pairs have been designed. We show some in Figure 17.9. In all these flange designs the gasket is plastically deformed to fill irregularities in the surface of the mating flange. An early seal design consists of an unconfined wire ring of 0.4 to 0.5-mm-diameter crosssection clamped between two flanges (17.9a). Indium, aluminum, copper, silver and gold are typical of the metals that have been used in wire seals. Gold works well but is expensive. A gold-sealed joint must be separated by force after a high temperature bake [1]. Aluminum wire cannot be used to seal stainless steel flanges baked to 450°C, because an iron-aluminum compound forms at the interface [28]. Stainless steel flanges have also been designed which seal by shearing a copper gasket [29]. The seal is formed by the surface shear resulting from the high frictional force. A confined soft metal gasket (17.9b) is similar to the confined O-ring in Fig. 17.8c. An aluminum wire is confined between floating stainless steel rings [30]. As the flanges are compressed, the harder rings compress inward and expand outward radially, while the softer aluminum ring is compressed axially. This seal cannot be used over 370°C without alloying to the flange.

The copper gasket has proven the most popular bakeable seal. It has been used in many forms, but the design which has proven the most reliable is the ConFlat [31] seal (17.9c). It consists of two symmetrical flanges each containing a work-hardened knife edge. The flanges are tightened until they touch and capture a section of the copper ring between the knife edge and the outer shoulder. This seal forces the copper to plastically flow into the surface irregularities with a pressure variously estimated to range from 75,000 lb/in^2 (500 MPa) [4] to 180,000 lb/in.^{2} (1200 MPa) [32]. The knife-edge seal has become widely used in both the U.S. and Europe. It is almost a standard in circular geometries up to 30 cm diameter. It can be repeatedly baked up to 450°C with excellent reliability. A few flanges are made up to 40 cm, but they cannot be

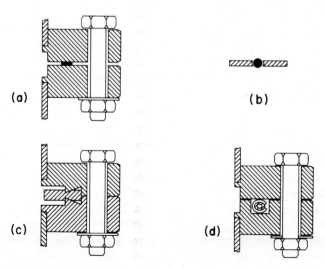

Fig. 17.9 Metal gasket seal geometries: (a) wire seal; (b) confined gasket; (c) ConFlat type knife edge seal; (d) Helicoflex seal.

reliably baked at temperatures above 200°C. Large flanges are difficult to fabricate from the knife-edge design, because alignment of the knife edges is difficult, and because it is difficult to tolerate mechanical motion introduced by differential thermal expansion in baking ovens. It is difficult to keep the ovens uniform enough to prevent leaking.

Edwards et al. [33] have shown that hydrogen firing of knife-edge flanges at 900°C for 4 h to remove the dissolved hydrogen, will not affect the reliability after 40 closures. The knife edge was seen to flatten out slightly with no loss in sealing ability. Other workers have observed that all flanges fabricated by this design are not identical in their hardness after firing. Those flanges fabricated from a high quality alloy containing additives to retard grain growth will loose only a small percentage of their initial hardness. Fuente [34] has developed a nickel gasket for use with small knife-edge flanges baked to 525°C.

The Helicoflex ring (17.9*d*) consists of an Inconel helical spring surrounded by an aluminum core [35-37]. Flemming et al. [38] have used the Helicoflex for large, non-circular seals. Non-circular seals pose additional problems because of differences in flange stiffness between straight and curved sections. Basic to this sealing problem is how the necessary elastic energy is stored. It was found that silver plated stainless steel covered Inconel spring made a reliable non-circular seal when the flanges were tightened against a spacer ring [38].

Two of the previously described seals, the wire seal and the shear seal differ from other metallic gasket seals in that a spring force is not necessary once the joint is formed. This is a result of the formation of the intermetallic (aluminum gasket), and the diffusion bond (gold gasket). In all the other seal designs an elastic restoring force is necessary. The knife-edge design is today the most universally used metal gasket seal for commercial demountable joints in UHV systems. The copper gaskets used in this flange should be compressed until the flange faces meet. If the gasket is only partially tightened, so that it can be re-used, the pressure between the copper and the steel is not great enough to withstand baking without leaking.

17.3 VALVES AND MOTION FEEDTHROUGHS

The components we use to control gas flow, connect or isolate chambers and pumps, and provide mechanical motion through a wall all use some form of an elastomer or metal seal in combination with a moveable vacuum wall, usually a metal bellows. The seals used in valves are adaptations of the demountable seals described in the previous section. Conductance, leak rate, vacuum range, baking temperature, radiation or corrosive gas exposure and the need for line-of-sight view in the open position are all variables which influence the choice of a valve and how it is designed. Transmitting linear or rotary motion through a vacuum wall is complicated. Dynamic sealing or flexible walls, and lubrication of the moving parts in vacuum are now required. The sealing techniques and materials are also dependent on the vacuum range, torque and the rotational speed at which the motion will be transmitted. Weston [39] has extensively reviewed valves and feedthroughs for ultrahigh vacuum applications.

In this section we review small and large valves, special purpose valves and motion feedthroughs. Basic design differences so divide our discussion.

17.3.1 Small Valves

We define a small valve as one which the throat diameter is less than, say, 6 cm, and which the sealing force is directly applied via a hand lever or screw or a pneumatically operated shaft. Figure 17.10 names the components of a simple, hand-operated, elastomer-sealed right-angle valve. Most small valves are right-angle, because that is a simple mechanical way of directly applying the sealing force in a direction normal to the valve seat. All valves, small or large, should be constructed from materials whose outgassing load is low enough so that it does not contaminate the process at the operating pressure. We desire the valve to have a maximum conductance for gas flow, and to form a leak-tight seal between the plate and the seat, and between the stem and the bonnet. We also desire the valve to be reliable and maintenance free and have a long operating life. The valve design illustrated in Fig. 17.10 is typical of an older, brass valve mainly used in roughing lines. Its basic problem, even if made from stainless steel, is its unreliable O-ring stem seal. Today this design is used only in valves designed to release a medium or high vacuum component to atmosphere. In that application the stem seal

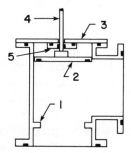

Fig. 17.10 Components of a simple elastomer-sealed valve: (1) valve seat; (2) valve plate; (3) bonnet; (4) stem; (5) stem seal.

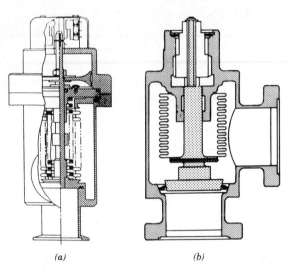

(a)   *(b)*

Fig. 17.11 Small right-angle valves: (a) elastomer sealed, pneumatically driven valve with a bellows shaft seal. Reproduced with permission from HPS division of MKS Corp., 5330 Sterling Drive, Boulder, CO 80301. (b) All metal valve with metal sealed seat. Reproduced with permission from VAT Inc. 600 W. Cummings Park, Woburn, MA 01801.

is never exposed to vacuum. If the stem seal side of the valve is evacuated, its use is limited to the low vacuum range.

In Fig. 17.11a we illustrate a right-angle valve in which the O-ring stem seal has been replaced with a stainless steel bellows. The linear motion needed to open and close the valve is transmitted to the valve plate via the bellows. The only elastomer seals remaining in the valve are static seals. The valve in this illustration is pneumatically driven and can be baked to 150°C in the closed position, and 200 °C in the open position. It is used on systems pumped to the 10^{-7} Pa range. Bellows-sealed valves of this type have lifetimes ranging from 20,000 to over 200,000 cycles, and valve plate leak rates of 10^{-12} to 10^{-10} Pa-L/s. A less expensive valve of this type is constructed from brass with a brass bellows and Buna-N seals for non-corrosive applications in the medium vacuum range. Butterfly valves, which operate by rotating a valve plate through 90° are made with an elastomer O-ring seal on the outer edge of the valve plate. They are generally available in hand-operated versions.

Elastomer seals prevent the use of this valve in UHV systems baked to temperatures over 200°C. An all-metal welded body and metal plate seals are necessary for UHV applications. The first all-metal valve was developed by Alpert [40] in the early 1950s. Today, several geometries are commercially available which most commonly use some form of copper or gold valve plate, and a stainless steel knife-edge seat. Designs with a sapphire valve plate and gold-plated stainless steel seats are also available. The design illustrated in Fig. 17.11b employs a coated steel valve plate and a seat fabricated from a steel with a high elastic limit in the shape of a conical Belleville washer. The body of this valve is welded and it is joined to the system by copper knife-edge flanges. Line-of-sight versions of valves such as those illustrated in Fig. 17.11 are available. These valves are constructed by mounting the valve seat at a 45° angle to the connecting tubing. Because of the geometry, a line-of-sight version requires a seat diameter over 1.5 times the pipe diameter. The distance between flanges reduces their conductance to a value less than a right-angle version.

17.3.2 Large Valves

Large valves differ from small valves in several ways. The total force required to close the valve plate scales with the seal area. For an elastomer-sealed valve this area is proportional to the O-ring chord diameter and the valve plate diameter. A 5-cm-diameter valve sealed with a 3.2-mm-diameter O-ring requires a total sealing force of 400 N. A 15-cm-diameter valve with a 6-mm O-ring requires about 2000 N, while a 30-cm valve with an 8-mm O-ring requires close to 5000 N. If the valve is to remain closed against atmospheric pressure the body must withstand an even greater force. A 30-cm valve must withstand 7500 N at atmospheric pressure. These forces place stringent design require-ments on the valve body. The valve seat must remain flat under these forces at room temperature and while being baked. Direct line-of-sight transmission through the valve is required for some applications. Evapo-ration sources are often isolated by a valve during substrate changing. Particle accelerators and storage rings require clear aperture valves in the beam lines. Multichamber, in-line thin-film deposition systems use large line-of-sight valves, not always circular, to pass wafers between isolated chambers in device processing. Other applications require an isolation valve to have a high open conductance, long life, ease of maintenance and an ability to be baked. These valves range in sizes up to 1.2 m diameter and include both high vacuum and UHV.

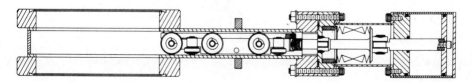

Fig. 17.12 Bellows sealed gate valve. Reproduced with permission from High Vacuum Apparatus, Inc., 1763 Sabre St. Hayward, CA 94545.

The most common large valve design is the gate valve. Gate valves are so common some think they are synonymous with large valves. Gate valves are made in sizes as small as 2 cm; however, they are not the only large valve. Figure 17.12 illustrates a common gate valve design. The gate travels horizontally until the valve plate is centered under the seat. Continued translational motion of the shaft moves the valve plate upward as the hinges are pivoted under it to a closed position past the vertical. In this way the valve is locked closed in the event of operation power loss. The bonnet seal is either a grease-packed double O-ring, or in better quality valves, a bellows. Inexpensive varieties of this design are made with cast aluminum bodies while higher quality versions are made with stainless steel.

A second gate valve design is illustrated in Fig. 17.13. In this design the shaft moves the valve plate in and aligns it with the seat. At the end of the valve plate travel, the shaft forces ball bearings out of detents and expands the valve plate against the seat and the backing plate against the valve housing. The use of multiple balls is one advantage of this design. By increasing the number of points at which the force is applied to the seal plate, the wear on each point can be reduced to a value below the design illustrated in Fig. 17.12. These two gate valves are commonly available in sizes up to 30 cm. However large elastomer sealed gate valves have been fabricated on special order to sizes up to 1 m. Metal-sealed gate valves are currently available in diameters up to 30 cm, or in some cases 40 cm. The reliability of such a valve drops drastically as the diameter increases beyond 30 cm. Baking of a large metal-sealed valve is limited to about 200°C. Large metal-sealed valves have the the same problems as large knife-edge flanges.

Other designs use a swinging plate shaped like a Belleville washer, which is forced into position by a wedge pushing on the center, or a coaxial bellows to seat the valve plate against a knife edge located in the opposite face.

Valves with aluminum bodies and grease packed shaft seals are acceptable on unbaked systems which pump to 10^{-5} Pa, while stainless steel

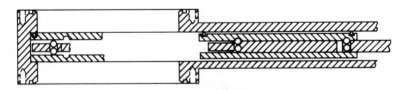

Fig. 17.13 Sealing concept used in the VAT gate valve. VAT Inc. 600 W. Cummings Park, Woburn, MA 01801.

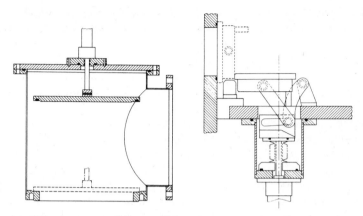

Fig. 17.14 Poppet (left) and (right) load lock valves. Reproduced with permission from CVC Products, Inc., 525 Lee Road, Rochester, NY.

valves with bellows shaft seals whose moving parts are lubricated with a low vapor pressure solid lubricant can be baked to 150°C with the valve closed or 200°C open, and can be used on UHV systems. Viton is the most common elastomer used in large valve seals, although others have experimented with polyimide [41]. In all gate valves, elastomer and metal sealed, the valve plate must be seated in a direction orthogonal to the drive motion with a force up to several thousand newtons. Moving parts need to be lubricated and the valve interior pumped. For these reasons the gate valve is the most complicated piece of machinery in a vacuum system. Stainless steel valves can have plate seal leak rates of order 2×10^{-8} to 2×10^{-9} Pa-L/s, with life times ranging from 10,000 to 50,000 cycles with the typical value on the low side of this range. They can fail by seal leaks, bellows leaks, and wear of moving parts. Bellows failure is the most common problem.

Elastomer-sealed poppet and load lock valves are made in large sizes. These styles are illustrated in Fig. 17.14. Poppet valves up to 1.2 m diameter are commercially available. Poppet valves use a simpler mechanism than gate valves, because the operating force is applied via a large compressed air piston in line with the valve plate motion. The load lock valve pivots about an axis parallel to the valve plate surface. These valves are much deeper than a gate valve and their conductance is less. Most large diffusion pumps require an elbow to connect them to a system so the poppet does not reduce the overall conductance as much as we first assume. In some small ion-pumped systems, the poppet valve serves both to isolate and baffle the pump entrance. In-line chambers generally cannot be connected closer than needed for a load lock valve. Any loss in conductance is more than offset by increased reliability and lower cost. Poppet valves are best used to seal a pump under vacuum while the chamber is vented. In this way the atmospheric pressure forces the valve plate closed. Flap valves are also constructed in rectangular shapes. For example a 5 × 30 cm valve might be used on an in-line sputtering system for isolating chambers and load locks. The plates are constructed from stainless steel, use elastomer gaskets, and are designed for operation in the high vacuum region. Butterfly valves are also made in large diameters.

Large elastomer and metal gasket valves are available for a number of applications, baked and unbaked. In the range of 15 to 30 cm diameter there are a number of off-the-shelf commercially available gate and poppet valves. Valves larger than this are custom made for individual projects. Some of these designs have been custom made with double-sided all metal seals [42]. Large valves can be purchased with contacts to indicate the full open or closed position. Large diameter, all-metal valves bakeable to temperatures over 300°C are not simple off-the-shelf items. Poppet valves are available to fit large diffusion pumps. Special purpose rectangular load lock valves are usually custom designed for a particular application. As the size of the valve increases, so does the cost, the probability of failure, maintenance, and the difficulty in baking. Ishimaru et al. [43] have devised a valve which can be baked and requires no elastomer or metal knife edge seal. It makes use of polished metal sealing surfaces which are forced together by compressed air in bellows. The region between is double pumped. See Fig. 17.15.

17.3.3   Special Purpose Valves

The valves described in the previous sections all serve to connect and isolate components of various diameters. There are special cases where we wish to control the conductance between the two chambers with a partially open valve, or control the rate with which gas enters a chamber.

Controlling gas flow in the pressure range 1 to 10 Pa is necessary to regulate the pressure in a sputtering or etching chamber. The pressure is regulated by putting a throttle valve in the pumping line, usually next to the pump. Typically the high vacuum pump will be operating at a maximum pressure of, say, 3×10^{-2} Pa, while the deposition or etch process may require 3 Pa. The throttle valve can be closed enough to provide this pressure drop. For a chamber using an argon flow of 100 Pa-L/s this corresponds to a valve conductance of 35 L/s. This compares with an open conductance of over 5000 L/s in the molecular flow region for a 15 cm valve.

The techniques used to throttle gas flow can be simple, for example a hole in the plate of a gate or butterfly valve. The valve is opened to pump to the base pressure and closed to place the desired throttling conductance in series. This does not allow for adjustment of the closed conductance. Several arrangements have been designed to allow control

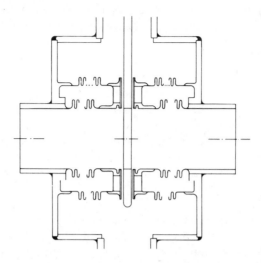

Fig. 17.15 Differentially pumped gate valve. The smooth metal mating surfaces are held together by compressed air. Reproduced with permission from *J. Vac. Sci. Technol. A,* **3**(3), p. 1703, H. Ishimaru et al. Copyright 1985, The American Vacuum Society.

of the closed conductance. A butterfly valve can close against an external shaft stop. This design and others which use venetian blinds, rotating pie-shaped segments or a small iris, are commercially available. Lehmann et al., [44] have designed a large iris valve which has a variable aperture and electronic control. 15-cm valves have throttled conductances which are adjustable in the range 1 to 200 L/s. Throttle valves are typically made with elastomer sealed flanges and are not bakable. The operating characteristics of a throttle valve are usually displayed as a plot of conductance versus pressure drop over the range of adjustment. The techniques used to control the closed position of the throttle range from a manual screw adjustment of the stop setting, to closed-loop pressure control systems in which the error signal from a pressure gauge operates a motor to control the closed position.

Gas flow regulator valves, sometimes called leak or metering valves are used to admit controlled quantities of gas from a source external to the vacuum. The external source may be at high pressure (100 to 150 kPa gauge), atmospheric pressure, or at a reduced vapor pressure of a supply gas. The simplest form of metering valve is a needle valve. When closed the needle is seated against a hollow tapered cone of soft metal. Another design used a cylinder in which a fine spiral groove is turned. A screw moves the cylinder inside a hollow mating piece and effectively changes the length of the capillary. These designs are not bakeable and are commercially available in sizes that span 1 to 4 decades of flow and cover the range 10^{-5} to 2000 atm.-cc/s. Leak valves in which a sapphire flat is pressed into a metal knife edge are designed to be baked. A sapphire-metal leak can control flows as low as 10^{-8} Pa-L/s.

17.3.4 Motion Feedthroughs

Vacuum systems would have few applications if there were no way of transmitting motion to the vacuum environment. Rotary and translational motion are necessary to operate pumps and valves, move samples and sources, open and close shutters, and perform many specialized tasks. Rotary and linear motion feedthroughs are characterized by the torque, the speed at which it is transmitted, and the operating pressure. Rotary and linear feedthroughs for the medium and high vacuum region usually use elastomer seals.

Figure 17.16 depicts two basic forms of dynamic elastomer seal. The rotary seal shown in 17.16*a* is one form of a simple hand or low speed rotary seal. In this sketch the O-ring groove is cut into the housing while the shaft contains a retaining ring to prevent translational motion of the shaft. Alternatively, the O-ring groove may be cut into the shaft. A better version of this feedthrough uses ball bearings on both sides of the O-ring. O-ring manufacturers can provide tables of groove dimensions for these designs using common elastomers. Groove dimensions are somewhat different than those used in static seals. Unlike static seals, dynamic O-ring seals need to be greased. The only exception is Teflon. The properties and selection of greases for use in vacuum is discussed in Chapter 18. An improved rotary seal, which can also be used for translation, is shown in Fig. 17.16*b*. It is a double-pumped seal and can be used in the high vacuum range. The fabrication cost is reduced by replacing machined grooves with sleeves.

These two elastomer feedthroughs are only two of a large number of elastomer seals. Today it is often easier to substitute a metal seal rather than design a high quality elastomer seal. One exception is for applica-

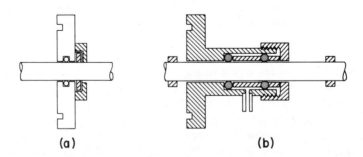

(a)   **(b)**

Fig. 17.16   (a) Rotary and (b) translational elastomer-sealed feedthroughs.

tions which require a high rotational speed and high torque in the very high vacuum or near UHV region where baking may be necessary. For this application a Teflon seal is superior to a greased elastomer. Teflon cold flows and must be spring loaded to avoid leaking. Seal gaskets which are C-shaped in cross section have been designed with a special spring located inside the C-ring [45-47]. These seals can be used singly, or in pairs which are differentially pumped and baked to 250°C. They can be rotated at high speeds and transmit high torques. Silverman [48] has designed a differentially pumped rotary flange seal using two large Teflon C-rings. This design allowed one 15-cm-diameter flange to rotate with respect to another with a dynamic leak rate of 10^{-6} Pa-L/s.

Two other techniques for transmitting motion through a wall which do not make use of an elastomer or all-metal seal are the differentially pumped seal and the magnetic liquid seal. The differentially pumped seal illustrated in Fig. 17.17, uses a closely spaced surface to create a high impedance between the vacuum chamber and the differentially pumped chamber. This is an expensive way to proceed, but if a pump is already in use for a load lock it is convenient. The inner parts of this seal can be baked to high temperature without any problems.

The magnetic liquid seal depicted in Fig. 17.18, allows rotary motion to be transmitted on a solid magnetically permeable shaft, by sealing the gap between the housing and the shaft with a magnetic liquid [49-51]. The magnetic liquid consists of a low vapor pressure fluid (e.g. polyphenyl ether), iron oxide (Fe_3O_4) and a surfactant. The magnetic lines of force concentrate the magnetic liquid at the ridges machined on the shaft. This seal allows high speed (5000 rpm) and high torques. It is useful in unbaked systems which are pumped to the 10^{-5}-Pa region.

Except for the differentially pumped feedthrough, none of the motion feedthroughs we have described are suitable for the UHV where baking over 250°C is required. For these conditions all-metal linear and rotary motion feedthroughs or the differentially pumped seal are required. All metal feedthroughs make use of flexible metal bellows.

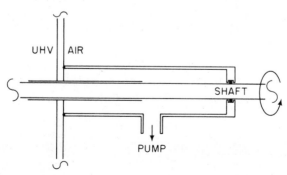

Fig. 17.17 Bakeable differentially pumped motion feedthrough. The inner seal is a closely fitting cylinder about a round shaft.

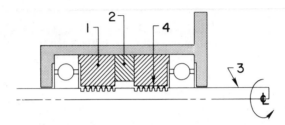

Fig. 17.18 Magnetic liquid seal: (1) pole piece; (2) magnet; (3) non-magnetic housing; (4) bearing; (5) magnetic liquid, (6) magnetic flux path. Reprinted with permission from *Vacuum*, **31**, p. 151, K. Raj and M. A. Grayson. Copyright 1981, Pergamon Press.

Crosssectional views of two types of bellows, hydroformed and welded bellows are shown in Fig. 17.19. Hydroformed bellows are made by hydraulically stretching rolled and welded thin-wall tubing, or deep-drawn thin-wall cups. Welded bellows are made by sequentially welding a series of thin-wall diaphragms. In both designs 304L, 316L and 321 are the most commonly used materials for vacuum applications. Welded bellows are more expensive and more flexible than hydroformed bellows. However fewer convolutions are needed. The limiting extension, compression and ultimate lifetime of a bellows depend on the stresses encountered at the ends of the stroke. These in turn are dependent on the ratio of inner-to-outer radii, material, and how much it has been cold worked, etc. A designer chooses the extension, compression and total pitched bellows length to keep the stresses in the bellows below a design value. This design value will be low when long life is a requirement. A few rough generalizations can be made: Bellows are often designed to work only in compression, but work well in both compression and extension. Stroke

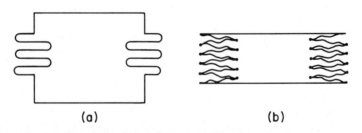

Fig. 17.19 (a) Hydroformed and (b) welded bellows.

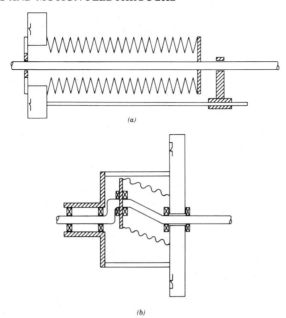

Fig. 17.20   (a) Translational and (b) rotary feedthroughs using metal bellows.

ratios ranging from 1/3 extension to 2/3 compression, to 1:1 extension-compression have been used in specific designs.   The total stroke length for a bellows used in a vacuum application ranges from 20% to 33% of the pitched length for hydroformed bellows, to 80% of the pitched length for edge-welded bellows.

Hydroformed bellows are adequate for many applications, especially where the bending or compression is small.   Edge-welded bellows find many applications where a long stroke is needed in a very short space. Bellows were traditionally furnace brazed onto a small pipe section for welding onto larger assemblies.   Mating pieces machined with heat relief grooves have resulted in elegant pipe joints and valve bodies.   See, for example, Fig. 17.1.

A long bellows section can be used for translational motion as described in Fig. 17.20a.   Another solution uses a rotating bellows to drive a lead screw to convert rotary motion outside the chamber into linear motion inside the chamber.

Using bellows to transmit rotary motion requires ingenuity.   Indeed, many solutions have been proposed.   One of these is shown in Fig. 17.20b.   This design illustrates the problems basic to all rotary bellows feedthroughs.   Metal-sealed bellows feedthroughs are adequate for low speed, light loads.   Since the shaft is not continuous, it cannot carry a heavy load.   Ball bearings are also required on both sides of the seal. Lubrication techniques are described in Chapter 18.   Alternatively, power may be coupled through a thin wall by placing a permanent magnet rotor of a small motor in vacuum and the field coils on the outside. The large air gap limits the torque available in this feedthrough. A variety of metal bellows rotary feedthroughs are commercially available and frequently used in UHV shutters and manipulators for precision movement of small parts, as well as rotation of shutters and substrate holders.

Bellows need to be carefully cleaned. Solvents are not recommended as they are difficult to remove. A vacuum, hydrogen or argon bake is effective. Baking temperatures should not exceed those of brazing, if it has been used for assembly.

REFERENCES

1. J. H. Singleton, *J. Vac. Sci. Technol. A*, **2**, 126 (1984).
2. A. Roth, *Vacuum Sealing Techniques*, Pergamon, Oxford, 1966.
3. A. Roth, *J. Vac Sci, Technol.*, **9**, 14, (1972).
4. A. Roth, *J. Vac Sci, Technol., A*, **1**, 211, (1983).
5. *Metals Handbook*, 8th ed., *Welding and Brazing*, Vol. 6, American Society for Metals, 1971, p. 113.
6. C. H. Rosendahl, *Sheet Metal Industries*, February 1970, p. 93.
7. C. Geyari, *Vacuum*, **26**, 287 (1976).
8. R. N. Peacock, *Vacuum Joining Techniques*, HPS Division of MKS Corporation, 5330 Sterling Drive, Boulder, CO, 80301.
9. W. H. Kohl, *Handbook of Materials and Techniques for Vacuum Devices*, Reinhold, New York, 1967.
10. American Welding Society, *Brazing Manual*, AWS.
11. Handy and Harmon, *Brazing Manual*, Handy and Harmon Co., 850 Third Ave., New York, NY 10022.
12. T. Takamori, *Treatise on Materials Science and Technology: Glass II,* Vol. 17, M. Tomozawa and R. H. Doremus, Eds., Academic, New York, 1979, p. 117.
13. J. E. Benbenek and R. E. Honig, *Rev. Sci. Instr.,* **31**, 460, 1960.
14. F. Rosebury, *Handbook of Electron Tube and Vacuum Techniques*, Adison-Wesley, Reading, 1965,
15. W. Espe, *Materials of High Vacuum Technology*, Vol. 2, Pergamon Press, New York, 1966.
16. R. N. Peacock, *J. Vac. Sci. Technol.*, **17**, 330 (1980).
17. B. Sessink and N. Verster, *Vacuum*, **23**, 319 (1973).
18. The relation between International Rubber Hardness Degrees and Young's Modulus is given in ASTM D1415-62T (American Society for Testing Materials, 1916 Race St., Philadelphia, PA 19103). Shore A degrees are approximately IRH degrees. Typical values of Young's modulus are E = 35, 54 and 68 kg/cm^2 corresponding to 60, 70 and 75 Shore A degrees.
19. L. de Csernatony, *Vacuum*, **16**, 13 (1966).
20. L. de Csernatony, *Vacuum*, **16**, 129 (1966).
21. L. Laurenson and N. T. M. Dennis *J. Vac. Sci. Technol. A*, **3**, 1707 (1985).
22. D. R. Wheeler and S. V. Pepper, *J. Vac. Sci. Technol.*, **20**, 226 (1982).

23. L. de Csernatony, *Proc. 7th Int. Vac. Congr. & 3rd Int. Conf. Solid Surf.*, R. Dobrozemsky, Ed., Vienna, 1977, p. 259.

24. L. de Csernatony, *Vacuum*, **27**, 605 (1977).

25. P. W. Hait, *Vacuum*, **17**, 547 (1967).

26. T. W. Edwards, J. R. Budge and W. Hauptli, *J. Vac. Sci. Technol.*, **14**, 740 (1977).

27. T. M. Miller and K. A. Geiger, *Trans. 9th Vac. Symp. (1962)*, Macmillan, 1963, p. 270.

28. L. Holland, *7th Nat. Symp. Vac. Techn. Trans. 1960*, Pergamon, Oxford, 1961, p. 168.

29. R. Brymmer and W. Steckelmacher, *J. Sci. Instrm.*, **36**, 278 (1959).

30. E. I. Seal, product of Pont-A-Mousson, S.A., Rue du Pont de Fer, 71105 Chalon-sur-Saone, France.

31. W. R. Wheeler and M. Carlson, *Trans. 8th Nat. Vac. Symp., and Proc. 2nd. Int. Congr. Vac. Sci. Technol., 1961,*, Pergamon, New York, 1962. p. 1309.

32. W. R. Wheeler, *Trans. 10th Nat. Vac. Symp., 1963*, New York, 1964. p. 159.

33. D. Edwards, Jr., D. McCafferty and L. Rios, *J. Vac. Sci. Technol.*, **16**, 2114 (1979).

34. A. O. Fuente, *J. Vac Sci. Technol A*, **1** 220 (1983).

35. H. Ishimaru, *J. Vac Sci Technol,* **15** 1853 (1978).

36. I. Sakai, H. Ishimaru and G. Horikoshi, *Vacuum*, **32**, 33 (1978).

37. A. P. Kaan, *Vacuum*, **31**, 85 (1981).

38. R. B. Flemming, R. W. Brocher, D. H. Mullaney and C. A. Knapp, *J. Vac. Sci. Technol.*, **17**, 337 (1980).

39. G. F. Weston, *Vacuum*, **34**, 619 (1984).

40. D. Alpert, *J. Appl. Physics*, **24**, 860 (1953).

41. K. Yokokura and M. Kazawa, *J. Vac. Soc. Japan*, **24**, 399 (1981).

42. C. L. Foerster and D. McCafferty, *J. Vac. Sci. Technol.*, **18**, 997 (1983).

43. H. Ishimaru, T. Kuroda, O. Kaneko, Y. Oka and K. Sakurai, *J. Vac. Sci. Technol. A*, **3**, 1703 (1985).

44. H. W. Lehmann, B. J. Curtis and R. Fehlman, *Vacuum*, **34**, 679 (1984).

45. BAL-Seal, BAL Engineering Corp., Boulder, CO.

46. Omni-Seal, Aeroquip, Los Angeles, CA.

47. Fluorocarbon Co., Los Angeles, CA.

48. P. J. Silverman, *J. Vac. Sci. Technol. A*, **2**, 76 (1984).

49. R. E. Rosensweig and R. Kaiser, Office of Adv. Res. and Tech., NASA CR 1407, August 1969.

50. K. Raj and M. A. Grayson, *Vacuum*, **31**, 151 (1981).

51. Ferofluidic Corporation, Burlington, MA.

PROBLEMS

17.1 †Why is tungsten inert gas welding the commonly used technique for fabricating leak-tight stainless steel or aluminum vacuum joints?

17.2 †When through welding (TIG) from the atmospheric side, is it necessary to use a shield gas on the vacuum side?

17.3 A 6-mm-diameter tubing is to be furnace brazed to a stainless steel flange with 60% copper–40% gold filler with a melting point of

1010°C. Calculate the diameter of the hole (including tolerance) to be milled in the flange so the parts will have a clearance of less than or equal to 0.05 mm at the brazing temperature.

17.4 A 0.5-mm-diameter metal wire with expansion coefficient $110 \times 10^{-7}/°C$ is sealed through a glass wall in a perpendicular direction and then cooled to room temperature. The glass expansion coefficient is $100 \times 10^{-7}/°C$. After cooling to room temperature, is the glass near the interface in compression or tension in the (a) radial direction, (b) longitudinal direction, and (c) tangential direction?

17.5 †What is one fundamental difference between ceramic-to-metal and glass-to-metal joints?

17.6 †Describe an effective procedure for cleaning an O-ring. Why not wash an O-ring in acetone?

17.7 A groove 4.76 mm wide by 2.38 mm deep has been cut in a flange for use with a Viton O-ring whose cord diameter is 3.53 mm at 20°C. (a) What is the groove filling factor for Viton at 20°C when the flange is closed? (b) What percentage of the grove is filled at 150°C? (c) Your manager wants to bake the apparatus at 275°C, and asks you to replace the Viton ring with a Kalrez ring of the same diameter. Is this request reasonable?

17.8 Why is it not possible to obtain a good seal on a groove or face which has been polished with abrasive papers, even though the depth of the scratches may be less than those left from machining the groove surface?

17.9 (a) †It is often observed that a gate valve is the most unreliable component of a vacuum system. Why is this so? (b) Which considerations determine whether a high vacuum gate valve is oriented with the seal plate facing toward or away from the chamber?

17.10 A stainless steel bell jar can be sealed to a metal base plate with either the O-ring gasket or the L-gasket sketched in Fig. 17.21. Discuss the differences in the two seals regarding (a) gas release, (b) seal compression and (c) operation and maintenance.

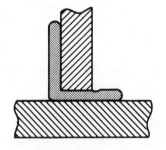 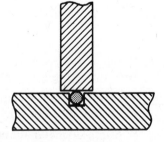

Fig. 17.21

CHAPTER 18

Lubrication

Lubrication is important to vacuum users in oil-lubricated pumps (vane, piston, lobe and turbomolecular), moving parts in vacuum (bearings and feedthroughs), and in peripheral applications, such as bolt lubricants. Not all of these applications involve high vacuum, and some take place outside the vacuum environment. Moving parts may be lubricated in vacuum with a liquid film, a grease, a dry film, or in certain cases, no film at all.

Buckley [1] defines a lubricant system to have three components; the materials to be lubricated, the lubricant, and the environment. These three components are interdependent. They are interdependent not only because surface adhesion and lubrication are characteristic of the material and the lubricant, but also how gases affect their interaction. The vacuum case is unique, because the gaseous environment has been removed. Lubricants are placed between moving surfaces to prevent contact. When surfaces contact, friction occurs and surfaces wear. Friction generates heat, while wear destroys material and produces debris. In an atmospheric environment it is the wear, not friction, which is responsible for most problems. In a vacuum environment friction is as much of a problem as wear, The absence of gas eliminates convection cooling, increases lubricant evaporation rates, reduces oxidation of the metals, hastens the onset of cold welding, and alters the size distribution of wear particles.

Lubrication processes have been reviewed by Booser [2] and Tipei [3]. Tabor [4] has reviewed lubrication and adhesion from a microscopic view while Buckley [1] and Friebel and Hinricks [5] discuss the problems of lubrication in vacuum. In this chapter we review basic lubrication processes, fluid rheology and techniques for vacuum lubrication. Many lubricating fluids are also working fluids for diffusion and turbomolecular pumps. In Chapter 13 we discussed their role as pump fluids, while here we examine other vacuum applications.

18.1 LUBRICATION PROCESSES

Full-film, elastohydrodynamic and boundary lubrication are three types of lubrication. Basic to the lubrication process is the ratio of the lubricant film thickness to the surface roughness. Lubricant thickness depends on the absolute viscosity η, relative surface velocity U, and the load L. Figure 18.1 relates these variables to the coefficient of friction f.

A combination of high absolute viscosity, high relative speed and low loading results in an oil film whose thickness is much greater than the roughness of either surface. The rapidly moving load rides on an oil wedge. This is called hydrodynamic or full-film lubrication. Viscosity is the most important lubricant property in this regime. Friction can be minimized for a given load or viscosity by the appropriate choice of viscosity.

At some value of low speed, low viscosity and high loading, irregularities on the two surfaces will contact. This is the boundary lubrication region [2], and can be encountered in pump bearings and vanes during starting and stopping, and between slowly sliding surfaces. Bowden and Tabor [6] derive a relation for the friction coefficient f, between surface irregularities:

$$f = s/p \tag{18.1}$$

where s is the shear strength and p is the yield pressure of the metal. Both quantities are related to the structure of the material. The coeffi-

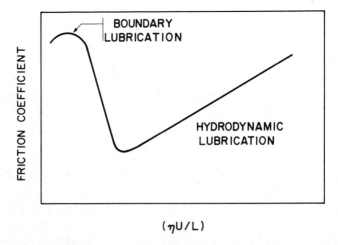

Fig. 18.1 Dependence of friction on viscosity η, relative velocity U, and load L.

cient of friction can be reduced by decreasing the shear stress s, or increasing the yield pressure p, by increasing the area over which the force is distributed. In the boundary region the coefficient of friction is determined not only by the manner which the adjacent surface peaks contact, but also by the additives which affect the chemistry of these contacting surfaces.

Intermediate between these two regimes is the elastohydrodynamic region. In this region the oil undergoes a tremendous pressure increase. Its viscosity increases exponentially, and it behaves like a low shear solid. Its viscosity is so great that it can force surface irregularities to deform without contact.

18.2   RHEOLOGY

Rheology is the study of fluid flow. Fluid flow is one physical property of a lubricant which we need to describe in order to determine its usefulness for either a vacuum or atmospheric application. Here we discuss absolute viscosity, kinematic viscosity, viscosity index and techniques by which each can be measured.

18.2.1   Absolute Viscosity

The definition of absolute viscosity was given in (2.21) and is repeated here in a slightly different form.

$$\tau = \eta s' \tag{18.2}$$

where τ is the shear stress and s' is the rate of shear. The viscosity of a Newtonian fluid is independent of shear rate. Not all fluids behave in this manner. The viscosity of emulsions like hand cream, grease and some synthetic fluids decreases with increasing shear rate. The decrease is small for the synthetic fluids we use so that η will be considered a constant. The viscosity of a grease drops rapidly with shear rate until it approaches the viscosity of the oil from which it is formulated.

The viscosity of a liquid is predominantly a result of cohesive forces between the molecules. Since cohesion decreases with temperature, the viscosity of a liquid decreases on heating. In SI dynamic viscosity η, has units of Pa-s. One mPa-s = 1 centipoise. Research instruments measure absolute viscosity of liquids by timing the flow through a long capillary tube under constant head pressure. Such instruments are required for measurement with ASTM and ISO standards. The volume flow rate through the tube (m^3/s) is related to the viscosity by the Poiseuille equation:

$$\frac{V}{t} = \frac{\pi d^4 \Delta P}{128 \eta l} \tag{18.3}$$

where ΔP is the pressure drop in the tube, d is the diameter and l is the length of the tube. This equation neglects end effects and is reasonably accurate for long tubes. Engineering instruments attempt to measure absolute viscosity by measuring the torque on an immersed rotating spindle or a rotating cone adjacent to a flat plate. Because the flow is not laminar for low viscosity liquids, corrections are needed to obtain absolute viscosity [7].

18.2.2 Kinematic Viscosity

The kinematic viscosity v, of a gas or liquid is simply the absolute viscosity divided by the density

$$v = \eta/\rho \tag{18.4}$$

In the cgs system kinematic viscosity is expressed in units of stokes. 1 S = 1 cm²/s. In SI the (unnamed) units are m²/s, but data are usually plotted in mm²/s because 1 cS = 1 mm²/s. Kinematic viscosity, like diffusivity and permeability, is a transport property—observe that all three quantities have dimensions of L^2/T. Kinematic viscosity is measured directly in research instruments by timing the (Poiseuille) flow of a liquid through a long capillary under its own head [8]. Engineering instruments use short tubes or orifices in which the flow is not always laminar, so it is necessary to tabulate factors by which their readings can be converted to kinematic viscosity. Saybolt and Redwood instruments measure the times required for a known quantity of oil with a falling head to flow through short tubes of specific dimensions. The Engler instrument measures the ratio of the times required by equal volumes of oil and water to flow through a particular tube. The Saybolt (U. S.), Redwood

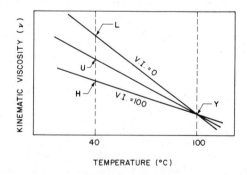

Fig. 18.2 Oil viscosity index

(English), and Engler (German) are the most common engineering instruments used for measuring the kinematic viscosity of oils. They are being replaced slowly by long tube instruments to meet SI standards. Appendix F.5 gives the factors for converting Saybolt universal seconds (SUS), Redwood No. 1 seconds, and Engler degrees to kinematic units.

The variation of kinematic viscosity with temperature for petroleum oils empirically fits the equation

$$\log (\log (\nu + 0.7)) = A + B \log T \qquad (18.5)$$

Measurement of ν at two temperatures, 40°C and 100°C is adequate for interpolation and extrapolation down to the pour point when plotted according to (18.5). This curve also fits most synthetics except chlorosiloxanes which show some curvature when plotted in this empirical way.

18.2.3   Viscosity Index

Viscosity index VI, is an empirical way of classifying how kinematic viscosity varies with temperature. It is an arbitrary and purely historical scheme. It recognizes the fact that the (highly paraffinic) Pennsylvania oils have a uniformly low change in viscosity with temperature. These oils have been arbitrarily assigned a VI of 100. Oils from the Gulf (highly naphthenic) with the greatest viscosity slope are assigned a VI of 0. VI is calculated from the 40°C and 100°C viscosities of the unknown oil (U), and the tabulated viscosities of oils of index 0 (L) and 100 (H). The unknown and the two standards must have the same viscosity at 100°C. See Fig. 18.2. The formulas for calculating VI are not stated here because they cannot be used without the extensive tables given in the ASTM standard [9]. Alternatively VI may be obtained from Fig. 18.3. This plot, generated from the ASTM data, greatly simplifies the procedure and yields sufficiently accurate results.

The VI system was originally intended for use with mineral oils, and was modified for use with synthetics. Some synthetics (e.g. chlorofluorocarbons) have a high viscosity-temperature slope and give a negative index. Hydrocarbons with viscosity improving additives, silicones and sebacate esters have indexes over 100%.

18.3   LUBRICATION TECHNIQUES

We can lubricate moving surfaces with liquids, e.g. petroleum oils or synthetic fluids, greases, or a variety of solid films such as silver and molybdenum disulfide. The use of these techniques in vacuum gives different results than when they are used in air. Each technique has its own problems when the surrounding gas is removed.

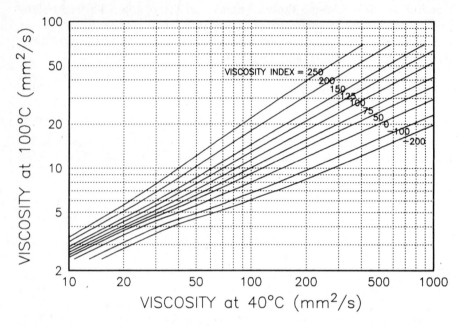

Fig. 18.3   Relation between 40 and 100°C viscosities for oils of varying viscosity index.

18.3.1   Liquid Lubrication

Petroleum and synthetic fluids are used to lubricate vacuum pumps and moving surfaces in vacuum. In the full film region fluid fills the spaces between moving surfaces and keeps them apart. The viscosity should be low enough to allow rotating or sliding motion at the lowest ambient temperature and remain sufficiently high at operating temperature (high viscosity index). The viscosity required for full-film lubrication depends on the operational temperature range and the load. From Fig. 18.1 we see that the ratio η/L should remain a constant. Therefore high viscosity fluids are required for heavy loads. Within a class of fluids vapor pressure is usually inversely proportional to viscosity. Lubrication in the full-film regime is dependent on the ability of the fluid to adhere to the metal surface and form a film of adequate shear strength which will support the bearing load.

Other necessary attributes are adhesion, stability and heat conductivity. The oil must wet the surface and form a film which is stable under high shear rate. Shear generates heat which must be removed to prevent degradation and further viscosity decrease.

Oil adhesion is determined by the strength of the liquid-metal bond. Liquids with unsaturated bonds strongly adhere to or wet metal surfaces.

Cohesion is a measure of intermolecular attraction. Oils that are very cohesive do not disperse or spread out rapidly on the surface of a metal. Certain highly cohesive oils such as polyphenyl ether adhere to metals strongly and also form a barrier that prevents creep. The outermost molecules align themselves in such a way that the exposed groups have a low liquid-metal adhesion. These fluids form a high contact angle with a metal surface and are termed autophobic. Fluids with good adhesion that do not have a resistance to creep will flow and wet a metal surface. Hydrocarbon oils have this property. Except for a fluid like pentaphenyl-tetramethyl trisiloxane, silicone fluids have high creep rates.

Sliding and rolling surfaces are easily lubricated with liquids for long lifetimes. The simplest liquid lubrication system is a wiped fluid coating. More complex systems can be designed to provide a continuous supply by migration or evaporation from a reservoir. If the fluid has been sufficiently outgassed, it can be used up to the temperature at which the vapor pressure of the fluid is intolerable. Vapor pressures of common fluids used in vacuum are given in Appendixes F.2 and F.3. Polyphenyl ether is a common choice for lubricating hand-operated sliding and rotating surfaces in vacuum. The mechanisms used to supply the lubricant can also cause organic contamination of the vacuum chamber. Contamination can be reduced by choosing an oil of sufficiently low vapor pressure, or by enclosing the lubricated area with a shroud. A creep barrier will prevent lubricant from creeping along the shaft where it passes through the shroud. PTFE is a hydrocarbon creep barrier, while nylon is a fluorocarbon creep barrier. It may be necessary to cryogenically cool the shroud in the area near the region where the sliding or rotating shaft enters the chamber.

Rolling friction is encountered with ball or roller bearings. Microscopically a rolling surface looks much like a sliding surface and the effects on viscosity, adhesion and heat conduction are the same. High speed ball bearings such as those found on turbomolecular pumps require an oil of lower viscosity than is used on slow speed bearings. Examination of Fig. 18.1 shows the product ηU needs to remain constant to minimize friction.

Boundary lubrication requires fluids with unique properties which are usually attained with additives. Sulfur, chlorine and lead make effective boundary lubricants for steel-on-steel because they form low shear stress compounds that are wiped or etched from the high spots. These chemicals function as etchants. Phosphorus functions by forming iron phosphide at hot spots that plastically flows and fills in the asperities. Plastic deformation, like etching, causes the load to be redistributed over an increased area [10]. In both cases the shear strength is reduced, the yield pressure is increased and the friction is reduced according to (18.1).

Oiliness, or lubricity, is an important property of boundary lubricants. It is an imprecise term and cannot be defined quantitatively. It refers to the ability of polar molecules to align themselves in double layers that

slide easily over adjacent double layers. Animal fats have a high degree
of oiliness, but they do not make acceptable hydrodynamic or boundary
lubricants because they are not chemically or thermally stable. Some
esters are stable and are used as oiliness additives.

18.3.2 Crease Lubrication

A grease is either a heavy petroleum distillate or, more commonly, a
thickened petroleum or synthetic liquid. Petroleum oil can be vacuum
distilled to yield a wax-like, high molecular weight grease (petrolatum)
which is uniform in composition. A grease also can be made by the
process of gelling or thickening a liquid. Various compounds such as
clay, esters, metal soaps, and powdered polytetrafluoroethylene (PTFE)
are used to increase the viscosity of an oil to the consistency of grease.
The oil is physically entrapped in the thickener, adsorbed on it or is held
in place by capillary action [1]. The typical starting material for a thickened grease is an oil whose room temperature viscosity is about 500
mm^2/s. The room temperature viscosity of a thickened grease is much
higher than the base oil when measured at low shear rates. However, it
drops by a factor of 30 at high shear rates.

Greases are characterized by their chemical type, method of thickening,
vapor pressure, service temperature range, reactivity, and consistency.
Characteristics of several greases are shown in Table 18.1. The vapor
pressure of a grease increases with temperature because of the properties
of both the oil and the thickener. Laurenson [11,12] has shown the
vacuum performance of a grease to be dependent on the vapor pressure
of the grease and its method of manufacture. He demonstrated that the
evaporation rate and quantity of a gelled grease made from hydrocarbon
oil or silicone fluid of moderate vapor pressure depended on the total
mass of the grease in vacuum, while the evaporation rate of a molecularly
distilled grease depended only on the surface area. Laurenson attributed
this to the fact that the molecularly distilled grease evaporated only from
the surface, while the surface oil molecules of a filled grease rapidly
evaporated from the large effective surface area of the gel. These molecules were replaced by migration from the bulk just as oil feeds a lamp
wick. He could not experimentally observe the mass dependence of
evaporation rate of filled greases made from very low vapor pressure
silicones or perfluoropolyethers (PFPE).

The ASTM penetration test [17] measures the depth in millimeters that
a particular cone will penetrate a grease in a given time. This is a measure of grease consistency. A No. 2 grease with a penetration of 265 to
295 mm is typically used for low speed bearings. Soft greases may be
needed for high speed ball bearings. Each of the grease types described
in Table 18.1 is made in several consistencies.

Table 18.1   Typical Properties of High Vacuum Greases

Trade Name	Chemical Type	Temperature Range (°C)	P_v at 25°C (Pa)	Sp. Gr. at 25°C	Penetration (mm)
CVC Celvacene Light[13]	ester-thickened hydrocarbon	-40 to 90	10^{-4}	–	150
Apiezon[14] AP 100 Grease	PTFE-thickened hydrocarbon	10 to 30	10^{-8}	1.042	–
Apiezon[14] L Grease	distilled hydrocarbon	10 to 30	10^{-8}	0.896	–
Dow Corning[15] High Vacuum	PTFE-thickened silicone	-40 to 260	10^{-7}	1.0	<260
Krytox[16] LVP L-10	PTFE-thickened fluorocarbon	-20 to 200	10^{-13}	1.94	280

[a] ASTM D-217

Grease lubrication is used for long life, low maintenance applications. Grease is often used in low speed ball bearings, and rotary feedthroughs and has been used in turbomolecular pump bearings [18]. Each grease type has unique characteristics. Hydrocarbon grease has a characteristically low maximum service temperature. It is rarely much above room temperature for distilled greases, but is higher for filled greases if the gelling agent is properly chosen. Silicone and PFPE fluids have high maximum service temperatures. They are typically formulated from a mixture of fluid and a gel or powdered PTFE. Silicone fluids are poor steel-on-steel lubricants, and greases made from them are also poor lubricants unless thickened with a lubricant like PTFE. If a large reservoir of grease is needed, a distilled hydrocarbon, or a filled grease made from very low vapor pressure silicone or PFPE fluid is recommended [11]. If a thin film is adequate, a gelled grease may be used. PFPE greases are preferred to silicones where electron bombardment can cause fragmentation. The reaction between greases and elastomers is similar to that for fluids and elastomers shown in Appendix F.1. The gelling agent may cause an additional reaction. Manufacturers can provide information on the reactivity of greases containing proprietary additives.

Bolt lubricants are often used around but not in vacuum systems. These greases are formulated for the extreme boundary lubrication region. They contain extreme pressure additives such as silver, copper, lead or molybdenum disulfide. Some base oils and additives used in these greases have high vapor pressures.

18.3.3 Dry Lubrication

The process of lubricating with a solid film is called dry lubrication. In (18.1) we saw that friction can be reduced in the boundary region by increasing the contact area and reducing the shear force of the lubricant. The main advantage of a dry lubricant is its low vapor pressure. Low vapor pressure reduces system contamination and allows operation at high temperatures. Solid film lubricants are useful when loads are extremely high, speeds are low, surface temperatures are extreme, and the design must be simplified for maintenance-free long life in vacuum. Dry lubrication is limited by its finite thickness and the debris generated by its sacrificial removal. Dry lubrication failure often results from a local defect and therefore the lifetime of a dry lubricated system is not as predictable as that of a liquid lubricated system in which the lubricant evaporates or migrates in a uniform manner [5].

The desirable properties of a dry lubricant are low vapor pressure, low shear strength, and good adhesion to the base metal. Many solids have been used as dry lubricants. Among them are graphite, the sulfides and selenides of molybdenum, tungsten, gold, silver, and PTFE. It was originally thought the lubricating ability of graphite and other layered solids resulted from their loosely bound layered structure. The weak binding between layers was thought to allow sliding with low shear. This theory was found wanting when it was discovered that graphite was not a good lubricant in vacuum or at high altitude, but only in the presence of water vapor [19]. MoS_2 is a good lubricant in vacuum [20]. The difference in the lubricating ability of these two materials is a result of their differing structure. Graphite consists of a layered structure with a high interlayer binding energy. Contamination from water vapor lowers the binding energy and allows motion with low shear. Molybdenum disulfide films are oriented layers of S-Mo-S with a low binding energy between the adjacent sulfur layers, so contamination is not necessary for low shear sliding [21].

Molybdenum disulfide and tungsten disulfide are the most widely used solid vacuum lubricants. Farr [22] has reviewed its structure and properties as they apply to lubrication. Mattey [23] described how these and other solid films have been used in space hardware. MoS_2 has a very low vapor pressure and can be applied by many techniques including sputtering and spraying. Spavins [24] first deposited MoS_2 by DC sputtering, and later by RF sputtering [25]. Sputter-deposited films give satisfactory lubrication but do generate debris [25]. Thomas [26] has studied the wear properties of sprayed films and concluded that a MoS_2-graphite-sodium silicate coating on steel gave satisfactory performance and long life. Wear rates of $< 3 \times 10^{-18}$ m^3/(N-m) were observed for spray-gun coated parts; the wear rates were 100 times larger for MoS_2 films applied in an aerosol. Kurilov [27] has studied the effect of vacuum on the friction and wear of MoS_2 lubricated steel surfaces. In the pressure range 10^5 to 10^4 Pa he observed the friction and the wear to decrease because of the removal of water vapor. Wear and friction remained constant in the pressure range 10^4 to 10 Pa. In the pressure range 10 to 0.1 Pa the friction continued to decrease while the wear increased greatly. The friction decrease was attributed to the removal of oxygen while the wear increased because heat could not be dissipated in vacuum. Below 0.1 Pa no further change in wear or friction was observed. Oxygen increases the friction by oxidation of MoS_2 to MoO_3 and MoO_2 [27,28].

Soft metals will also lubricate siliding surfaces. Kirby et al. [29] studied the lubrication of gears and bearing surfaces in UHV. They observed pair hardness, ductility and redeposition to be important in designing a low friction, long life system. Their best results were obtained when the material pairs were as dissimilar in hardness as was practical. They concluded the lubricant should be non-brittle to prevent flaking (e.g. MoS_2 coatings on steel, and silver on aluminum), and should redeposit during running to prevent cold welding. Their successful gear pair was a silver plated aluminum gear running against an MoS_2-graphite-sodium silicate coated steel gear.

Polytetrafluoroethelyne coatings have also been used for vacuum applications. PTFE transfers and recoats [4]. Tabor [3] has shown that the long PTFE molecules are oriented parallel to the direction of sliding.

Almost anything placed between two moving surfaces will reduce friction and wear. The environment affects the adhesion of a lubricant, its evaporation, oxidation and intercrystalline forces. If we remove all foreign materials, similar metals will instantly cold weld. However, the cold welding of adjacent surfaces can be eliminated by use of materials of dissimilar lattice constants such as sapphire balls in stainless steel races [5]. The friction coefficient of this system is higher than a lubricated system.

Maeba et al. [30] have developed magnetically levitated bearings for atmospheric-to-vauum rotary motion feedthroughs, that produce no wear particles.

REFERENCES

1. E. R. Booser, in *Kirk-Othmer Encyclopedia of Chemical Technology, 3rd ed.,* Vol. 14, M. Grayson and D. Eckroth, Eds., Wiley, New York, 1980, p. 484.

2. N. Tipei, *Theory of Lubrication*, Stanford Univ. Press, 1962, p. 11.

3. D. Tabor in *Microscopic Aspects of Adhesion and Lubrication,* J. M. Georges, ed., *Tribology Series*, Vol. 7, Elsevier, Amsterdam, 1982.

4. D. H. Buckley, *Proc. 6th Int'l. Vac. Congr. 1974, Japan J. Appl. Phys., Suppl. 2, Pt. 1.,* 1974, p. 297.

5. V. R. Friebel and J. T. Hinricks, *J. Vac. Sci. Technol.,* **12**, 551 (1975).

6. F. D. Bowden and D. Tabor, *The Friction and Lubrication of Solids, Part II*, Clarendon Press, 1964.

7. J. R. Van Wazer, J. W. Lyons, K. Y. Kim and R. E. Colwell, *Viscosity and Flow Measurement*, Interscience, New York, 1963.

8. J. F. Swindells, R. Ullman and H. Mark in *Technique of Organic Chemistry, 3rd ed.,,* A. Weissberger, Ed., Vol. 1, *Physical Methods of Organic Chemistry, Part 4,* Interscience, New York, 1959, p. 689.

9. ASTM D-2270, *1981 Annual Book of ASTM Standards*, Part 24, American Society for Testing Materials, Philadelphia, 1981, p. 277.

10. R. E. Hatton in *Synthetic Lubricants*, R. C. Gunderson and A. W. Hart Eds., Reinhold, New York, 1962, p. 402.

11. L. Laurenson, *Vacuum*, **27**, 431 (1977).

12. L. Laurenson, *Vacuum*, **30**, 275 (1980).

13. CVC Products Inc., 525 Lee Road, Rochester, NY 14603.

14. Edwards High Vacuum Corp., 3279 Grand Island Blvd., Grand Island, NY 14072.

15. Dow Corning Co, Inc. 2030 Dow Center, Midland, MI 48640.

16. Du Pont and Co. Chemicals and Pigments Department, Wilmington, DE 19898.

17. ASTM D-217, *1981 Annual Book of ASTM Standards*, Part 23, American Society for Testing Materials, Philadelphia, 1981.

18. G. Osterstrom and T. Knecht, *J. Vac. Sci. Technol.,* **16**, 746 (1979).

19. R. H. Savage, *J. Appl. Phys.,* **19**, 1 (1948).

20. A. J. Haltner, *Wear*, **7**, 102 (1964).

21. P. J. Bryant, P. L. Gutshall and L. H. Taylor, *Wear*, **7**, 118 (1964).

22. J. P. G. Farr, *Wear*, **35**, 1 (1975).

23. R. A. Mattey, *Lubr. Engng. (ASLE)*, **34**, 79 (1978).

24. T. Spavins and J. S. Przybyszewski, *Deposition of Sputtered Molybdenum Disulfide Films and Friction Characteristics of Such Films in Vacuum*, NASA TDN-4269, December 1969.

25. C. E. Vest, *Lubr. Engng. (ASLE)*, **34**, 31 (1978).

26. A. Thomas, *The Friction and Wear Properties of Some Proprietary Molybdenum Disulphide Spray Lubricants in Sliding Contact with Steel*. Report No. ESA CR(P) 1537, European Space Tribology Laboratory, Risley, England, January 1982.

27. G. V. Kurilov, *Soviet Materials Science*, **15**, 381 (1979).

28. M. Matsunaga and T. Nakagawa, *Trans. ASLE*, **19**, 216 (1976).

29. R. E. Kirby, G. J. Collet, and E. L. Garwin, *Proc. 8th Intl. Vac. Congr.*, **II**, J. P. Langeron and L. Maurice, Eds., Cannes, France, Sept., 1980, p. 437.

30. Y. Maeba, Y. Minamigawa and H. Yamakawa, *Proc. 3rd. Symp. Automatic Integrated Circuits Manufacturing,*, **88-13**, J. B. Anthony, Ed., The Electrochemical Society, Pennington, NJ, 1987, p. 91.

PROBLEMS

18.1 The viscosity of a gas increases with increasing temperature. Why does the viscosity of a liquid decrease with increasing temperature?

18.2 A kinetic viscometer gives values of 40 and 36 mm^2/s for two liquids, while an absolute viscometer gives a reading of 0.04 Pa-s for both. Explain.

18.3 A 10-cm-diameter shaft revolves in a 3-cm-long sleeve bearing at 400 rpm with a radial clearance of 0.01 cm. Find the torque required to overcome the resistance of the oil (density 0.88 g/cc, viscosity of 50 mm^2/s) used to lubricate the sleeve.

18.4 The viscosity of a substance was found to be 52, 9.2, 3.2, and 1.8 Pa-s, when measured at shear rates of 10, 100, 1000 and 10,000 s^{-1}. It has a density of 0.91 g/cc. Plot the kinematic viscosity vs shear rate. What is the material?

18.5 How will the vacuum properties of a grease change if the grease is left exposed to the atmosphere?

18.6 Why should grease not be applied with the fingers?

18.7 What are three advantages of a synthetic lubricant over a paraffin based oil?

18.8 Why use a solid lubricant instead of a liquid or grease?

18.9 Describe the sliding behavior of molybdenum plates which are placed in vacuum after the following separate treatments: (a) high temperature hydrogen firing, (b) high temperature oxidation, and (c) high temperature baking in hydrogen sulfide.

18.10 Describe the lubrication environment in the following applications: (a) shaft bearing in a mechanical pump, (b) a turbomolecular pump bearing, (c) rotating planetary dome roller bearings in an un-baked routine aluminum metallizing system used at 10^{-4} Pa, (d) rotating substrate holder bearing (1 rpm) in a molecular beam epitaxy system. What kind of a metal-lubricant system would you choose for each and why?

Systems

This group of four chapters focuses on three types of vacuum systems used in the thin-film and semiconductor industries: systems that will pump to the high vacuum and ultrahigh vacuum range and systems that will pump a large gas flow in the medium and low vacuum ranges. One way of viewing them is to observe that the end use places emphasis on different facets of the technology. In the high vacuum systems discussed in Chapter 19 the large process gas load and the need for line-of-sight motion of molecules from the hearth to the substrate are paramount in determining pump size and type and chamber design.

The ultrahigh vacuum systems described in Chapter 20 are used to keep surfaces free of monomolecular layers of gas contaminants for long periods. The achievement of ultrahigh vacuum pressures is dependent on the careful selection of materials, joining and cleaning techniques, and the choice of pump. Ultrahigh vacuum systems will find their way into production lines, as the need for dense semiconductor chips demands improvements in processing purity.

The systems discussed in Chapter 21 are used when high gas flows are needed in the medium or low vacuum range. Because the gas flow–pressure range is outside that of high vacuum systems, they are throttled or replaced by mechanical pumping systems.

Chapter 22 briefly reviews leak detection. Although the leak detector is primarily a measuring instrument, it is used in conjunction with completed systems more than in testing individual components. Its mode of operation is dependent on the pumping system.

CHAPTER 19

High Vacuum Systems

Pumping systems capable of producing a high vacuum environment are used for a variety of applications, for example, the evaporation and condensation of materials to form thin-film layers. The high vacuum pumping packages reviewed in this chapter encompass those that will produce a base pressure of an order of 10^{-6} Pa, and a process pressure less than 10^{-4} Pa. The base pressure of such a system is determined by the pumping speed and by the outgassing of the chamber. The pressure will be higher during the process because of the gas load evolved from the source and other heated surfaces. The pumping packages used on high vacuum chambers must have a high enough pumping speed not only to reach the base pressure, but also to maintain the process pressure as well. In addition, systems that are used for production applications should be able to reach this base pressure quickly. These pump packages must pump water rapidly, because it is the most difficult gas to remove from a chamber filled with air.

In this chapter we discuss systems with these objectives in mind. We describe the operation of diffusion-, turbomolecular-, ion-, and cryogenic-pumped systems, respectively. Included is a procedure for starting, stopping, cycling, and fault protecting each kind of pumping system. We describe problems commonly encountered when using these pumps, and conclude with a discussion of the high vacuum chamber.

19.1 DIFFUSION-PUMPED SYSTEMS

The layout of a small diffusion pumped system is sketched in Fig. 19.1. Several accessory items are shown as well. Not every system has or needs all of these additional items but we show them here in order to discuss their proper placement and operation. The main components of this system, the diffusion pump, trap, gate valve, and mechanical pump, together form the single most common high vacuum system.

In less critical applications only a water baffle is necessary. Hablanian [1] measured a backstreaming rate of 5 × 10^{-6} (mg/cm^2)/min for DC-705 fluid over a 6-in. diffusion pump and simple liquid nitrogen trap, while a value of 8 × 10^{-6} (mg/cm^2)/min over a 4-in. diffusion baffled with a chevron cooled to 15°C may be calculated from data given by Holland [2]. For many applications the cold-water baffle alone will perform adequately, provided a pentaphenyl ether or pentaphenyl silicone fluid is used in the pump. As Singleton [3] points out, these two fluids are valuable in systems which are trapped with a less than elegant trap. The simplest way to reduce backstreaming in a low quality system is to use a high quality fluid.

In the most critical applications a liquid nitrogen trap, a water- or conduction-cooled cap, and possibly a partial water baffle are used. The liquid nitrogen reservoir is a trap for water vapor and pump fluid fragments. A 6-in. diffusion pump stack like the one sketched in Fig. 19.1 will have a net pumping speed of 1000 L/s for air at the baseplate, while a matching liquid nitrogen trap can pump water vapor at speeds up to 4000 L/s. Diffusion pumps, fluids and traps of modern design are capable of reducing backstreaming rates to a level such that organics from other sources may be more prevalent in the chamber.

The arrangement shown here for automatically filling the liquid nitrogen trap is convenient when there is a long vacuum-jacketed supply line that must be chilled each time the trap needs filling. This does not eliminate the need for phase separation on large distribution systems. Without this arrangement the incoming liquid boils, the gas quickly warms the trap, and releases condensed vapors. To avoid trap warming, a valve is added. When the low level sensor is activated, the vent valve (3 in Fig. 19.1) is opened and the warm gas is vented to the atmosphere. A second sensor is placed in the supply line near the vent valve. When cooled by the liquid, it closes the vent valve and opens the fill valve. The controller then performs its normal function. A commercial controller with two-level sense elements (low level and empty) may be modified to perform this function. If the trap is filled from a dewar by a short length of tubing, this gas bypass operation is not be required. Alternatively, we could use a triaxial delivery pipe in which the gas bubbles are automatically vented [4]. If the liquid nitrogen is pressurized when it fills the trap, it will be warmer than 77 K.

In smaller systems one mechanical pump is used alternately to rough pump the chamber and back the diffusion pump. During roughing pressure will build up in the valved-off foreline but it is generally not significant if the roughing cycle is shorter than 15 min. Large systems, for example, with 16-in. or larger diffusion pumps, frequently use a Roots

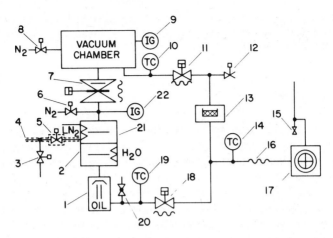

Fig. 19.1 Diffusion pump stack: (1) diffusion pump, (2) partial water baffle, (3) LN₂ vent valve, (4) LN₂ inlet, (5) LN₂ fill valve, (6) port for gas purging diffusion pump, (7) bellows sealed high vacuum valve, (8) chamber bleed valve, (9) chamber ionization gauge, (10) chamber thermal conductivity gauge, (11) roughing valve, (12) mechanical pump vent, (13) roughing line trap, (14) roughing line thermal conductivity gauge, (15) sump for collecting condensable vapors, (16) bellows for vibration isolation, (17) mechanical pump, (18) foreline valve, (19) foreline thermal conductivity gauge, (20) leak testing port, (21) liquid nitrogen trap, (22) diffusion pump ionization gauge.

blower backed by an oil-sealed piston pump for chamber roughing. A smaller, separate "holding" pump can back the diffusion pump during roughing. At crossover to high vacuum the Roots pump is switched over to back the diffusion pump to provide maximum throughput during this period of high gas flow.

Thermal conductivity gauges are located in the work chamber and in the foreline for control and fault protection and in the roughing line to check the mechanical pump blank-off pressure. One ionization gauge is located in the chamber, or on the spool piece adjacent to it, and a second is positioned between the cold trap and the gate valve. Many systems are designed and constructed without a gauge at the latter point; however, it is extremely useful. The most straightforward diagnostic measurement on a diffusion pump is the blank-off pressure, and that measurement can be made only with a gauge located beneath the main valve. Some larger traps are fabricated with an ion gauge port, while gate valves with extra piping or ports on the diffusion pump side facilitate easy mounting of the lower ion gauge tube. In general, it is advisable to install the gate valve with its seal plate facing upward to keep the valve interior under vacuum at all times. This reduces the volume and surface area exposed to the atmosphere in each cycle and thus minimizes the pump-down time. All gauge tubes, both ion and thermal conductivity, should be positioned with their entrances facing downward or to the side to prevent them from becoming traps for particulates.

Some other items shown in Fig. 19.1 are not often needed or found on standard systems. A leak detection port located in the foreline provides the best sensitivity and speed of response for leak detecting the chamber, trap, and diffusion pump. By use of this port a leak detector may be attached while the system is operating. A flow restriction placed on the inlet side of the diffusion pump cooling water coil will save water and increase the temperature of the oil ejector stage. Roughing traps are often used in diffusion pump systems and when properly maintained can reduce the transfer of mechanical pump oil to the chamber. In this system the trap is used only as a roughing trap. When placed in the roughing line (Fig. 19.1), it does not become contaminated with diffusion pump fluid. The mechanical pump vent should be located above the trap to permit the oil vapors to be blown away from the chamber each time the mechanical pump is vented.

19.1.1   System Operation

Let us review the operation of a diffusion-pumped system by studying the sequence of chamber pumping and venting and system start-up and shut-down. Assume that the system is operating and that the work chamber is at atmospheric pressure. The high vacuum and roughing valves are closed. When the roughing sequence begins, the foreline valve will be closed and the roughing valve opened. In small pumping systems only a single roughing valve is required. In large chambers it is a multistep process. First, the chamber may be rough-pumped through a small diameter orifice or by pass line that chokes the flow and prevents turbulence from stirring up particulates on the floor of the work chamber. Particulates may become electrostatically attached to insulating substrates. Second, when the pressure has been reduced to a value at which turbulent flow is no longer possible, the main roughing valve is opened. Third, if a Roots blower is used, it will be bypassed or allowed to free-wheel until the pressure drop between its inlet and outlet is reduced to a safe operating value, say 1000 Pa, at which time it begins to pump in series with the rotary vane or piston pump. Turbulent-free roughing is discussed in Section 19.5. Single set point pressure switches are used to control these functions. Roughing will continue until the pressure reaches 15 Pa. At this pressure the roughing valve will be closed and following a time delay the high vacuum valve will be opened. Never allow the roughing pump to exhaust the system below a pressure of 15 Pa or gross backstreaming of mechanical pump oil will occur.

The pressure at which crossover from rough to diffusion pumping occurs is sensed by a thermal conductivity gauge that operates a relay. The time constant of a 0 to 150 Pa thermocouple gauge is typically 2 s

[5]. In a small system which has a system roughing time constant that is much less than the gauge time constant the pressure will reach a level less than that set on the gauge. This problem is easily corrected by adjusting the set point to a pressure higher than actually desired or by using a pressure switch with a faster time constant. The fact that the gauge reading lags the system pressure can be used effectively by the controller to look for leaks or excessive outgassing in the chamber. When the gauge reaches the 15-Pa set point, the roughing valve closes, and a timer delays the opening of the high vacuum valve. The thermal conductivity gauge will reach a pressure minimum that is below the set point and will drift upward at a rate determined by the real or virtual leak. If the pressure increases beyond the set point pressure before the timer opens the valve, the sequence may be programmed to abort; if the pressure is below the set point at the end of the timed interval, the high vacuum valve will be programmed to open. The diffusion pump and liquid nitrogen trap will then pump the system to its base pressure. The time required to reach the base pressure is a function of the speed of the diffusion pump and the liquid nitrogen trap, the chamber volume, surface area, and cleanliness. The main vapor species is water vapor and a liquid nitrogen trap pumps water vapor efficiently. Diffusion pumps with expanded tops and oversized liquid nitrogen traps have high pumping speeds for water vapor.

System shut-down begins by closing the high vacuum valve and warming the cryogenic trap as the diffusion pump removes the evolved gases and vapors. If the trap is allowed to equilibrate with its surroundings naturally, it will take between 4 and 20 h, depending on the trap design. This time can be shortened considerably by flushing dry nitrogen gas through the liquid nitrogen reservoir. When the trap temperature reaches 0°C, the power to the diffusion pump will be turned off. When the pump will be cooled to 50°C, the foreline valve is closed. The mechanical pump is then shut down and vented. Venting of the diffusion pump should be done with the valve located above the trap (6 in Fig. 19.1). Valve 20 in Fig. 19.1 may be used to vent the pump, but the pump should be cooled to 50°C or lower and venting should proceed slowly; otherwise the fluid and jet assembly may be forced upward. The diffusion pump should not be vented by opening the foreline valve and allowing air to flow back through the roughing trap, roughing line and into the diffusion pump. This procedure would allow excessive mechanical pump oil vapor to be transported into the diffusion pump. Last, cooling water in the diffusion pump should be turned off. If the pump interior is exposed to ambient air when cooled below the dew point, water vapor will condense and contaminate the pump fluid.

Start-up of a diffusion-pumped system begins by flowing cooling water through the diffusion pump jacket and starting the mechanical pump. After the gas in the roughing line is exhausted to a pressure of, say, 100

Pa the foreline valve may be opened and the diffusion pump heater activated. On a system operated by an automatic controller, this function is done by a preset gauge. Warm-up time for a diffusion pump varies from 15 min for a 4-in. or 6-in. pump to 45 min for a 35-in. pump. When starting a diffusion pump, precautions may be taken to prevent the initial backstreaming transient from contaminating the region above the trap before it is cooled. This can be accomplished by precooling the trap or by gas flushing. Not all system controllers activate the liquid nitrogen trap at the same time in the start-up cycle. In fact, some controllers leave this as a step to be manually controlled by the operator because a partly cooled trap is preferred when the pump begins to function.

If economy dictates that the pump is to be started and shut down each day. Santeler's [6] gas purge technique should be used to prevent transient backstreaming from contaminating the system. When systems are run continuously, the total transient contamination from an occasional start-up or shut-down is so small that it can be ignored, at least for the kind of system described in this chapter. Daily shut-down multiplies this effect enormously. Pump fluid fractions desorbing from the trap will backstream into the region between the trap and the high vacuum valve, even though the pump is operating. In a similar manner the burst of fluid vapor emitted during the collapse of the top jet [7,8] will diffuse above the trap and travel to the work chamber the next time the high vacuum valve is opened. These contaminants can be flushed from the system by use of a gas purge.

19.1.2   Operating Concerns

The operations sketched in the preceding section did not deal in depth with some of the real and potential problems in diffusion-pumped stacks—problems such as reduction of backstreaming, operation of the trap, and fault protection. Let us consider these here.

Backstreaming is a subject for which many vacuum system users have a limited appreciation. Unfortunately the definition that states that backstreaming is the transfer of diffusion pump fluid from the pump to the chamber is for many a general definition. It is still incorrect when restricted to diffusion pump systems, because mechanical pump oil is the largest organic contaminant in the work chamber. It need not be, because techniques are available to prevent mechanical pump oil vapors from reaching the chamber.

In a diffusion pumped system backstreaming is controlled by three factors: (1) the diffusion pump, traps, and baffles; (2) the roughing pump, traps and piping; (3) the system operating procedures. Most important is that the backstreaming from all three sources must be reduced to a level that can be tolerated. It is a meaningless exercise to eliminate totally backstreaming from diffusion pump fluid while operating the system in a manner that allows gross contamination from mechanical pump oil. In this section we review ways of reducing backstreaming from each of the above three sources to a level that will not affect most processes. Procedures for further reduction of backstreaming to make diffusion pumps suitable for ultrahigh vacuum work are discussed in Chapter 21.

Methods of reducing fluid contamination from diffusion pumps were discussed in Chapter 12. The diffusion pump should use a high-quality fluid and be baffled by a water-cooled or conduction-cooled cap and a liquid nitrogen trap.

Contamination from the roughing pump may be reduced by use of a liquid nitrogen [9] or ambient temperature trap and a low vapor-pressure oil. The liquid nitrogen trap is the most effective but the most difficult to maintain and consequently is not frequently used. Ambient temperature traps do not require constant refrigeration but do saturate at some point. Zeolite [10,11], alumina [2,11,12], and bronze or copper wool have been used for this purpose. Water vapor will soon saturate a zeolite trap [3,13] and slow the roughing cycle. However, zeolite traps can remove more than 99% of the contamination [10] from the fact that particulates can drift into valve seats and into the mechanical pump, thus hastening wear of the valve seat and pump interior. All traps except those that are liquid-nitrogen-cooled saturate in at least three months under normal use. If they are not rigorously maintained, they will lose their value. Experience has shown that they are almost never maintained and so give a sense of protection that is false.

Kendall [14] designed a thermoelectrically cooled (−40°C) zeolite adsorption trap which included a bypass valve to avoid water vapor saturation during the initial portion of the pumping cycle. More recently Buhl [15] has demonstrated a catalytic trap [16] which works on the principle of oxidation and reduction of copper oxide. Hydrocarbons are oxidized to CO_2 and H_2O in a heated catalyst. The catalyst is regenerated when it is exposed to air.

Oil backstreaming from the mechanical pump can be reduced by changing to an oil with a lower vapor pressure [6]. The use of a low vapor pressure oil, such as Convoil-20 or Invoil-20, can reduce the backstreaming. Friction will eventually degrade the oil [2,17] and periodic changing will be required. When the rotary pump is backing the diffusion pump, the foreline is in free molecular flow and mechanical pump oil can freely flow toward the diffusion pump. This is no problem if the diffusion pump is the fractionating variety because it is able to selectively direct high vapor fractions to the foreline where they are ejected [3]. Attempts have been made to eliminate the effect of mechanical pump fluid backstreaming to the diffusion pump by use of a pure silicone diffusion pump fluid in the mechanical pump [18]. The non-polar silicone molecules do not wet and therefore do not form a lubricating or sealing film.

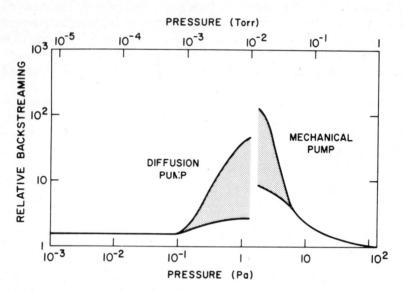

Fig. 19.2 Pumping and lubrication fluid backstreaming in the transition zone between the diffusion and mechanical pump operating regions. Reprinted with permission from *Japan. J. Appl Phys.*, Suppl. 2, Pt. 1, p. 25, M. H. Hablanian. Copyright 1974, The Japanese Journal of Applied Physics.

System operation, the third component in the recipe for low back-streaming, requires more emphasis than it is usually given. Improper operation can cause gross contamination of the work chamber, even in a well-designed system.   Figure 19.2 from Hablanian [19] qualitatively outlines the problem of backstreaming from a diffusion and a mechanical pump. The pressure region between 10^{-2} and 15 Pa is of most concern. Hablanian and Steinherz [20] and Rettinghaus and Huber [8] discuss the rise in backstreaming of diffusion pump fluids at high pressures due to oil-gas scattering; Baker, Holland, and Stanton [13] measured the pressure dependent backstreaming of oil vapor from a rotary mechanical pump.   The substance of Fig. 19.2 is that both diffusion and roughing pumps have maximum backstreaming in adjacent regions. These effects are of serious concern when roughing the chamber and crossing over to the diffusion pump.

The usual technique for roughing the system is to pump to 15 Pa, close the roughing valve, open the high vacuum valve, and pump to below 10^{-1} or 10^{-2} Pa as quickly as possible. When the high vacuum valve is initially opened at 15 Pa, the viscous gas flow will reduce the backstreaming [8, 21], even though the top jet may be momentarily overloaded. As the pressure is reduced, the backstreaming will peak in the transition flow region, below which the backstreaming is small and linearly proportional to pressure. Rettinghaus and Huber [8] found that the maximum back-streaming rate over a cold trap occurred at a pressure of 5×10^{-2} Pa for one 6-in. diffusion-pumped system. The pressure at which peak backstreaming occurs is geometry dependent. Hablanian [19] describes a controlled-opening high vacuum gate valve which is designed to avoid this kind of backstreaming by keeping the pump out of the overload region. Another scheme accomplishes the same thing by opening the valve in two steps.

The contamination introduced by roughing to the free-molecular-flow region is illustrated in Fig. 19.3. Baker, Holland, and Stanton [13] measured the backstreaming of oil through a 300-mm-long × 25-mm-dia. roughing line connected to a 4.5-m^3/h rotary vane mechanical pump. They found that the backstreaming was small ($\sim$10^{-4} mg/min) at high pressures because of the viscous flushing action of the flowing air, but at pressures below approximately 15 Pa the viscous flushing action was diminished until at a pressure of 1.3 Pa the backstreaming was 7×10^{-3} mg/min, or 70 times greater than at high pressures. The data in Fig. 19.3 show that static air, that is gas admitted at the pump ballast port, did not reduce the backstreaming as much as air that was allowed to flow through the length of the tube as it would during roughing. Figure 19.4b illustrates the total integrated backstreaming calculated for the roughing line described in Fig. 19.2, assuming the hypothetical pump-down curve given in Fig. 19.4a. At a backstreaming rate of 10^{-4} mg/min from atmos-

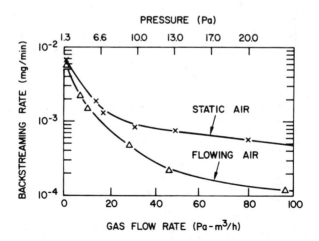

Fig. 19.3 Backstreaming rate as a function of gas flow. The flowing air is traveling the length of the tube. The static air is admitted at the ballast port. Gas pressure is a mean value for flow in the pipe. Edwards 4.5-m^3/h rotary vane pump, 300-mm-long, 25-mm-diameter pipe. Reprinted with permission from *J. Vac. Sci. Technol.*, **9**, p. 412, M. A. Baker, L. Holland, and D. Stanton. Copyright 1972, The American Vacuum Society.

phere to 15 Pa the total backstreaming during the 4-min pump down would be 4×10^{-4} mg. If the pump had been permitted to exhaust the chamber to 1 Pa in the next 11 min, the total backstreaming would have increased from 4×10^{-4} mg to 5×10^{-2} mg. This illustration should make the danger of roughing below 15 Pa adequately clear. Horikoshi and Yamaguchi [22] have measured and calculated the diffusion of light and heavy gases against a nitrogen counter flow. Their data for He, Ar and Xe diffusing against a flow of nitrogen shows that viscous flushing is extremely efficient at reducing counter-flow diffusion. However we should note that the backstreaming measurements shown in Fig. 19.3 for oil vapor, do not decrease as rapidly as are predicted from the above calculations or from the calculations of Jones and Tsonis [21]. This may be due to wall creep, as the measurements of Baker et al. seem to be pressure independent at high pressures.

Comparison of the steady-state backstreaming from a diffusion pump to the backstreaming from the roughing pump cycle reveals that the roughing pump backstreaming need not dominate the system contamination. A 6-in. diffusion pumped system has a roughing line with an inside diameter of approximately 35 mm, or twice the cross sectional area of the pump used in the example in Fig. 19.3. If we assume a backstreaming rate of 2×10^{-4} mg/min (twice that of the preceding example) for the roughing line on the 6-in. system and a pump-down time of 5 min to a pressure of 15 Pa, the total amount of oil backstreamed into the system would be 10^{-3} mg per roughing cycle. The lowest value of vapor backstreaming quoted by Hablanian [1] would result in a total backstreaming of 3×10^{-5} mg/min for a 6-in. pump. For the worst case the roughing cycle could contribute as much backstreaming as the diffusion pump does in 30 min of operation at high vacuum. Even for a diffusion pump operating on a 1 h cycle, contamination from the roughing pump is less

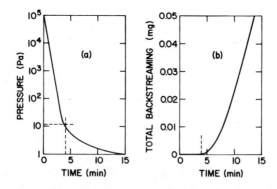

Fig. 19.4 Chamber pressure (a) and total backstreaming (b) as a function of pumping time for the backstreaming rate (flowing air) given in Fig. 19.3.

**Table 19.1   Methods of Reducing Pump Fluid Contamination
in a Diffusion-Pumped High Vacuum System**

Diffusion Pump	Mechanical Pump	System Operation
Cold cap	Simple trap	Crossover at $P > 15$ Pa
LN_2 trap	Good oil	Minimize time dp is in overload region
Fractionating dp		Mild bake
Pentaphenyl ether or pentaphenyl silicone		Gas flush during start/stop[a]

[a] Only if the system is shut down daily.

than that from the diffusion pump, provided that the roughing cycle is always terminated at approximately 15 Pa.

The two most important system-operation concepts for minimizing backstreaming are to stop roughing at 15 or 20 Pa, and to pump through crossover as quickly as possible or use a conductance-limited high vacuum valve. Table 19.1 summarizes the results of this discussion of methods of reducing contamination in a diffusion-pumped, high vacuum system.

The dominant gas load in most high vacuum systems is water vapor. A liquid-nitrogen-cooled surface will provide a high pumping speed (14.5 L/(s-cm^2)) for this gas load. The vapor pressure of water at this temperature is about 10^{-19} Pa; therefore the pumping speed is neither a function of the operating pressure nor sensitive to small trap surface temperature changes. Several gases, however, can reevaporate at low pressures or during trap temperature fluctuations. Carbon dioxide is partly condensed; its vapor pressure is 10^{-6} Pa at 77 K. Methane and CO have high vapor pressures and are not condensed. They are adsorbed to some degree on liquid-nitrogen-cooled surfaces.

Hengevoss and Huber [23] have shown how such gases evolve from a liquid nitrogen trap as it slowly warms up: CH_4 is first, followed by CO and CO_2. Santeler [6] has shown that the reevaporation of CO_2 can be eliminated by delaying full cool-down of the trap until the system pressure falls below 10^{-3} Pa or by momentarily heating the trap to 135 to 150 K after the system pressure falls below 10^{-6} Pa. Because CO_2 is partly condensed on the trap, it can cause problems. Siebert and Omori [24] have shown how certain liquid nitrogen cooling coil designs with temperature variations of only 1 K allow the release of enough CO_2 to cause total pressure variations of 20% in a system operating in the 10^{-4}-Pa range.

In any vacuum system faults may occur that could potentially affect equipment performance or product yield. Misoperation of any one of several valves could cause diffusion pump fluid to backstream to the work chamber and contaminate the chamber and product. A leak in the foreline, an inadequate mechanical pump oil level, or a foreline valve failure could cause the forepressure to exceed the critical value. When the forepressure exceeds the critical value, pump fluid will rapidly backstream into the chamber. In addition, many diffusion pump system faults could cause harm to the pumping equipment. Cooling water failure or an inadequate diffusion pump fluid level will result in excessive fluid temperatures, accompanied by decomposition of some fluids. Certain decomposition products are vapors, while others form tar-like substances on the jets and pump body.

Modern diffusion pumped systems can be adequately equipped with sensors that will remove power from the system or place it in a standby condition in case of loss of utilities, a leak, loss of cryogen, or loss of diffusion pump fluid. Pneumatic and solenoid valves, which leave the system in a safe position during a utility failure, are used on automatically controlled systems. When the compressed air or electrical power fails, all valves except the roughing pump vent should close. The roughing pump vent should open to prevent mechanical pump oil from pushing upward into the roughing line. A flow meter may be installed at the outlet (not the inlet) of the diffusion pump cooling water line. The signal from the flow meter is used by the system controller to remove power to the diffusion pump and to close both high vacuum and foreline valves. The use of a water filter on the inlet to the diffusion pump cooling water line will prevent deposits from clogging the line and causing untimely repairs. A thermostatic switch mounted on the outside of the diffusion pump casing performs these tasks much more simply, and a thermoswitch mounted on the boiler can be used to detect a low fluid level. The set points on the thermocouple or Pirani gauge controllers are used to signal an automatic controller to close the high vacuum valve in case of a chamber leak and to isolate and remove the power from a diffusion pump with excessively high forepressure. Many liquid nitrogen controllers have low level alarm sensors which close the high vacuum valve when the cryogen is exhausted. If not, an insulated thermoswitch can be attached to the trap vent line; the boil-off keeps the vent-line temperature below 0°C. Table 19.2 lists some of the concerns with which the operator should be familiar.

This section raised several issues in regard to the operation and fault protection of a diffusion-pumped system. In many cases fluid backstreaming could result from a utility failure or improper procedure. The easiest way to address these concerns is by fully automatic operation of

Table 19.2   Diffusion–Pumped System Operating Concerns

Do	Do Not
Check mp oil level	Exceed dp critical forepressure
Check mp belts	Overload dp in steady state
Measure dp heater power when system is new	Leave cooling water flowing when system is shut down
Clean roughing trap periodically	Vent cold LN_2 trap to air
Pump oxygen with proper fluids	Rough pump below 15 Pa

the system. With automatic control and scheduled maintenance complete fault protection can be provided and backstreaming minimized to a point at which it is only a small source of organic contamination.

19.2   TURBOMOLECULAR-PUMPED SYSTEMS

Turbomolecular-pumped systems are configured in two ways: they are assembled with a separate roughing line and foreline or as a system without valves in which the chamber is roughed through the turbomolecular pump. Diffusion-pumped systems can be configured that way also. Figure 19.5 illustrates the valved system with a separate roughing line. This system is much like a conventional diffusion-pumped system. The entire roughing and foreline section apparently is identical to that in a diffusion-pumped system. The criteria for sizing the forepump, however, are not the same as for a diffusion pump. Proper diffusion pump operation requires that the forepump speed be large enough to keep the forepressure below the critical forepressure at maximum throughput. In a turbomolecular pumped system a forepump of sufficient capacity should be chosen to keep the blades nearest the foreline in molecular flow or just in transition flow at maximum throughput. For most pumps the maximum steady-state inlet pressure P_i is between 0.5 and 1.0 Pa and the maximum forepressure P_f is about 30 Pa. Because the throughput is constant, $P_i S_i = P_f S_f$, or

$$\frac{P_f}{P_i} = \frac{S_i}{S_f} \tag{19.1}$$

Equation (19.1) states that the staging ratio, or ratio of turbomolecular pump speed to forepump speed, S_i/S_f, should be in the range 30 to 60. In Chapter 11 a more restrictive condition was placed on forepump size for turbomolecular pumps with low hydrogen compression ratios. A staging ratio of 20:1 was found to be necessary for pumps with maximum compression ratios for H_2 less than 500 in order to pump H_2 with the same speed as heavier gases. See Fig. 11.8.

No conventional liquid nitrogen trap is required on a turbomolecular pump to stop bearing or mechanical pump fluid backstreaming. The compression ratios for all but hydrogen, the lightest gas, are high enough so that none will backstream from the foreline side to the high vacuum side, provided that the pump is rotating at rated angular velocity. A liquid-nitrogen-cooled surface will not trap the small amount of hydrogen that does backstream into the work chamber because of its low compression ratio. Liquid nitrogen may be used to increase the system pumping speed for water vapor. Incorrect as it may seem on first consideration, the optimum place to locate this liquid-nitrogen-cooled water vapor pump behind the high vacuum valve is directly over the throat of the turbomolecular pump. Alternatively one could conceive of placing a tee behind the high vacuum valve and attaching the turbomolecular pump to one port and a liquid nitrogen-cooled surface to the other port. The conductance between any two ports of the tee is less, however, than the conductance of a liquid nitrogen trap. A 200-mm-diameter tee whose length is

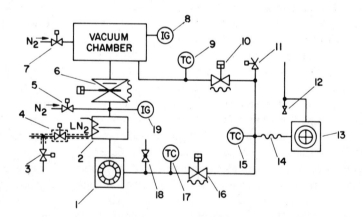

Fig. 19.5 Turbomolecular pump stack with a separate roughing line: (1) turbomolecular pump, (2) liquid nitrogen trap, (3) LN_2 vent valve, (4) LN_2 fill valve, (5) turbomolecular pump vent valve, (6) bellows-sealed high vacuum valve, (7) chamber bleed valve, (8) chamber ionization gauge, (9) chamber thermal conductivity gauge, (10) roughing valve, (11) mechanical pump vent valve, (12) sump for collecting condensable vapors, (13) mechanical pump, (14) bellows for vibration isolation, (15) mechanical pump thermal conductivity gauge, (16) foreline valve, (17) foreline thermal conductivity gauge, (18) leak testing port, (19) turbomolecular pump ionization gauge.

equal to its inside diameter has an air conductance between any two ports of 1275 L/s, while a low-profile liquid nitrogen trap designed for a 6-in. diffusion pump has a conductance of 1700 L/s. Clearly, the optimum arrangement for pumping both water and air, each at their highest speed, is a liquid nitrogen trap located between the turbomolecular pump and the high vacuum valve, even though it may convey the erroneous impression that the trap is present to prevent hydrocarbon backstreaming. In some processes use of a Meissner trap in the chamber is required, just as it would be in a diffusion pumped system.

Neither a roughing trap nor a foreline trap is needed. High crossover pressures prevent oil transfer to the chamber via the roughing line and a high compression ratio for hydrocarbon fragments prevents back diffusion from the foreline through the pump to the chamber. Buhl [15] demonstrated the ability of the catalytic trap to prevent mechanical pump vapors from mixing with the turbopump oil. He measured a 100-fold decrease in the partial pressure of $C_2H_3^+$ in a UHV pumping unit equipped with a catalyzer trap. Valve 18 in Fig. 19.5 is used for leak testing, not pump venting. Venting the pump at this location would drive oil vapors toward the high vacuum side after the rotor loses about 60% of its speed.

Figure 19.6 shows an unvalved system in which the chamber is roughed directly through the pump. Because there is no roughing line, the problems associated with roughing line contamination and improper crossover are eliminated. Physically the system becomes simpler. There is no roughing valve, piping, or trap. No high vacuum valve is needed and only one ion gauge and thermal conductivity gauge are required. This system must be completely shut down and restarted each time the chamber is

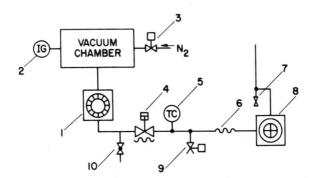

Fig. 19.6 Valveless turbomolecular pump stack: (1) Turbomolecular pump, (2) chamber ionization gauge, (3) chamber pump and vent valve, (4) mechanical pump isolation valve, (5) foreline and mechanical pump thermal conductivity gauge, (6) bellows for vibration isolation, (7) sump for condensable vapors, (8) mechanical pump, (9) mechanical pump vent.

opened to the atmosphere. This makes the use of liquid-nitrogen-cooled surfaces awkward because they must be warmed each time the system is shut down.

Roughing through the turbomolecular pump places a size restriction on the mechanical pump if the turbomolecular and mechanical pumps are to be started simultaneously. If the mechanical pump is small, it will not exhaust the chamber to the transition region before the turbomolecular pump reaches maximum rotational speed. When this happens, the motor over-current protection circuit will shut down the turbomolecular pump. A properly sized and operated mechanical pump will exhaust the chamber to 20 to 200 Pa by the time the turbomolecular pump has reached, say, 75% of its rated rotational speed and prevent the backstreaming of the mechanical pump oil. Most turbomolecular pumps are designed with an acceleration time of 5 to 10 min.

19.2.1 System Operation

The operation of the turbomolecular pumped system with a separate roughing line shown in Fig. 19.5 is much like that of the diffusion-pumped system described in Fig. 19.1. Before rough pumping is begun, the high vacuum valve and the roughing valve are closed, while the foreline valve is open. Chamber pump-down begins by closing the foreline valve and opening the roughing valve. As in a diffusion or any other high vacuum pump the rough pumping hardware varies in complexity with the chamber size. For 500-L/s or smaller turbomolecular pumps a two-stage rotary vane pump is used; turbomolecular pumps larger than 1000 L/s use a Roots pump backed by a rotary piston pump. At a pressure of 100 to 150 Pa the roughing valve is closed and the chamber is crossed over to the turbomolecular pump. In some pumps this may cause a slight, but delayed, momentary speed reduction which will have no effect on the pump-down. Here, as for a diffusion pumped system, the dominant species is water vapor and the pump-down time will be controlled by the speed of the liquid nitrogen trap. If no liquid nitrogen trap is used, this system will pump water vapor somewhat more slowly than an untrapped diffusion pump of the same speed. The large unbaked internal surface area of the pump adsorbs water during the early stages of the high vacuum pump-down cycle and then reemits it at lower pressures. This effect is more noticeable in a valveless system because it potentially can adsorb more water vapor than a valved system that is not exposed to ambient air at a pressure greater than 150 Pa.

System shut-down begins by closing the high vacuum valve and warming the liquid nitrogen trap, if used, as described in Section 19.1.1. When the trap has equilibrated, the foreline valve is closed and the power to the turbomolecular pump motor is removed. The rotor will now decelerate. Typically, it should take 10 min or more for the rotor to come to a complete stop, but if that were to happen hydrocarbons from the foreline would rapidly diffuse to the region above the pump inlet. To prevent backstreaming of mechanical pump oil vapors and turbomolecular pump lubricating oil vapors the pump is vented with a reverse flow of dry gas. Argon or nitrogen for example, should be admitted at a point above the pump inlet or part way up the rotor stack when the rotor speed has decreased to approximately 50% of maximum rotational speed. The flow should continue until the pump is at atmospheric pressure. This can be properly accomplished by admitting gas through valve 5 (Fig. 19.5). Turbomolecular pumps should not be routinely blasted with atmospheric pressure gas while running at rated speed. It is not good for long-term bearing life. At any time after the foreline valve is closed the mechanical pump system can be shut off and vented by valve 11 in Fig. 19.5 or if so equipped, an internal pump vent. The cooling water should be promptly shut off to prevent internal condensation. Condensation that may form on outer portions of the pump body during normal operation may be eliminated by tempering the water to just above the dew point.

System start-up begins by initiating cooling water flow, opening the foreline valve, and simultaneously starting the mechanical and turbomolecular pumps. After the pump accelerates to rated rotational speed, typically within 5 to 10 min, the liquid nitrogen trap may be filled. At this point the chamber may be pumped as described in the preceding section.

Operation of the valveless system illustrated in Fig. 19.6 is considerably simpler than the valved system. Operation begins by opening the cooling water and foreline valves and starting the mechanical and turbomolecular pumps simultaneously. If the roughing pump has been properly chosen to make the chamber roughing cycle equal to the acceleration time, the system will exhaust the chamber to its base pressure without backstreaming pump oil vapors.

The valveless system is vented and shut down by first closing the foreline valve, waiting for the rotor speed to decrease to 50% of maximum rotational speed, and admitting dry gas above the pump throat (valve 3 in Fig. 19.6). The gas vent valve must be shut off when the system reaches atmospheric pressure or it will over pressurize the work chamber. The mechanical pump is shut down as described and the cooling water flow is stopped.

19.2.2 Operating Concerns

Turbomolecular-pumped systems share a few problems with diffusion-pumped systems and present some that are unique. All are easily soluble. Backstreaming of mechanical pump oil vapors during roughing or turbomolecular pump shut-down, electrical interference, and damage from utility failure are all potential problems in a turbomolecular-pumped system.

Backstreaming of mechanical pump oil to the chamber via the roughing line is even less important than in a diffusion-pumped system because the crossover pressure is 100 to 150 Pa and not 15 Pa, as it is in a diffusion pump. If proper procedures are followed, backstreaming by this path is of no concern. The roughing pump in the valveless system (Fig. 19.6) cannot contaminate the chamber because the turbomolecular pump will have reached full rotational speed before the foreline is in molecular flow.

The compression ratio for hydrocarbon vapors at full rotor speed is high enough in turbomolecular pump to prevent any backstreamed fore-pump oil vapors from reaching the work chamber. This is not so when the rotor is stopped or operating at reduced speed. Nesseldreher [25] has studied the partial pressures of several gases at the inlet as a function of the rotor speed and found that heavy hydrocarbons were observed at the inlet for rotor speeds of 40% maximum speed and less for the pump under scrutiny (Balzers TPU-400). Stopping the pump with the fore-pump operating and the work chamber under vacuum will result in rapid backstreaming of oil vapors from the foreline to the clean side of the pump. To prevent this backstreaming the valved and valveless turbomolecular pumped systems are always vented during shut-down. In particular, it is most important that the pump be vented with a dry gas and that the vent gas be admitted in such a way that it will flow to the foreline through at least a portion of the rotor and stator assembly. Oil vapors in the foreline are then flushed away from the high vacuum chamber. The pump must never be vented from the foreline or the forechamber, because oil vapors will be forced back toward the pump inlet and into the high vacuum chamber.

Proper venting procedures should be followed even after a utility failure. Automatic venting is a necessity when running the pumping system unattended. The simplest electronic solution delays the opening of the vent valve until the pump coasts to about 50% of rated speed. A dc solenoid valve is driven by a power supply with a large capacitive output; the capacitor provides sufficient energy to keep the valve closed for a fixed time until the rotor loses speed. Another method uses a battery-operated circuit to delay the opening of the vent until the speed decreases and then closes it after the pump has reached atmospheric pressure. Both circuits allow the pump to coast through momentary power failures without initiating venting. Speed sensing can be used to vent the pump automatically as a result of a catastrophic leak like a broken ion gauge tube.

Nearby amplifiers may pick up mechanical or electrical noise emanating from turbomolecular pump power supplies. Mechanical vibration may be reduced significantly by use of bellows between the pump and chamber. Improper connection of earth and neutral in three-phase supplies may also generate noise. Other electrical noise can be eliminated by connecting the ground of each piece of equipment, including the pump, to the ground terminal on the most sensitive amplifier stage and then grounding to earth. This is most efficiently done with solid copper strips.

Turbomolecular pumps must be protected against mechanical damage as well as against the loss of cooling water because the pump is a high speed device with considerable stored energy. If a large, solid particle enters the rotor or a bearing seizes, serious damage may be done to the pump. The expense of repair could easily exceed that of scraping the varnish from a diffusion pump. Such catastrophes need not occur and do not occur if the most elementary precautions are taken. A splinter shield located at the pump throat adequately protects the rotors and stators from physical damage at some loss in pumping speed, and some pumps are available with side entrance ports. Water cooling is used in oil-lubricated or grease-packed bearings. Proper cooling is necessary to remove heat from the bearings and to extend bearing life. Most pumps are manufactured with an internal cooling water thermal sensor so it is generally not necessary to add an external flow sensor. A water flow restriction may be added to conserve cooling water. Any device with internal cooling water passages requires a clean water supply. Even though filters are used, it is advisable to reverse-flush the pump water lines once or twice a year to remove material that has passed through the filter. If the water supply is unreliable, a recirculating water cooler is in order. No protection is needed for loss of liquid nitrogen because it does not serve a protective or baffling function; it merely pumps water vapor.

Turbomolecular pumps will give reliable trouble-free operation if the oil is changed at recommended intervals and they are protected against cooling water failure, power failure, mechanical damage, and excessive torque. This protection is easily and routinely provided with the available

Table 19.3 Turbomolecular Pumped System Operating Concerns

Do	Do Not
Periodically change oils	Use undersized forepump
Vent mp to outside exhaust	Vent cold LN_2 trap to air
Vent tmp from high vacuum side	Flow cooling water when vented
Keep forechamber in molecular flow	Run tmp without cooling water fault protection
Check mp oil level	Stop tmp while under vacuum
Check mp belts	Run tmp below maximum rotational speed

technology. Table 19.3 lists some of the needs for proper turbomolecular pump operation.

19.3   ION-PUMPED SYSTEMS

The sputter-ion pump is the most common of all ion pumps; it is used with a titanium sublimation pump and cryobaffle to form a high vacuum pumping package that is easy to operate and free of heavy hydrocarbon contamination. Invariably it uses a sorption roughing module that is also free of hydrocarbons.

A typical, small, sputter-ion pumped system is shown in Fig. 19.7. The TSP and cryocondensation surface may be in the chamber on which sputter-ion pump modules are peripherally located or they may be in separate units as sketched here. Titanium is most effectively sublimed on a water-cooled rather than a liquid nitrogen-cooled surface in a system that is frequently vented to atmosphere. If liquid nitrogen cooling were used, the film on the cooled surface would be composed of alternate layers of titanium compounds and water vapor. Because water vapor is the dominant condensable vapor in rapid-cycle systems, titanium flaking and pressure bursts would frequently result. In rapid-cycle systems it is best to separate the two functions and condense water vapor on a liquid nitrogen-cooled surface and titanium on a water-cooled surface.

A two- or three-stage sorption pump is used to rough the system. Alternatively, a gas aspirator or carbon vane pump may be used to first exhaust the chamber to about 15,000 Pa before sorption pumping. The gas aspirator requires a high mass flow of nitrogen at high pressure and is noisier and less practical than a carbon vane pump. Neither pump is necessary, but their use does allow more sorption cycles between baking.

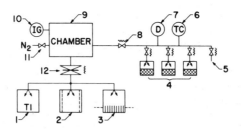

Fig. 19.7 Components of a small sputter-ion pumped system: (1) titanium sublimation pump, (2) sputter-ion pump, (3) liquid nitrogen-cooled array for pumping condensable gases, (4) sorption roughing pumps. (5) port for the attachment of a gas aspirator or carbon vane roughing pump, (6) thermocouple gauge, (7) diaphragm gauge, (8) roughing valve, (9) work chamber, (10) ionization gauge, (11) chamber release valve, (12) high vacuum valve.

High capacity modules equivalent to about 10 to 15 small pumps (10 cm diam., 25 cm high) are available for roughing large volumes.

If the system is to be cycled frequently, a gate valve should be installed between the high vacuum pumps and the chamber to minimize operation of the ion pump at high pressures. An ion gauge is not needed on the pump side of the gate valve because the ion current is a measure of pressure. A gas release valve is provided for chamber release to atmosphere. Nitrogen is used for this function because it is easily pumped by the sorption and ion pumps.

19.3.1 System Operation

Evacuation of a small, sputter-ion pumped system begins with sorption pump chilling. Small commercial pumps will equilibrate in 15 min; a chill time of 30 min is not unreasonable. If the sorption pumps have been saturated by prior use, they must be baked at a temperature of 250°C for at least 5 h before they are ready to chill. Pumping on a sorption pump with an oil-sealed mechanical pump is not a good way to speed the outgassing of water vapor, because backstreamed oil vapors contaminate the sieve and reduce its pumping capacity. The first stage of a two-stage pump manifold is used to rough to 1000 Pa; the pump is then *quickly* valved from the manifold and the second stage is used to pump to 0.4 to 0.2 Pa. If a three-stage manifold is used, a pressure of 3000 to 5000 Pa is obtained in the first stage, 15 Pa in the second, and 0.1 Pa in the third. Staged pumping traps the neon that entered the first stage in viscous flow and also reduces the quantity of gas to be pumped by the last stage. Both effects reduce the ultimate pressure. At a chamber pressure of approximately 0.5 Pa continuous sublimation may begin. The sputter-ion pump is started when the chamber pressure reaches 0.05 Pa. The roughing line can be valved from the system, if the chamber and pump are clean. When the ion pump voltage increases to about 2000 V, its pumping speed will rapidly increase until it reaches its maximum value at about 10^{-3} Pa. Below 10^{-5} Pa continuous operation of the TSP is not necessary because the sublimation rate of the titanium exceeds the gas flux. As the system pressure decreases, the interval between successive titanium depositions may be increased. When the layer is saturated, the pressure will rise as explained in Fig. 12.4. Timing circuits are available to control the sublimation time and interval between depositions.

A sputter-ion pump which has been exposed to atmosphere before its operation will not pump gas from the chamber as quickly as if it were clean and under vacuum. Operation of an exposed system begins with chilling the sorption pumps and flowing cooling water to the TSP. The sorption pumps and TSP are used to rough as described for an operating system. When the sputter-ion pump is turned on, the system pressure will rise because of electrode outgassing. The solution to this dilemma is

Table 19.4   Sputter–Ion Pump Operating Concerns

Do	Do Not
Operate s-i pump below 10^{-4} Pa in steady state to extend life	Obstruct the safety vent on the sorption pump
Valve s-i pump when releasing chamber to atmosphere	Pump a large quantity of H_2 at high pressure
Clean Ti flakes when replacing filaments	Operate the pump with the high current switch in the start mode
Adequately bake sorption pumps	Start the pump at high pressure
Sequentially operate sorption pumps	Leave polystyrene dewars on pumps while baking

to continue pumping with the sorption and TSP until the outgassing load is pumped. If the outgassing is not removed in a short time, power to the ion pump should be pumped to avoid overheating. The outgassing should be pumped after switching the sputter-ion pump on and off a few times for about 5 min each time. When the pump voltage reaches about 2000 V, the roughing line may be valved from the system and operation continued in the normal manner.

The shut-down procedure for an ion pumped system is the simplest of any vacuum system. The high vacuum valve, if any, is closed and the power to the ion pump is removed. The entire system may remain under vacuum until it is needed again. The TSP cooling water should be disconnected if the system is to be vented to atmosphere to prevent condensation on the interior of the system.

19.3.2   Operating Concerns

One advantage of an ion pump is that no fault protection equipment is needed to prevent damage from a utility failure. Loss of electrical power, cooling water, or liquid nitrogen will not harm the pump. Pumping will simply cease, gases will be desorbed from the walls and cryobaffle, and the pressure will rise. If the pressure does not exceed 10^{-1} Pa, the pump can be restarted by applying power to the ion pump; rough pumping is not required.

Of the severest concern to ion pumps in these applications is that they regurgitate previously pumped gases, in particular hydrogen, and do not pump large gas loads well. Essentially all of the pumping above 10^{-3} Pa is done by the TSP and at that pressure filament life is short. Because of their slow pumping between 10^{-1} and 10^{-3} Pa, ion pumps are not commonly used for routine, unbaked, rapid-cycle systems.   Table 19.4 lists some of the needs for proper operation of an ion-pumped system.

19.4   CRYOGENIC-PUMPED SYSTEMS

The layout of a typical cryogenic-pumped system driven by a helium gas refrigerator is sketched in Fig. 19.8.   As in Fig. 19.1, more valves and other parts are shown than may be necessary.   The system requires no forepump, and mechanical pump operation is required only during roughing.   A liquid nitrogen trap is not needed for the prevention of backstreaming from the cryogenic pump, but a Meissner trap, cold-water baffle, or room-temperature baffle, none of which is shown in the sketch, may be necessary in the chamber to baffle the process heat load.   These baffles also reduce the overall system pumping speed. Most cryogenic pumps include a hydrogen vapor pressure gauge for monitoring the temperature of the second stage.   The coarseness of the gauge makes it difficult to read extremely low temperatures.   A silicon diode or gold-germanium thermocouple are more accurate than a hydrogen vapor pressure bulb at cold stage temperatures.

The chamber may be safely crossed over to the high vacuum pump in a range of pressures.   The lowest permissible pressure is governed by the

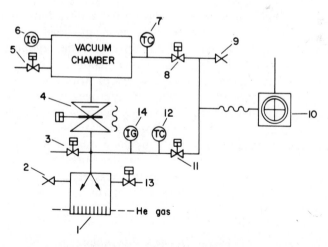

Fig. 19.8 Components of a helium gas refrigerator cryogenic pump stack: (1) cryopumping surfaces, (2) pressure relief valve, (3) flush gas inlet valve, (4) high vacuum valve, (5) chamber vent valve, (6) ion gauge, (7) thermal conductivity gauge, (8) roughing valve, (9) mechanical pump vent, (10) mechanical pump, (11) roughing valve, (12) thermal conductivity gauge, (13) vent valve, (14) ion gauge.

desire to prevent oil backstreaming from the mechanical pump. For a 4-
to 6-cm-diameter roughing line the minimum pressure is of an order of 10
to 20 Pa. For very large roughing lines it will be less; its value is greater
than that which gives a Knudsen number of 0.01. The highest crossover
pressure is determined by the cryogenic geometry, the heat capacity of
the gas and the refrigeration capacity of the expander. The cryopumping
surfaces have a reasonably large heat capacity and can accept a "burst"
of gas without irreversibly warming. The maximum quantity of a gas PV_i
admissible to the pump in a burst is a constant and is available from the
manufacturer. The quantity of gas instantly admitted when the high
vacuum valve is opened is given by $PV_i = P_cV$, where P_c is the crossover
pressure and V is the chamber plus pump volume. From this we see that
the maximum crossover pressure is

$$P_c(\text{max}) = \frac{PV_i}{V} \tag{19.2}$$

The crossover pressure range for rough pumping air in a typical roughing
line of a small pump is

$$15 \text{ Pa} \leq P_c \leq \frac{PV_i}{V} \tag{19.3}$$

If the value obtained from (19.2) is low enough to allow oil backstreaming before crossover, the cryogenic pump will be too small in relation to
the size of the chamber.

Equation (19.2) implies that the crossover pressure could be several
orders of magnitude higher than is typical for a diffusion pump. Bentley
[26] states a rule of thumb for the maximum gas load is $PV_i/W_2 \leq 4$
Pa-L/W where W_2 is the refrigeration capacity of the cryopump's second
stage. If the gas burst is very large, water vapor in the gas burst could
reach the adsorbent stage during viscous or transition flow, coat or
saturate it, and hinder the pumping of hydrogen and helium. Water
vapor in an atmospheric gas load actually reduces the time required to
pump the gas burst. Water vapor has a higher thermal velocity and a
lower heat capacity than nitrogen. Thus the crossover pressure for an air
load can be higher than that used for dry nitrogen [27]. In most
cryogenic-pumped systems the crossover pressure will be 50 to 70 Pa.
To minimize oil backstreaming it is desirable to use a crossover pressure
as as large as possible.

19.4.1 System Operation

A cryogenic-pumped chamber is evacuated by a roughing pump until the
crossover pressure is reached. At that time the roughing valve is closed
and the high vacuum valve is opened. The time required to reach the

system base pressure is a function of the history of the cryogenic pump and its radiation loading, as well as other characteristics of the chamber which affect all pumps. The chamber is cycled to atmosphere in the usual way by closing the high vacuum valve and venting with dry nitrogen gas.

System shut-down commences by closing the high vacuum valve, removing power to the compressor, and equilibrating any liquid nitrogen trap in the chamber or pump with nitrogen gas. When the pump is restarted it will need to be regenerated.

19.4.2 Regeneraton

The regeneration procedures that are recommended or are available from automatic controllers make use of various combinations of external heat, rough pumping, and gas flushing. The object is to remove the captured gases from the pump after the power has been removed. Not all procedures clean the sorbent and pumping surfaces properly. For example, removing the power and warming the refrigerated surfaces will allow water vapor to transfer from the warm stage to the cold stage, and eventually to form a puddle in the bottom of the pump housing. Cryopumps contain a safety relief, but desorbed vapors should not simply be discharged through it when it opens slightly above atmospheric pressure. The regeneration gas load should be pumped away in a mechanical pump, but not at pressures less than 10 Pa. Loading the charcoal with water vapor only increases the time required to clean it. Procedures which advocate heat and uncontrolled mechanical pumping or flushing nitrogen "through the cryopump body" are to be avoided.

Longsworth [28] has compared several regeneration cycles using different combinations of gas purge, heat, purge location and exhaust pressure. He concluded that the shortest regeneration time, consistent with sorbent cleanliness (as measured by helium pumping speed) was a cycle consisting of heating the outer body, venting the pump with ambient temperature nitrogen gas until a pressure of 1 atm. At this point the heat and purge gas were removed, and the mechanical pump was used to evacuate the chamber to 15 Pa at which time the mechanical pump was removed and the cryopump started. Using this procedure, Longsworth was able to regenerate a 200-mm-diam cryopump in 3 h and retain 90% of the rated helium pumping speed. He did not heat the nitrogen gas as it contains only 1.4 W/std liter/min at 100°C. He tested a slightly longer procedure (4.5 h) which used dry nitrogen purge gas directed at the cold stage and recovered 100% of rated helium speed.

One common effective procedure consists of directing warm nitrogen purge gas into the cold stage and pumping the exhaust gases with a mechanical pump. The nitrogen gas flow quantity is adjusted to a large enough value to prevent the mechanical pumping line from ever going below a pressure of 10 to 20 Pa. It is important to remove the heat from the nitrogen purge when the elements reach room temperature or it will delay cooling [29]. Using this technique the charcoal is first warmed followed by the second stage. In so doing the water desorbed from the second stage is not condensed on the first stage. It is important that the gas flow be directed at the first stage [27-29] or regeneration times will be considerably lengthened. Many industrial pumping systems use this procedure and perform regeneration on nights or weekends. For high vacuum processes, regeneration resulting from sorbent saturation or ice deposits will not cause frequent regeneration. Regeneration is required when the power fails or dips for a brief time. Momentary loss of power is a serious concern, especially if the sorbent is saturated with helium. If power is lost for a short time, the helium will be released from the sorbent and will conduct a large amount of heat from the chamber walls to the pumping surfaces. It will serve no purpose to rough the pump to 20 Pa because that is not sufficient to prevent continued conductive heat transfer. Stated another way, a pump will wipe out after a burst of helium and require complete regeneration. If the power is off for a longer time, say 10 to 20 min, long enough for water vapor to release from the first stage and deposit on the second stage, the sorbent will saturate and regeneration also will be required. Experience has shown that the mean time between regeneration in pumps operated at high vacuum is determined by external events, rather than by normal saturation.

19.4.3 Operating Concerns

The most important concern is for the magnitude of the heat load on the first stage. In addition to the 300-K radiation from nearby chamber walls, the first stage is subject to thermal radiation from any source such as an electron beam hearth, heater lamp, or sputtering discharge. Heat loads up to 100 to 150 W, which are possible in many processes, easily exceed the capacity of the large 35 to 40 W expander stages. To reduce the incident flux on the first stage some form of baffling is necessary. The simplest is a reflective, ambient-temperature baffle. If that is insufficient, a cooled chevron array may be required. Water or liquid nitrogen may be required for baffle cooling. In many instances the manufacturer is unaware of the details of the process and so cannot provide the correct baffling. It is the user's responsibility to ensure that the pump is not thermally overloaded by the process.

Loss of pumping by power failure or gas overload will not harm the pump. A malfunctioning over pressure relief is about the only way that damage can be done to the operator and pump. It will delay operation and may destroy a partially completed experiment, but it will not damage the pump.

Table 19.5   Cryogenic Pumping Concerns

Do	Do Not
Cross over at high pressure to avoid backstreaming	Obstruct safety release
Periodically change oil adsorption cartridge	Allow oil to collect on sorbent stage
Regenerate completely with gas purging	Operate an ion gauge within the pump housing.
Baffle pumping arrays from heated sources	Concentrate dangerous gases

Cryogenic pumps do not handle all gases equally well. The capacities for pumping helium and hydrogen are much less than for other gases. Some gases are easily pumped but they do present safety problems. Because cryosurfaces condense vapors, they can accumulate significant deposits which when warmed are able to react with one another or with the atmosphere. Some combinations can ignite so hot filament gauges should never be used within the pump body. the danger of an explosion or reaction is not lessened by venting flammable or hazardous gases.

High neutron or gamma radiation will subject internal polymeric parts to degradation. It is not recommended that they be used in such applications [30].

High purity helium (99.999%) is required to fill the compressors. Neon is the most common impurity in helium and may condense on the low temperature stage and lead to seal wear. Table 19.5 summarizes some of the problems in cryogenic pump operation.

Cryogenic pumps have their own unique features. When they are understood, they can be reliable and safely applied to many pumping situations.

19.5   HIGH VACUUM CHAMBERS

All-metal high vacuum chambers are almost exclusively constructed from TIG-welded 300 series stainless steel with elastomer-sealed flanges. Viton is the preferred gasket material for high quality systems because of its low outgassing rate and permeability to atmospheric gases. Buna-N is used when cost is a factor and silicone is used in certain high temperature applications. Any metal, glass, or ceramic whose outgassing rate and vapor pressure are adequately low can be used in the chamber, assuming that it is compatible with other materials and with the thermal cycle. The use of elastomers in the chamber should be approached with more caution. Extreme heat, as well as excited molecules from glow discharge cleaning, sputtering, or ion etching, will decompose materials like polytetrafluoroethylene. Whenever possible, a ceramic or glass should be used for electrical insulation. Valves with bellows and O-ring stem seals are used on high vacuum chambers. The bellows stem seal is used in locations such as the high vacuum line, foreline, and roughing line, where a vacuum exists on both sides of the seat. Both sides of this valve are leak tight but the stem side has a larger internal surface area than the seat side. Valves with O-ring stem seals have a high leak rate and are used only for applications like the chamber air release, where the seat side faces the vacuum. Rubber hose roughing lines or vibration isolation sections should be avoided. Rubber hose deteriorates.

19.5.1   Chamber Evacuation

The pumping rate and base pressure of high vacuum systems are limited by the surface (water) desorption rate sketched in Fig. 4.6. The pumping time is not only dependent on the pump size, chamber volume, and internal surface area but also on the state of surface cleanliness. This is especially true when the interior surfaces become covered with deposition residues, some of which are hygroscopic. Consider an electron beam deposition system capable of coating a 3000-cm^2 substrate area. The 0.4-m^3 chamber is pumped at the rate of 2000 L/s at the high vacuum gate valve. The chamber wall area up to the gate valve is 3.9 m^2. Internal fixturing accounts for 1.5 m^2 of surface area and an additional 6.6 m^2 of stainless steel removable liner plates brings the total internal surface area to 12 m^2. The chamber door, feedthroughs, liquid nitrogen trap, and gate valve are sealed with 6 m of Viton gaskets. The equation that describes the time dependence of the pressure in this problem is

$$SP - Q = -V\frac{dP}{dt} \qquad (19.4)$$

The first term on the left is the quantity of gas exiting to the pump per unit time, the second term is the amount of gas per unit time entering the chamber from outgassing, permeation, and leaks, and $-VdP/dt$ is the net rate at which gas is removed from the system. If Q is constant, or at least changes much more slowly with time than the system constant (V/S), as it does for outgassing, we may write the approximate solution as

$$P = P_o e^{-St/V} + \frac{Q}{S} \qquad (19.5)$$

The first term in the solution represents the time dependence of the pressure that is due to the initial gas concentration; the second term represents the contribution of other gas sources. If this term represents outgassing, it is a slowly varying function and is the largest of the two terms after the initial pumping period has passed. After some time the pressure is given by

$$P = \frac{Q_1}{S} t^\alpha \tag{19.6}$$

where Q_1 is the value of Q at some initial time, usually 1 h, and α is often -1. From this equation the pump-down curve in Fig. 19.9 can be constructed. The effect on the base pressure of electropolishing and chemically cleaning the stainless steel and prebaking the O-rings is clearly visible. The variable effects, such as the hygroscopic nature of the film residue and the gas used to release the chamber, cannot be made any more explicit in this example. The amount of water vapor adsorbed by clean stainless steel will rapidly increase with ambient exposure time for the first several hours, after which it will saturate [31]. The saturation time and quantity are a function of the surface cleanliness, nature and temperature. The quantity adsorbed can be reduced by heating the walls and flushing with dry gas while the chamber is open. The gas inlet should be located to allow the gas to flow across the chamber toward the open access port.

Chambers used for routine evaporation are subjected to minimal or no baking. Many chambers are designed with an exterior chamber tubulation that can be heated with (50°C) water to assist in degassing the

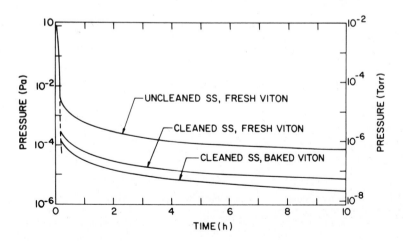

Fig. 19.9 Pumping time for a 0.4-m^3 stainless steel chamber for various steel and elastomer precleaning conditions.

chamber. Partial outgassing may be achieved by the use of interior heating lamps and glow discharge cleaning, but neither can effectively cover 100% of the interior surface area. A surface cooled by liquid nitrogen or an auxiliary cryogenic pump is often necessary to achieve the necessary base pressure in a time commensurate with an efficient production schedule.

After establishing an operating procedure the serious user may choose to record an initial or "clean and dry" pump-down curve for the system and, if possible, record a residual gas spectrum after the system reaches its base pressure. Preservation of these data in a log book will assist in problem solving and requalifying the equipment after routine maintenance. These and some other general operating comments are tabulated in Table 19.6.

19.5.2 Soft Roughing and Venting

Turbulence caused by rapidly removing the air from a chamber or venting to atmosphere too rapidly has been known for some time to be an incredibly large source of particulate contamination. As long as 40 years ago this was observed but not documented [32]. Hoh [33] studied the effect of soft roughing and venting in a silicon monoxide evaporator and observed that a flow corresponding to a Reynolds number less than 1000 did not distribute large amounts of debris. Maeba et al. [34] studied the effect of varying the roughing speed and found that if the time constant (V/S) were kept longer than 400 s, particulates larger than 0.2 μm were not transported to the test wafer. Other measurements have verified particulate transport and have also observed the formation of water vapor droplets upon suddenly exhausting the chamber. At pressures greater than 1 Pa, observed

Table 19.6   General System Operating Concerns

Do	Do Not
Keep a system log	Vent the system with air
Record initial pump-down and rate of rise data	Vent system while Meissner trap is cold
Record clean RGA spectrum	Outgas ion gauge at P > 10^{-3} Pa
Operate ion gauge at reduced emission current	Handle fixturing without clean gloves
Vent continuously with dry gas while chamber is open	Apply vacuum grease to O-rings with bare hands

particles are most likely water droplets. This is an important aspect of maintaining cleanliness in the chamber, but needs further study. Soft roughing and venting combined with the use of load locks on the entrance and exit of the process chamber are two very important mechanisms for maintaining cleanliness.

REFERENCES

1. M. H. Hablanian, *J. Vac. Sci. Technol.*, **6**, 265 (1969).
2. L. Holland, *Vacuum*, **21**, 45 (1971).
3. J. H. Singleton, *J. Phys. E.*, **6**, 685 (1973).
4. Vacuum Barrier Corp., Woburn, MA 01810.
5. J. M. Benson, *Trans. 8th Nat. Vac. Symp. (1961)*, Vol. 1, Pergamon, New York, 1962, p. 489.
6. D. J. Santeler, *J. Vac. Sci. Technol.*, **8**, 299 (1971).
7. B. D. Power and D. J. Crawley, *Vacuum*, **4**, 415 (1954).
8. G. Rettinghaus and W. K. Huber, *Vacuum*, **24**, 249 (1974).
9. For example, see P. M. Danielson and F. C. Mrazek, *J. Vac. Sci. Technol.*, **6**, 423 (1969).
10. M. J. Fulker, *Vacuum*, **18**, 445 (1968).
11. R. D. Craig, *Vacuum*, **20**, 139 (1970).
12. M. A. Baker and G. H. Staniforth, *Vacuum*, **18**, 17 (1968).
13. M. A. Baker, L. Holland, and D. A. G. Stanton, *J. Vac. Sci. Technol.*, **9**, 412 (1972).
14. B. R. F. Kendall, *Vacuum*, *18*, 275 (1968).
15. R. Buhl, *Vacuum-Technik*, **30**, 166, (1981).
16. T. Kraus, F. R. G. Patent No. 1,022,349, Aug. 28, 1956.
17. N. S. Harris, *Vacuum*, **28**, 261 (1978).
18. M. J. Fulker, M. A. Baker and L. Laurenson, *Vacuum*, **19**, 555 (1969).
19. M. H. Hablanian, *Proc. 6th Int. Vac. Congr. (1974)*, Japan. *J. App. Phys.* Suppl. 2, 1974, p. 25.
20. M. H. Hablanian and H. A. Steinherz, *8th Nat. Vac. Symp. (1961)*, Vol. 1, Pergamon, New York, 1962, p. 333.
21. D. W. Jones and C. A. Tsonis, *J. Vac. Sci. Technol.*, **1**, 19 (1964).
22. G. Horikoshi and H. Yamaguchi, *J. Vac. Soc. Japan*, **25**, 161 (1982).
23. J. Hengevoss and W. K. Huber, *Vacuum*, **13**, 1 (1963).
24. J. F. Seibert and M. Omori, *J. Vac. Sci. Technol.*, **14**, 1307 (1977).
25. W. Nesseldreher, *Vacuum*, **26**, 281 (1976).
26. P. D. Bentley, *Vacuum*, **30**, 145 (1980).
27. M. Bridwell and J. G. Rodes, *J. Vac. Sci. Technol. A*, **3**, 472 (1985).
28. R. A. Longsworth and G. E. Bonney, *J. Vac. Sci. Technol.*, **21**, 1022 (1982).
29. R. A. Scholl, *Solid State Technol.*, December, 187 (1983).
30. K. M. Welch, *An Introduction to the Elements of Cryopumping*, K. M. Welch, Ed., American Vacuum Society, New York, 1975, p. III-20.
31. H. Galron, *Vacuum*, **23**, 177 (1973).
32. J. F. O'Hanlon, *J. Vac. Sci. Technol. A*, **5**(4), 2067 (1987).

33. P. D. Hoh, *J. Vac. Sci. Technol. A,* **2,** 198 (1984).

34. Y. Maeba, H. Yoshikawa, H. Yonagida, H. Yamakawa and S. Komiya, *Electrochem. Soc. Exten. Abstr.,* **85-1,** The Electrochemical Society, Pennington, NJ, 1987, p. 158.

PROBLEMS

19.1 †A diffusion-pumped vacuum system uses electropneumatic valves and is operated by an automatic controller. What operations should the controller be capable of performing if each of the following accidents occurred individually while the system is operating: (a) The ionization gauge tube is carelessly broken, (b) the liquid nitrogen supply is inadvertently valved off at its source, or the dewar runs dry, (c) during an electrical storm the line voltage drops by 30% or fails completely, (d) a water main breaks, (e) the compressed air supply fails, (f) the mechanical pump belt breaks, and (g) a water-cooled baffle located directly over a diffusion pump develops a massive leak as a soft-soldered joint in the baffle fails. (h) What transducers should be located on the system to sense these abnormal conditions?

19.2 †Chromium films are deposited in a diffusion pumped system. The procedure calls for loading the cleaned glass plates in the chamber, roughing to 2.5 Pa, closing the roughing valve, and opening the gate valve to the liquid-nitrogen-trapped 20-cm-diameter diffusion pump. The chamber is pumped to 5×10^{-5} Pa and the chrome source is outgassed, and chromium is deposited at a rate of 5 nm/s. The diffusion pump is charged with tetraphenyltrisiloxane and the mechanical pump with mineral oil. It is observed that the chromium is not adhering to the entire substrate surface. It is also observed that the condition is worse after the system has been opened to atmosphere for routine cleaning. Propose a solution to the problem.

19.3 †Water flow sensors are often used to monitor the presence of cooling water flow in turbomolecular pumps. Why should the water flow sensor be installed at the outlet rather than the inlet of a pump? What other sensor could we use instead of a flow sensor?

19.4 The 25-cm turbo pump has a 20-cm-diam valve, whose thickness is 15 cm, in the line between the pump and the chamber. The valve is connected to the pump by means of a 20-cm-diameter tube 15 cm long. An engineer wants to increase the pumping speed at the chamber entrance by replacing the valve and the tube, and enlarge the chamber opening so the pumping path is 25 cm diameter everywhere. The manager doesn't want to spend the money. You are asked to provide a technical evaluation. What do you recommend?

19.5 A 10,000-L/s turbopump is connected to the 500-m³/h lobe blower and 100 m³/h rotary vane pump shown in Fig. 10.9. Is this an adequate combination to back the turbopump for turbo inlet pressures of 0.5 Pa?

19.6 Carbon vane pumps are sometimes used to rough pump chambers before starting the sorption pump to reduce the load on the molecular sieve. They will pump to about 10,000 Pa. Given a 50-L chamber and a sorption pump with a capacity of 10^7 Pa-L of air, how many cycles can we obtain from the sorption pump with and without the use of the carbon vane pump.

19.7 †A diffusion-pumped system has been retrofitted with a helium gas cryogenic pump. Discuss any problems which might ensue.

19.8 A large cryogenically pumped electron beam lithography system is being started. The steel chamber is 2 m × 2 m × 0.5 m. The internal surface area of the tooling contains an additional surface area of 8 m². A 3500-L/s cryogenic pump is appended directly to the chamber with a flange of ratio $l/d=0.05$, and an internal poppet valve. The rough pumping system consists of a 100-cfm lobe blower backed by a 35-cfm rotary vane pump. In order to ensure proper cryopump operation the pumping stages are regenerated by pumping overnight with the mechanical pump running and with the lobe blower stopped. In the morning the compressor is started. After a few hours of operation the ion gauge reads 1.7 × 10^{-4} Pa. A rate of rise measurement with the poppet valve closed shows the pressure to rise from 0.01 Pa to 0.035 Pa in 100 s. The hydrogen vapor pressure thermometer reads 14 K with the poppet valve closed and it rises to 20 K when the valve is opened. You might begin by examining the rate of rise and trying to determine whether there is a leak or this is a normal outgassing load. The operator is concerned because the cold stage temperature is increasing when the load is applied. Is his concern legitimate?

19.9 A small clean stainless steel vacuum system contains a 100-L/s pump and 25 O-rings of 100 mm major diameter × 6 mm chord diameter. Assume a reasonable percentage of the O-ring faces the interior of the system, and calculate a typical pressure in the system after 1 h of pumping because of outgassing from the exposed O-ring surface.

19.10 You are asked to calculate the high vacuum pumping speeds required for a large production roll coater designed for deposition of aluminum on large rolls of plastic film. The rolls of film are 1 m wide and have a total outgassing rate of 1600 Pa-L/s after a short pumping cycle. Proper aluminum adhesion requires a glow discharge cleaning step which adds another 1600 Pa-L/s of gas desorption at the proposed winding rates. A desorption rate of 15 Pa-L/s is measured from a 1-m x 0.5-m sample of the film which has been glow discharge cleaned and is exposed to the lower chamber. The hearth will outgas at the rate of 7.8 Pa-L/s. The film quality is unacceptable if the pressure in the deposition chamber is greater than 1×10^{-2} Pa. The proposed system design is sketched in Fig. 19.10. The mild steel chamber is 2.5 m diameter and 1.5 m long. It is divided into an upper chamber and a lower chamber. Consider the upper and lower parts to have equal internal surface area. The upper half contains a feed roll, a take-up roll, and a glow discharge cleaning apparatus. The lower half contains the hearth. The chambers are divided by the coating roller and the shelf in the middle. The slits separating the shelf and the drum are 75 mm long, have a 1.5-mm gap and extend a total distance of 2.5 m along each side and across the back end of the roll. The slits which separate the front of the coater, containing the three rolls and shelf, from the outer walls are 10 mm long, have a 1-mm gap, and extend a total of 4 m around the two sides and the back of the chamber.

(a) Tabulate the outgassing loads in the upper and lower chambers. Which are dominant? Does the outgassing load of the wall have to be known with great accuracy?

(b) Solve for the high vacuum pumping speeds required at the upper and lower chamber entrances in order to pump each chamber to its required pressure.

(c) What would happen to the pumping speed requirement in the lower chamber, if one of the slit widths were twice as wide as assumed?

(d) What chamber high vacuum pumping speed would be needed if the process were to be done in one chamber (no dividing shelf)?

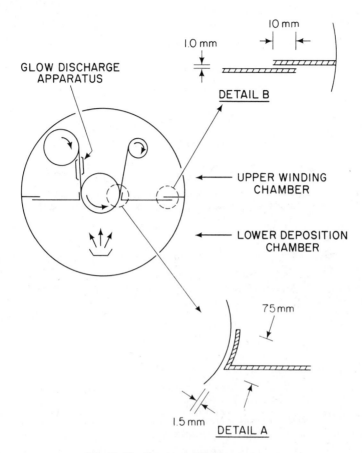

Fig. 19.10   Proposed roll coating system.

CHAPTER 20

Ultrahigh Vacuum Systems

Several fields of science and technology have made remarkable contributions to our understanding of nature, because of ultrahigh vacuum technology. One branch of science that has reaped significant benefits is surface science [1] because it required an environment for the preparation and preservation of atomically clean surfaces. At a pressure of 10^{-9} Pa it takes 50 h to form a monolayer of surface contamination. A technology that will require UHV technique is high density semiconductor device fabrication. Sub-micron structures will require cleanliness levels in processing only available in these systems.

The realization of an ultrahigh vacuum technology necessitated the development of pumps, gauges, materials, and fabrication and sealing techniques appropriate to this region. The first important development was the invention of a practical gauge by Bayard and Alpert [2] which reduced the x-ray limit below 10^{-6} Pa. Before the development of gauges with a reduced x-ray limit it was not possible to determine directly if ultrahigh vacuum pressures had indeed been reached. Development of the B-A gauge was followed by that of the ion pump, the cryogenic pump, and improved materials, components, and system designs.

Today there are commercially available ion-, cryogenic-, turbomolecular-, and diffusion-pumped systems that will attain pressures in the ultrahigh vacuum range. There is also a variety of gauges that can be used in this pressure range, some of which were referenced in Chapter 5. Many of the elements of an ultrahigh vacuum technology were included in more general discussions in earlier chapters.

The most important attribute that distinguishes an ultrahigh vacuum system from any other is cleanliness, that is, the elimination of contamination from all sources. There are many sources and types of contamination in any vacuum system. Gas can desorb from the chamber walls or

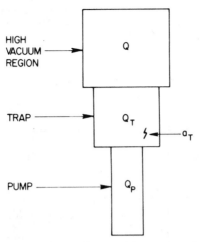

Fig. 20.1 High vacuum pumping system (schematic). Reprinted with permission from *J. Vac. Sci. Technol.*, **8**, p. 299, D. J. Santeler. Copyright 1971, The American Vacuum Society.

evolve from a pump. Gases from either source can be pumped by a cold trap with a probability of being released at a later time. We have chosen to view them after the manner of Santeler [3]. Figure 20.1 shows a generalized vacuum system. The total pressure in the system from vacuum chamber sources is

$$P = \sum_i \frac{Q_i}{S_i}$$

▶ (20.1)

The partial pressure of each gas or vapor desorbing from the chamber is equal to the rate of desorption of that species divided by its pumping speed.

Gases that originate in the pump will flow to the chamber through a trap (if used) in which a fraction of them will be trapped with trapping coefficient a_T. The quantity of gas $Q_p(1-a_T)$ will backstream to the pump where it will be pumped by the pump or the trap, with a resultant pressure contribution to the chamber of

$$P = \sum_i \frac{Q_{pi}\,(1-a_{Ti})}{S_i}$$

(20.2)

If a trap is added between the chamber and the pump, it will be a source of gases that may become partially trapped on other surfaces or flow to the chamber. This contribution to the pressure in the chamber may be expressed as

$$P = \sum_i \frac{K_{ai} Q_{Ti}}{S_i} \tag{20.3}$$

where

K_a depends on the trapping coefficient a_T and the location of the gas molecules in the trap.

The total system pressure results from all of the above sources. It is given by

$$P = \sum_i \frac{Q_i}{S_i} + \frac{Q_{pi}(1-a_{Ti})}{S_i} + \frac{K_{ai} Q_{Ti}}{S_i} \tag{20.4}$$

The essence of Santeler's argument is that the first term behaves in a normal manner (i.e., as the pumping speed increases, the pressure contribution decreases), but the second and third terms do not behave in a like manner. For example, if the trap surface and the pump size are increased, the backstreamed gas load and the pumping speed are increased in proportion. The result is an effective base pressure P_o for a particular pump that is affected only by the trapping efficiency. This base pressure will still exist even if the system does not contain a trap between the pump and the chamber; for example, more hydrogen will backstream from a large ion pump than from a small pump. By this argument (20.4) may be simplified to read

$$P = \sum_i \frac{Q_i}{S_i} + P_o \qquad \blacktriangleright (20.5)$$

The production of ultrahigh vacuum is therefore concerned with the pump and chamber. The contamination from the chamber may be reduced by the intelligent choice of materials, fabrication techniques, and operating procedures; the gas load originating in the pump may be reduced by trapping or processing procedures.

This chapter discusses the respective roles of the chamber and diffusion, turbomolecular, ion, and cryogenic pumps in achieving ultrahigh vacuum. The role of the chamber becomes paramount because the clean and reasonable way to reduce the pressure contribution from the chamber is to reduce Q and not to increase S. Recall that increasing S will not decrease P_o. The selection of the pump is important. Each pump type has its characteristic advantages and disadvantages that make it suitable for some applications and not suitable for others.

20.1 UHV CHAMBERS

The rate limiting steps during the pump-down of a vacuum chamber at ambient temperature were sketched in Fig. 4.6. In a vacuum chamber the volume gas is removed first, followed by surface desorption, out-diffusion from the solid, and last, permeation through the solid wall. All of these processes except volume gas removal are greatly temperature dependent. At room temperature the pressure decreases so slowly that the permeation limit can never be reached on a practical time scale. High-temperature processing is a necessity if ultrahigh vacuum is to be attained after a reasonable pumping time. The materials, fabrication techniques, and seals used for construction of the chamber walls and internal fixturing must be compatible with this thermal cycling.

In the early days, when ultrahigh vacuum was a laboratory phenomenon systems were constructed from glass because it could be easily baked clean. As the complexity and size of ultrahigh vacuum experimental work grew and fabrication techniques improved stainless steel became the preferred construction material. AISI grades 304, 304L, 316, and 316L or a stabilized grade such as 321 or 347 are the most frequently used. The overwhelming majority of the systems fabricated in this country are made from 304, while in England a grade equivalent to 347 is more readily available than 304. In Europe a grade equivalent to 316 is often used. Grade 304L, 321 and 347 steels have the advantage of reducing carbide precipitation which can occur in 304 when welded or subjected to an extended high temperature bake. All joints are made by TIG welding or metal gasketed-flanges; O-rings are not used anywhere in the high vacuum portion of the system, including the high vacuum connection to the pump or trap.

Inside the chamber stainless steel and other metals are used for the fixturing. With the exception of those metals with very high vapor pressures, most metals are suitable as long as they can be outgassed. As discussed in Chapters 5 and 8, some filament materials generate carbon oxides. High density alumina and sapphire are frequently used as insulators because they are stronger than quartz or machinable glass ceramic. The properties of these and other materials are discussed in Chapter 16.

To reduce the outgassing load, the components and subassemblies are first chemically cleaned in the same manner as an unbaked high vacuum system and further heat treated to reduce the outgassing to levels acceptable for these low pressure ranges. Several heat treatments are discussed in Chapter 16. For ultrahigh vacuum use the most thorough of these treatments is selected. The two initial treatments that are the most effective on stainless steel are a vacuum bake at 800 to 950°C and a 400°C air bake. The vacuum bake should be performed at a pressure of 10^{-6} Pa to keep the hydrogen concentration in the metal as low as possible. Nuvolone [4] suggests that the 400°C bake be performed in 2700

Pa of pure oxygen rather than air to avoid any complications that may result from the presence of water vapor during the bake. Metal-to-glass or metal-to-ceramic seals should not be subjected to temperatures higher than 400°C, and sealed copper-gasketed flanges should not be baked at temperatures higher than 450°C. Unsealed flanges, however, will tolerate higher temperatures.

The flange and knife edge will suffer some loss of temper during an 800 to 950°C bake. The exact amount depends on the grade of stainless steels and the fabrication steps used in flange construction. Most commercially available copper-gasketed flanges are forged from grade 304 and some contain trace additives that will retard grain growth during heat treating. Small grain size is necessary to prevent significant loss of hardness. Not all flanges are forged in the same manner or have the same properties. The extremely large project, such as a particle accelerator installation, can afford to write its own specifications and inspect incoming materials, but the average user must analyze and test a few samples for composition and hardness, before and after heat treatment, or rely on the data provided by the manufacturer.

Subassemblies that have been prefired by one of the two aforementioned techniques, can be stored until final assembly for periods of months [4]. It is not recommended that parts be stored in boxes or plastic bags and rubber bands because they will quickly contaminate the clean parts with organics [5]. Aluminum foil is frequently used to cover open flange faces or to wrap cleaned parts until final assembly. Aluminum foil that has been specifically degreased for vacuum use will not contaminate the cleaned parts. Ordinary household aluminum foil can coat the parts with residues of (typically sterate) rolling oils. Clean aluminum foil is commercially available. When the system has been completely assembled, it can be baked under vacuum at 150 to 250°C for 24 h to remove the surface gas [4,6]. This same bake cycle can be used each time the system is exposed to ambient air. If the system is released with dry nitrogen or argon, each succeeding pumping cycle will be shorter than if it had been released with air. It is also advantageous to open the smallest possible port and continue the dry gas purge until the flange is closed. Samuel [7] described an alternative procedure for baking a system in which the components were given an 850 to 900°C vacuum bake. The chamber was heated in atmospheric air at a temperature not exceeding 200°C for 2 h, after which the system was rough pumped. When the system reached 10^{-3} Pa, the temperature was reduced to 150°C until the gauges were outgassed; the heat was then removed.

If the system has been properly cleaned and operated, hydrogen will be the dominant residual gas at the ultimate pressure of a stainless steel system and helium will be the dominant residual gas in a glass system. Usually the situation is somewhat more involved than we have described. A system containing any amount of internal fixturing will have gas trapped between flat surfaces and in blind spots (e.g., around screw threads). Slots are usually cut in the screws and in flat mating surfaces to hasten the exit of trapped gas. A system that makes extensive use of stainless steel bellows may have its ultimate pressure limited by hydrogen permeation through the thin walls. Contaminants on the surfaces of copper gaskets can cause oxidation of the copper after repeated baking and open up minute leak paths through an otherwise rugged ultrahigh vacuum seal.

Unbaked areas of the system, which are usually located near the pump entrance, can be responsible for the largest quantity of residual gas in the system. Consider a system in which 99% of the chamber region can be baked and the remainder only chemically cleaned. A typical baked region has a desorption rate of 10^{-11} W/m^2, while an unbaked region has a desorption rate of 10^{-8} W/m^2. The total gas flux from the unbaked metal surfaces will be 10 times higher than that from the entire baked region.

The area of the system that cannot be baked depends largely on the choice of high vacuum pump. An ion pump and a liquid cryogenic pump can be completely baked. A turbomolecular pump and a gas refrigerator cryogenic pump can be baked at approximately 100°C; a diffusion pump cannot be baked at all. With the exception of the first two pumps, all systems will have some surface that cannot be subjected to a bake at a temperature in excess of 100°C.

Current commercial ultrahigh vacuum systems are limited to the 10^{-9}-Pa range by virtue of the baking and construction techniques. Only systems that are completely immersible in liquid helium can be routinely pumped to the 10^{-13}-Pa range [8] and, interestingly enough, in these low-temperature systems outgassing from the walls and cleaning procedures are less important because everything sticks to the walls. Bills [9] has provided an interesting discussion of the problems that prevent classical ultrahigh vacuum systems from advancing to the 10^{-14}-Pa range. Today UHV technique is finding its way into production lines as a method of reducing contamination. Processes which normally operate at higher pressures, such as sputtering, chemical vapor deposition and reactive ion etching are being performed in systems constructed according to UHV practice with entrance and exit from the process via a load lock. The load lock provides the isolation necessary to introduce and remove samples from the processing environment without introducing contamination, most notably water vapor.

20.2 PUMPING TECHNIQUES

Selection of the pumps for an ultrahigh vacuum system is important. Equation (20.5) demonstrates that an ultrahigh vacuum pump should act to reduce the pressure by providing high pumping speed (low Q/S) and to reduce the backstreaming from the pump to the chamber (low P_o). Here backstreaming is defined as the flow of *any* gas or vapor from the pump to the system. The goal of reducing contamination from the pump is the same for an ultrahigh vacuum system as it is for any other vacuum system. Only the ultimate pressure and permissible level of contamination are lower in these systems. To reduce the contaminant levels rigorous adherence to exacting pumping procedures is necessary. A liquid nitrogen trap that has been warmed for only a few minutes will backstream condensed vapors to the clean region above the trap. Although rechilling will prevent further contamination, it cannot remove the vapors that have accidentally diffused into the clean region. An accident that would be minor in the high vacuum region can destroy the validity of an experiment taking place at low pressures.

Each pump type has its own characteristic class of gases and vapors that it will not pump and that it will generate as impurities. Consequently, pumps and traps are used in combination to provide adequate pumping speed for all gases while minimizing backstreaming. Diffusion pumps are often combined with titanium sublimation traps, cryocondensation and sublimation pumps are used with ion pumps, and turbomolecular pumps are often assisted by sublimators. In general, no two ultrahigh vacuum systems will use the same combination of pumps. Even though commercial pumping systems are available, the design and selection of pumps for ultrahigh vacuum applications is usually done on an individual basis. Today, ultrahigh vacuum pump selection remains a cottage art with many "best" solutions. The remainder of this section reviews the selection and clean operation of some of the many possible configurations of diffusion, turbomolecular, ion, and cryogenic pumps that are suitable for use in the ultrahigh vacuum region.

20.2.1 Diffusion Pumps

Diffusion pumps are traditionally thought of as sources of hydrocarbon contamination in a vacuum system, but they can pump a chamber to the ultrahigh vacuum region when proper traps and procedures are selected. Diffusion pumps do not show a preference among gases and vapors; the exception is hydrogen or helium at low pressures in some pumps. The light gas compression ratio is not the same in all diffusion pumps. A pump with a hydrogen compression ratio of 10^6 and a hydrogen forepressure of 10^{-4} Pa will have an ultimate hydrogen pressure of 10^{-10} Pa. If that value is too great, the diffusion pump must be replaced, backed by another diffusion pump, or the special trapping techniques discussed in this section must be used. This backstreaming, if present, can easily be observed by admitting helium to the mechanical pump exhaust while watching the output of an RGA tuned to helium.

The major contaminants from a diffusion pumped system operating in the ultrahigh vacuum region result from backstreaming mechanical pump fluid through the roughing line, backstreaming diffusion pump oil, and backstreaming the permanent gases (H_2, CH_4, and C_2H_4) that are formed as a result of pump fluid degradation.

Backstreaming mechanical pump fluid vapors can be prevented by any one of several techniques. A molecular sieve trap will reduce but not totally eliminate mechanical pump fluid back diffusion. It will also transfer particulates to the mechanical pump during roughing or during purge pumping and therefore is not recommended. A liquid nitrogen trap designed for the roughing line is the most effective trap. If a liquid nitrogen trap warms, it will allow oil vapors to backstream when the gas is in the molecular flow region. Purge pumping is even more practical than trapping. Santeler [3] has described the technique of purge gas protection illustrated in Fig. 20.2. A purge gas valve is located on the chamber side of the roughing line trap. Whenever the trap is warm, dry nitrogen gas is admitted through the leak. This flow of gas prevents mechanical pump oil from backstreaming through the trap and cleans the trap. The gas purge can be activated by a thermoswitch in the trap. As we noted in Chapter 19, a reverse purge can contaminate the region above the trap equally well. This will happen if a mechanical pump vent is located between the trap and the mechanical pump. The vent should never be located in that position but always between the trap and the chamber.

Purging can also be used in place of a liquid nitrogen trap by admitting the purge gas flow in the chamber, as shown in Fig. 20.2. The leak valve is set so that the mechanical pump cannot exhaust the chamber below 15 to 30 Pa. At crossover the chamber purge valve is closed, the high vacuum valve is opened, and the roughing line purge is begun. In this manner neither the chamber nor the roughing line is in the molecular flow region when the mechanical pump is operating. The third technique for the prevention of roughing line backstreaming eliminates the roughing line altogether and roughs through the diffusion pump as it is starting. Here the concerns are the same as those described for the valveless turbomolecular pump in Chapter 19. A gas purge through the chamber, trap, and diffusion pump will prevent oil from backstreaming until the trap and diffusion pump become operative.

Mechanical pumps are traditionally used to rough out diffusion pump systems, but they are in no way a requirement. One of the simplest and most straightforward techniques for roughing out an ultrahigh vacuum chamber uses the two-stage sorption pumping system described in Chapter 19.

Although a fractionating diffusion pump with a side ejector stage will stop mechanical pump oil backstreaming, it will not stop its working fluid from entering the chamber. High vacuum trapping techniques were

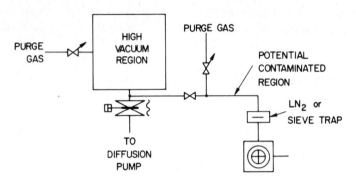

Fig. 20.2 Prevention of hydrocarbon contamination from the roughing pump by use of a purge gas. Adapted with permission from *J. Vac. Sci. Technol.*, **8**. p. 299, D. J. Santeler. Copyright 1971, The American Vacuum Society.

discussed in Chapter 12, but more elaborate techniques are necessary for ultrahigh vacuum. A minimum two-bounce array will reduce oil-gas and oil-oil collisional backstreaming. Recall from Chapter 12 that oil–gas backstreaming has a peak at $\sim 5 \times 10^{-2}$ Pa for the dimensions typical of a 6-in. trap. Because this oil transmission is characteristic of the transition flow regime trap dimensions and pressure, it can be reduced by a trap with two totally different sets of dimensions [10]. The two regions of the trap are in series and have backstreaming peaks at different pressures. At lower pressures each oil molecule makes two collisions and the transmission is proportional to $(1 - \alpha)^2$, where α is the accommodation coefficient for oil on the cooled surface. Minimum two-bounce traps can maintain the oil partial pressure in the system at less than that which can be detected [3,10], which is of an order of 10^{-13} Pa.

Permanent gases that evolve from the pump or trap are also contaminants. In Chapter 19 methods of handling CO_2 reemission from the trap were discussed. Hydrogen, methane, and ethane generated by the decomposition of small amounts of pump fluid in the boiler are permanent gases that are not pumped in the usual sense by a liquid nitrogen trap. Their vapor pressures are so high at 77 K that less than a monolayer of methane, little ethane, and no hydrogen will stick. Some improvement can be obtained by a liquid-nitrogen-cooled titanium sublimation trap between the liquid nitrogen trap and the system [11]. To produce an effective trap, the designer should have a clear conception of the difference between a sublimation trap and a sublimation pump. A pump is simply a large surface with a high pumping speed. To trap these gases effectively, this surface should form a two-bounce array and include a creep barrier [3]. Even with these elaborate precautions the elimination of all H_2 and CH_4 is not possible because their accommodation coefficients on titanium at 77 K are low. Table 20.1 summarizes contaminants from an oil diffusion-pumped system, how they can be eliminated, and some possible resulting pressures.

Despite the ability of diffusion pumps to achieve ultrahigh vacuum pressures, they are not an overwhelming favorite with users of small systems. Their operation demands perfection to remain free of hydrocarbon contamination. Purge pumping can clean organics from the stainless steel walls of an empty chamber [3] but it will not easily remove them from many of the materials under study or used in the construction of experimental fixturing.

20.2.2   Turbomolecular Pumps

The most significant contaminant from a turbomolecular pump is hydrogen. The amount of hydrogen that backstreams from or is not pumped by these pumps is affected by the design of the pump, the kind of trap, the backing pump, and the operating procedures. The ultimate pressure of a high-compression turbomolecular pump will be 5×10^{-9} Pa when backed by a two-stage rotary pump, $\sim 10^{-9}$ Pa when backed by a diffusion

Table 20.1   Contaminants From an Oil Diffusion Pump and How They Are Eliminated[a]

Contaminant	Typical Pressure (Pa)	How Eliminated	Resulting Pressure (Pa)
D. P. Oil	10^{-6}-10^{-7}	Good LN_2 trap	$<10^{-13}$
		Sublimation trap	0
M. P. Oil	10^{-4}-10^{-5}	Change oil	10^{-6}-10^{-7}
		Gas purge or valveless system	$<10^{-13}$
		Sorption pump	0
H_2 from D. P.	10^{-7}-10^{-8}	Sublimation trap, pre bake of pump	$<10^{-11}$
CH_4, C_2H_6 from D.P.	10^{-7}-10^{-8}	Sublimation trap	$<10^{-11}$

Source. Adapted with permission from *J. Vac. Sci. Technol.*, **8**, p. 299, D. J. Santeler. Copyright 1971, The American Vacuum Society.

[a] The pressure values given are representative of a large majority of diffusion-pumped systems but do not cover all possible situations.

pump, and $<10^{-9}$ Pa when a titanium sublimation trap is installed between the pump and the chamber.

For ultrahigh vacuum pumping the ultimate hydrogen partial pressure is limited by the compression ratio of the pump and the hydrogen partial pressure in the foreline. Modern pumps are available with compression ratios between 1500 and 5000. The hydrogen pressure in the foreline can be reduced by the use of a high quality mechanical pump oil or a diffusion backing pump. Even with a diffusion backing pump there will be a finite hydrogen partial pressure at the inlet because hydrocarbon oils lubricate the bearings in the forechamber and elastomers are used for seals between the forechamber and atmosphere. Lange [12] observed random H_2 pressure bursts in a turbomolecular pump when the cooling water temperature was raised above 14°C. The residual hydrogen can be partially trapped in a titanium sublimation trap.

Turbomolecular pumps have a large internal surface area and consequently must be thoroughly baked for the pump to reach its ultimate pressure. Unfortunately the construction of the pump is such that it is not possible to bake it much over 100°C. This bake cycle is necessary to remove the water from the interior of the stator and rotor surfaces.

The roughing cycle can contaminate a turbomolecular-pumped system just the same as any other system. For ultrahigh vacuum the simplest and safest procedure starts the turbomolecular pump and mechanical pump simultaneously and roughs through the turbopump. Alternatively, gas flushing, a sorption pump, or a second small turbomechanical pump set can be used for clean roughing. When properly roughed and baked the turbomolecular pump can pump to the 10^{-9}-Pa range without causing organic contamination.

20.2.3 Sputter–Ion Pumps

Sputter-ion pumps are frequently combined with TSPs to produce a compact, bakeable system. Structurally the pumps are identical to those used in high vacuum systems with the exception of the TSP which must be liquid nitrogen cooled rather than water cooled and the pump is provided with a baking mantle or oven.

The main requirement for use of the sputter-ion pump in the ultrahigh vacuum region is thorough baking. Assuming that the chamber and components have received an appropriate high temperature prebake, a 250°C bake of the entire system, including the pump, is adequate. The temperature may be increased slowly to 250°C after the pressure reaches 10^{-5} or 10^{-6} Pa. A baking time of 10 to 20 h is typical. The ion pump will continue to operate provided that its pressure is below 10^{-4} Pa. If the system contains an additional pump, for example a turbomolecular pump for differentially pumping an RGA, it can be used to reduce the gas load to the sputter-ion pump during baking. After baking is complete, the TSP liquid nitrogen surface may be cooled. The water vapor load in a baked ultrahigh vacuum system is much smaller than it is in an unbaked rapid-cycle system; therefore the sublimed titanium film will not entrap enough water to make titanium flaking a problem.

The main background gas present in a sputter-ion pumped system at low pressures is hydrogen, but other gases may be observed. Previously pumped gases may be reemitted and the presence of impurities in the titanium cathodes and TSP filaments can cause the generation of carbon oxides, methane, and ethane.

Sputter-ion-pumped systems will pump routinely to the 10^{-9}-Pa range; they are most useful for applications with small gas loads, but are not suitable for applications that require a constant pumping speed over a wide pressure range or in gas sampling systems in which reemitted gases can hopelessly confuse the measurements.

20.2.4 Cryogenic Pumps

Both liquid- and gas-refrigerator-cooled pumps are able to reach the ultrahigh vacuum region. Liquid-helium and liquid-nitrogen-cooled pumps have attained pressures of 10^{-10} to 10^{-12} Pa [8,13,14], while gas refrigerator pumps can reach the mid-to-low 10^{-8} Pa range.

The adsorption isotherms of helium and hydrogen on molecular sieve are strongly temperature dependent in the ultrahigh vacuum region. Because the second stage in a liquid cooled pump operates at 4.2 rather than 10 to 15 K, it has a much lower ultimate pressure than a gas refrigerator pump. A most important consideration in reaching very low pressures is the isolation of thermal and optical radiation from the cold stage. Liquid nitrogen-cooled baffles have been designed to maximize molecular transmission and minimize photon transmission [13,14]. Intermediate baffles located between the 4.2- and 77-K stages, which are cooled to 20 K by the liquid helium boil-off, have also been used [15].

The bonded molecular sieve construction [16,17] used in the 4.2-K stage allows liquid cooled pumps to be baked to temperatures as high as 250°C. Gas refrigerator cooled pumps cannot be baked above 70 to 100°C because their construction makes use of indium gaskets. In both cases there is oil-free rough pumping, and the pumps are exhausted by auxiliary pumps during baking. Sputter-ion or turbomolecular pumps are used for the latter purpose.

Cryogenic pumps easily pump large amounts of hydrogen, have no high voltages, and generate no hydrocarbon, metal film flakes, or other contaminants of their own. Gas refrigerators pump all gases well, with the exception of helium. Helium is pumped well only if the adsorbent is cooled to the 4 to 8 K range, a job best done by a liquid helium pump. Liquid pumps are completely free of magnetic fields and vibration. Gas refrigerator pumps have some vibration that results from the displacer motion. They must be carefully damped for use with sensitive surface analysis equipment such as ESCA or SIMS. Gas refrigerator pumps have the advantage of not requiring liquid cryogens. Becker [18] discusses the operation of a gas refrigerator cryogenic pump in an ultrahigh vacuum molecular beam epitaxy system. He has shown that it is a viable alternative to conventional pumping systems.

REFERENCES

1. P. A. Redhead, *J. Vac. Sci. Technol.*, **13**, 5, (1976).

2. R. T. Bayard and D. A. Alpert, *Rev. Sci. Instrum.*, **21**, 571 (1950).

3. D. J. Santeler, *J. Vac. Sci. Technol.*, **8**, 299 (1971).

4. R. Nuvolone, *J. Vac. Sci. Technol.*, **14**, 1210 (1977).

5. T. Sigmond, *Vacuum*, **25**, 239 (1975).

6. J. R. Young, *J. Vac., Sci. Technol.*, **6**, 398 (1969).

7. R. L. Samuel, *Vacuum*, **20**, 295 (1970).

8. W. Thompson and S. Hanrahan, *J. Vac. Sci. Technol.*, **14**, 643 (1977).

9. D. G. Bills, *J. Vac. Sci. Technol.*, **6**, 166 (1969).

10. N. Milleron, *Trans. 5th Nat. Vac. Symp. (1958)*, Pergamon, New York, 1959, p. 140.

11. R. D. Gretz, *J. Vac. Sci. Technol.*, **5**, 49 (1968).

12. W. J. Lange and J. H. Singleton, *J. Vac. Sci. Technol.*, **15**, 1189 (1978).

13. C. Benvenuti and D. Blechschmidt, *Japan. J. Appl. Phys.*, Suppl. 2, Pt. 1, 77 (1974).

14. H. J. Halama, and J. R. Aggus, *J. Vac. Sci. Technol.*, **12**, 532 (1975).

15. R. J. Powers and R. M. Chambers, *J. Vac. Sci. Technol.*, **8**. 319 (1971).

16. G. E. Greiner and S. A. Stern, *J. Vac. Sci. Technol.*, **3**, 334 (1966).

17. P. J. Gareis and S. A. Stern, *Bulletin l'Institut International du Froid*, Annexe 1966-5, p. 429.

18. G. Becker, *J. Vac. Sci. Technol.*, **14**, 640 (1977).

PROBLEMS

20.1 Table 1.1 classifies the low, medium, high, very high and ultrahigh vacuum ranges. Describe the suitability of the following materials in each range: brass, cadmium, nickel-plated copper, household aluminum foil used to wrap cleaned parts, aluminum or copper castings, copper sheet, aluminum sheet, quartz, alumia ceramic, Buna-N, nylon, Teflon, polyimide, and Viton. Assume materials used in the construction of ultrahigh and extreme high vacuum systems are baked to temperatures as high as 400°C, and those used in the construction of high vacuum chambers may be baked to 70 to 100°C.

20.2 A small vacuum chamber with an internal surface of 0.1 m² is pumped to its ultimate pressure by a pump of speed 100 L/s. The net outgassing rate of the inner surface is reported to be 3×10^{-11} W/m^2. (a) What is the ultimate pressure? (b) If the sticking coefficient of the surface is 0.1, what is the true outgassing rate of the surface?

20.3 Carefully estimate the outgassing rate of a fingerprint.

20.4 We are admitting oxygen into an ultrahigh vacuum system. The RGA shows a small oxygen peak and a large CO peak. The oxygen peak is much smaller than it should be. We suspect conversion on the hot filaments and turn off the ion gauge and reduce the emission in the RGA. The CO peak is still large and the oxygen peak is still small. What is happening?

20.5 Oxygen is admitted to an ion-pumped ultrahigh vacuum system. The pump current increases by a factor of 2 and the pressure as indicated by a Bayard-Alpert gauge increases by a factor of 10. What is happening?

20.6 †After a rapid system backfill with argon, flakes from a titanium sublimation pump have been blown into the ion pump and the power supply is shorted. What do you do?

20.7 †After exhausting a chamber with one sorption pump, the pressure stabilizes at 10 Pa. The system has been leak checked and found to be leak free. What is causing the pump to stop at this pressure?

20.8 Why should a pressure gauge be the last item to be cooled when baking a vacuum system?

20.9 (a) Calculate the rate of pressure rise, due to helium in the atmosphere, in a sealed 7740 Pyrex glass system with a wall area of $1\ m^2$ and a wall thickness of 1 mm which is initially pumped to 10^{-10} Pa? Assume the permeation constant for 7740 glass is $10^{-14}\ m^2/s$ at $25°C$. (b) Calculate the pumping speed necessary to hold the chamber at 10^{-10} Pa, if the permeating helium is the only gas source.

20.10 In a surface analysis system samples are introduced through a load lock on a long cylindrical rod. The rod passing through the load lock is isolated at the atmospheric side by an O-ring seal and at the UHV side by a long closely fitting cylinder which passes through the vacuum wall. Assume the rod is 1 cm diameter, the gap between the rod and the cylinder is 3×10^{-3} cm, and the cylinder is 8 cm long. (a) Calculate the gas leak into the main chamber through this load lock seal if the base pressure is 1×10^{-8} Pa in the main chamber and 1×10^{-4} Pa in the load lock. (b) If the main system pump has a speed of 400 L/s, what fraction of its gas load is entering via the concentric seal at the above base pressure?

CHAPTER 21

High Flow Systems

Not all thin-film deposition processes require high or ultrahigh vacuum environments. In fact, some of the most interesting processes take place in the medium and low vacuum range and in addition require a high gas flow. The pressure–pumping speed ranges for several processes as they are currently commercially practiced are shown in Fig. 21.1.

Sputter deposition is done in the 0.5 to 10 Pa pressure range. For certain materials sputtering is the preferred deposition technique. Various plasma processes are performed in the 5 to 500 Pa range. Plasma-deposited films are formed from the reaction of chemical vapors in the glow discharge, and plasma etching and reactive-ion etching are of current interest. Polymer films are formed from the glow discharge polymerization of a monomer such as styrene. Plasma etching is a simple isotropic chemical etching process that uses chemically active neutrals in the discharge; for example, the plasma decomposes CF_4 and creates fluorine atoms that react with a silicon surface to form SiF_4, a volatile product that is pumped away. Reactive ion etching is a directional process useful in fabricating semiconductor microstructures. Its directionality is due to high energy ions that are accelerated through a potential gradient toward the surface on which it is believed that, by some mechanism, they increase the reactivity of the chemically active neutral species with the unmasked portions of a thin film. Because the ion-stimulated neutral reaction proceeds at a rate tenfold faster than simple plasma etching, the thin film etches downward much faster than laterally with little undercutting. This allows the etching of fine lines. Reactive-ion etching can be performed over the entire range of pressures used for sputtering and plasma etching. Any differences attributed to pressure are differences in nomenclature rather than theory. Low pressure chemical vapor deposition (LPCVD) and reduced pressure epitaxy are thermal processes that

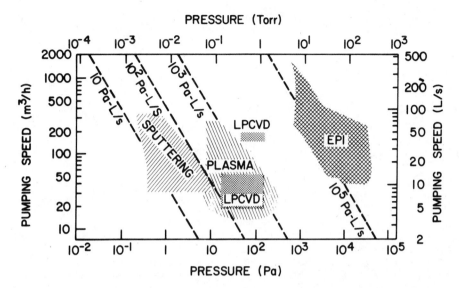

Fig. 21.1 Pressure–speed ranges for some thin-film growth, deposition, and etching processes that require medium to low vacuum and gas flow. Adapted with permission from M. T. Wauk, Applied Materials Inc., 3050 Bowers Avenue, Santa Clara, CA 95051.

take place at low pressures. The thermal energy is typically provided by induction heating. LPCVD, which is done in the 50 to 100 Pa range, has attracted wide attention. The high diffusivity of the thermally active species at low pressure improves the transport of the vapor throughout the reactor and allows the growth of more uniform films on a greater number of larger wafers than is possible with atmospheric pressure CVD. Reduced pressure epitaxy takes place in the 500-Pa to atmospheric pressure range. Epitaxial films grown at reduced pressure are higher in quality and have less autodoping than films grown at atmospheric pressure.

Many of the processes done in the medium or low vacuum range also require the use of vapors that are toxic, hazardous, or corrosive. Special precautions must be taken in the design, operation, and maintenance of these systems to ensure operator safety and equipment protection. This is a subject that is advancing so rapidly as to make it unsuitable for inclusion in a text. An idea of the current state of the art can be obtained from the AVS Recommended Practice for Pumping Hazardous Gases [1].

These thin-film deposition and etching processes span a pressure and gas flow range that far exceeds the capability of any one pump. The pressure range of each process is dictated by the physics of the process. Sputtering, for example, cannot commence until the pressure is high

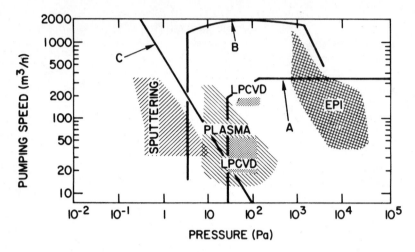

Fig. 21.2 Useful pressure–speed ranges for some pumping systems: (*a*) rotary mechanical pumps; (*b*) Roots pump backed by a rotary pump; (*c*) throttled high vacuum pump. Adapted with permission from M. T. Wauk, Applied Materials, Inc., 3050 Bowers Avenue, Santa Clara, CA 95051.

enough to initiate a self-sustained glow discharge, but the pressure must be low enough for the sputtered material to reach the anode without suffering a large number of gas collisions. Gas flows ranging from 10^1 to 10^6 Pa-L/s are needed for a different purpose in each of these processes. In some processes the high flow dilutes or replenishes the reactant species and simultaneously flushes away the products of reaction and other impurities; in others it mainly serves to flush away impurities.

The pressure–pumping speed operating ranges of the mechanical pump, Roots pump, and throttled high vacuum pump are quite different. Rotary vane or piston pumps can pump to the sputtering pressure range and still retain some pumping speed, but below 15 Pa a typical small roughing line is in free molecular flow and backstreaming a large amount of oil vapor is allowed. Rotary vane or piston mechanical pumps are economical up to speeds of 200 to 300 m³/h and provide effective pumping for all process-es in the region delineated by speeds lower than this value and pressures greater than 15 to 20 Pa. This region is bounded by curve A in Fig. 21.2. For speeds greater than 200 to 300 m³/h a Roots pump backed by a rotary piston pump is necessary. Again oil backstreaming limits the lowest pressure of operation. At low pressures the roughing line will be in free molecular flow and will allow mechanical pump oil to back diffuse to the Roots pump outlet, creep around the interior surfaces, and enter the process chamber. A molecular sieve trap could be used to reduce the oil backstreaming from the mechanical pump but it generates particulates that contaminate the mechanical pump; it also traps water that will react with certain process gases.   Oil contamination may be eliminated by bleeding gas into the roughing line to maintain the Roots pump outlet at a pressure greater than, say, 15 to 20 Pa.   This limits the inlet pressure of most Roots–rotary pump combinations to about 3 to 5 Pa [see Fig. 10.8 and the discussion following (10.1) and (10.2)].   The Roots pump has an upper pressure limit of ~1000 Pa which yields the useful operating region outlined by curve B in Fig. 21.2.   The only pump systems that will operate at pressures below a few pascals without backstreaming oil vapors are throttled high vacuum pumps.   Diffusion, turbomolecular, or cryogenic pumps will maintain chambers at sputtering pressures when the inlet to each is throttled to a pressure below its respective critical inlet pressure. Curve C, in Fig. 21.1 sketches the upper throughput limit of a typical small, throttled high vacuum pump.

As we noted in Chapter 20, cleanliness requirements for low contamination of semiconductor and magneto-optical disk production force these systems to be designed and constructed according to UHV practice.   To reduce overall contamination to the part per billion level, the ratio of background or chamber contamination to process gas pressure must be kept extremely small.   Entrance and exit load locks are used to isolate the environment, and gases of high purity will be delivered to the system through gas distribution systems that will be UHV leak tight.   Many of the gases used in these processes are corrosive, explosive or poisonous. Designs which focus on UHV construction techniques are resulting in cleaner systems with reduced maintenance.   For example, the reduction of moisture levels to the part per billion level reduces the corrosion that results from reaction with anhydrous gases such as chlorine.

This chapter reviews throttled high vacuum systems and unthrottled medium and low vacuum systems.   Throttled high vacuum pumps are used mainly for sputtering and ion etching.   Medium and low vacuum systems are used for ion etching, plasma deposition, LPCVD, and reduced pressure epitaxy.

21.1   THROTTLED HIGH VACUUM SYSTEMS

The pressure range encompassed by sputtering and other plasma processes is above the operating range of all high vacuum pumps.   These pumps can be used in the 0.5 to 10 Pa range by placing a throttle valve between the pump and the work chamber.   This throttle valve with its low conductance allows gas to flow from the high pressure chamber to the pump while keeping the pressure at the pump entrance below its maximum or critical inlet pressure.   A typical sputtering chamber for a 150 to 200-mm-diameter cathode is 500 mm in diameter and 250 mm high.   The traditional pumping plant contained a 6-in. diffusion pump, but cryogenic or turbomolecular pumps are now common.   For pumps of this size the maximum throughput will be limited to about 100 to 200 Pa-L/s; larger pumps, although more expensive, are capable of removing gas at a faster rate.

Residual gases pose a greater problem in a sputtering system than in a high vacuum evaporation system because of enhanced plasma desorption of impurities from the walls and because typical sputter deposition rates are much lower than typical evaporation rates. Even when the two processes yield rates of the same order the sputtered films have a greater exposure to residual gas impurities than have films condensed from evaporating sources. Electron and ion impact desorption are efficient at releasing gases from the chamber walls. They are even more effective than mild baking and in the plasma the desorbed gases are likely to exist in the atomic state, where they can easily react with the sputtered film. If hydrogen, for example, is not removed from an argon discharge, the sputtering rate will be reduced [2] and hydrogen will become incorporated in the film [3]. Sputtering discharges with argon or other noble gases can be kept clean by operating the discharge in a static mode with selective pumping or by flowing a large amount of argon through the chamber during sputtering. A static discharge is maintained by exhausting the chamber and refilling it with argon to the operating pressure while other gases are selectively removed by an auxiliary pump located within the chamber. One problem is that the ideal auxiliary pump does not exist. The closest approximation, the titanium sublimator, generates some methane and does not pump noble gas impurities. For these reasons static discharges are not frequently used for sputtering. Here, as for plasma processes with reactive gases, the one reliable technique for impurity removal is a continuously flowing clean source gas which will flush away gases evolved from walls and the process.

Viscous flushing will work only if the gas flows through the active sputtering region and chamber and if the arrival rate of impurities from the gas source is much less than the desorption rate from the chamber walls. Lamont [4] has pointed out that high throughput alone does not guarantee adequate flushing; it is necessary that the gas stream velocity be large in the region in which the cleaning action is desired. Contamination originating from within the chamber can be reduced with high gas flow. In the high flow limit the lowest possible level of contamination attainable is that of the source gas. In critical applications the source gas is scrubbed by passing it through a titanium sublimation pump. Because of this, there is little point in flushing a chamber of the size described above at a rate greater than a few hundred pascal-liters per second with the purest available source gas [5]. Both the source gas cleanliness and the gas flow rate are important to the maintenance of conditions suitable for deposition of pure films.

Gas flushing alone will not adequately clean all the residual gases from the chamber. Contamination-free sputtering requires high vacuum pumping to a suitable base pressure [6] followed by presputter cleaning with the discharge operating and with the shutter covering the samples on which a film is to be deposited. Shirn and Patterson [7] have noted that the flushing time of a typical 6-in. diffusion pumped system is so small (1/7 s) that the system can be cleaned by simply pumping to the process pressure and initiating the glow discharge without pumping to high vacuum. Unfortunately the glow does not clean all surfaces adequately, nor do the surfaces outgas that rapidly. The continued evolution of water vapor from surfaces not exposed to the glow can cause oxygen or hydrogen contamination of deposited films. Most importantly, the only routine way to check for minute leaks in the system without the use of an RGA is to pump to the same low base pressure each time before opening the leak to argon flow. Presputtering cleans by several techniques. If the sputtered material is a getter, it may be allowed to deposit on the chamber walls where it is an effective getter pump [8]. The sputtered material also covers adsorbed gases, whereas the discharge cleans the cathode and other surfaces exposed to the glow.

The value of base pressure and the length of presputtering time are process, equipment, and material dependent. No general observations can be made; for example d'Heurle [9] showed that for aluminum films a 10-min presputter cleaning was sufficient, while Blachman [10] required a minimum of 1 h cleaning time for molybdenum. Some material properties are so critically dependent on film purity that the value of base pressure and minimum presputtering time are of crucial importance in the repeatable fabrication of uniformly high quality films. In these systems, the use of entrance and exit load locks is mandatory.

The conclusion of this discussion is that high vacuum pumps are needed to establish the initial cleanliness, while throttled, high compression pumps are required for pre sputter cleaning and contaminant removal from the chamber without permitting backstreaming of hydrogen or a pump fluid. These two requirements can be satisfied by the same pump. Two completely different pumping systems could be used, a high vacuum, low-throughput system for initial cleaning and a medium vacuum system for high gas flow [4], but it would be difficult to design a medium vacuum system with the required hydrogen compression ratio.

The remainder of this section discusses the configuration and operation of diffusion, turbomolecular, and cryogenic pumps for sputtering applications. Ion pumps are not considered because of their inability to handle high gas loads.

21.1.1 Diffusion Pump

The maximum throughput of a diffusion pump is the product of its critical inlet pressure and its pumping speed at this pressure. The critical inlet pressure is determined by the pressure at which the top jet begins to fail. Typically this happens near 0.1 Pa in a 6-in. diffusion pump. If the pump inlet is maintained at a higher pressure, the pumping action will become unstable and pressure control difficult. Backstreaming may also increase in the region of jet instability. At 0.1 Pa a 6-in. pump has a speed of about 2000 L/s. Its maximum throughput is therefore 200 Pa-L/s. See, for example curve C, Fig. 21.2. A 10-in. diffusion pump with a speed of 5000 L/s at 0.1 Pa has a maximum throughput of 500 Pa-L/s. The maximum throughput is a property of the pump, and is affected by the amount of heat supplied to the boiler by the heater and by the type of fluid. Figure 12.5 illustrates the throughput dependence of the heater power. The maximum throughput will be reduced if the heater power is decreased or if a section of the heater becomes open-circuited. It can be exceeded if the forepump has inadequate capacity. The heater power was chosen by the pump designer to heat a selected pump fluid to a temperature that will produce the desired maximum throughput in the boiler. Use of a fluid other than that for which the pump was designed will cause the critical forepressure to change; for example, a pump that was designed to run with DC-704 but used with DC-705 or Santovac 5 will have a reduced critical forepressure unless the heater power is increased. The pump manufacturer should be consulted to determine the critical forepressure for the fluid and pump in question. This is a concern only in systems that are operated near the forepressure limit. For this reason oversized forepumps are often used to back diffusion pumps used for high gas flow applications.

The maximum throughput of the pump is not determined by the series conductance of the trap, high vacuum valve, and throttle valve; it is a property of the pump. The inlet gas leak valve controls the throughput, whereas the throttle valve controls the chamber pressure. Because the throughput at any point in a series flow path is constant, the pumping speed at the chamber under throttled conditions can be calculated as

$$S_c \; = \; \frac{P_p S_p}{P_c} \; = \; \frac{Q_p}{P_c} \qquad (21.1)$$

A 2000-L/s diffusion pump will have a throttled pumping speed, measured at the base plate of $\sim$ 100 L/s for a chamber pressure of 2.0 Pa. The maximum speed of the pumping system is naturally affected by the series conductances of the trap and opened valves and the pump can operate at its maximum speed only when the inlet pressure is less than the critical inlet pressure. Diffusion pumps with expanded inlets will have greater speed in the high vacuum region but cannot pump any larger gas quantity than a straight-sided pump of the same boiler diameter. The increased speed of the expanded inlet pump will decrease the time required to reach the base pressure only slightly, because the slow surface release of gas molecules limits the removal rate of the outgassing species.

The one question that has not been adequately resolved is the optimum location of the throttle valve in the pump stack. The two possible locations of the throttle are illustrated in Fig. 21.3. It may be placed below the liquid nitrogen trap directly over the throat of the diffusion pump (Fig. 21.3a) or it may be located upstream from the liquid nitrogen trap, between the liquid nitrogen trap and the chamber (Fig. 21.3b). The arrangement shown in (a) allows the trap to pump water vapor. At sputtering pressures the trap is in transition, or perhaps viscous flow, and its conductance for pumping water vapor is considerably more than it is in the free-molecular-flow range, although much of the water vapor does not strike the cooled surface because of vapor-gas collisions. Even though this is not the most efficient location, the trap pumping speed for water vapor is maximized when placed over the throttle.

The trap should capture oil vapors at a greater rate when placed below the throttle because the probability of an oil-trap collision is greater at low pressures where the mean free path is longer than it is at high pressures. The sweeping action of the gas flow from the sputtering chamber is constant and cannot exceed the maximum throughput; therefore as the trap pressure is raised the trap's effectiveness as an oil baffle is reduced by the short mean free path but not enhanced by any increased gas streaming. From a backstreaming consideration the location of the throttle over the liquid nitrogen trap will yield a lower net backstreaming rate provided that the added baffling action of the throttle is the same in both cases. This is not normally so, especially when the throttle is not cooled. If an ambient temperature throttle is located over the trap, any oil vapor condensed on the throttle may escape into the chamber, whereas for a throttle located between the trap and the pump the evaporation

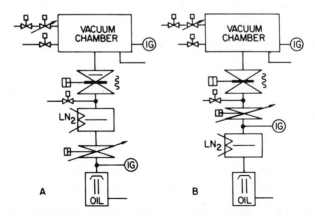

Fig. 21.3 Throttle valve location in a high throughput diffusion pump stack: (a) throttle located in between the liquid nitrogen trap and the pump, (b) throttle located between the liquid nitrogen trap and the chamber.

may be small compared with the primary backstreaming rate [11]. Rettinghaus and Huber [12] have suggested that the pump may be effectively throttled by a cooled throttle plate located near the cooled cap of the top jet, where it can intercept the primary backstreaming.

No general conclusions in regard to the best arrangement for minimizing backstreaming can be drawn from these considerations. The circumstances of each application must be considered. In many sputtering applications, however, a large water-vapor pumping speed is essential and takes precedence over the differences between two already small backstreaming rates in choosing to locate the throttle below the liquid nitrogen trap.

Occasionally throttled diffusion pumps will be used to pump reactive gases. In those situations a silicone or perfluoropolyether fluid should be used because hydrocarbons will react with many gases. Certain gases will condense on the liquid nitrogen trap and either affect the process or become a safety hazard.

21.1.2 Turbomolecular Pump

A turbomolecular-pumped system suitable for high flow applications must attain an adequate base pressure as well as exhaust a high gas flow at medium vacuum pressures. The two requirements are different but not conflicting. In Chapter 19 we discussed the requirements of a turbomolecular pump for good high vacuum pumping: high compression ratio for light gases, high pumping speed for all gases, and an LN_2-cooled trap for increasing the water-vapor pumping speed. The selection of a pump for a particular application is then made on the basis of pumping speed and compression ratio data like those in Figs. 11.4 and 11.6. These data describe the speed of the pump and the compression ratio for each gas with zero flow. These data, necessary for characterizing the pump at high vacuum, are inadequate for describing the pump's performance during high gas flow.

High gas flow imposes other constraints on the pump in addition to the obvious one of pumping a considerable quantity of process gas, as in sputtering. The pump must also maintain a reasonably high compression ratio and pumping speed for hydrogen while pumping a high throughput of a heavy gas such as argon. A flow of argon that is large enough to cause viscous flow in the blades or to reduce the rotational velocity will alter the compression and speed characteristics of the pump. Few data are available to show how the design or the operation of the pump are affected by high gas flow.

Visser [5] has measured the compression characteristics of methane in a turbomolecular pump as a function of the quantity of the argon gas flow. His results, shown in Fig. 21.4, demonstrated that a flow of argon from 2 to 8 Pa-L/s increased the compression ratio for methane in a 250-L/s pump. Data at higher argon flow rates have not been obtained.

Figure 21.5 shows the pumping speed for hydrogen and the argon flow rate as a function of argon gas pressure [13]. The argon flow is the product of the argon pumping speed and the inlet pressure. The pumping speed for hydrogen remained constant for argon inlet pressures up to 0.9 Pa, above which it dropped precipitously. The sudden drop in hydrogen pumping speed corresponded to a similar sharp decrease in the rotational speed of the turbomolecular pump blades in this pump. The exact pressure at which the rotational speed begins to decrease is a function of the pumping speed of the forepump. As the inlet pressure increased, the gas flow in the blades closest to the forechamber changed from molecular to transition and then to viscous. Near the onset of viscous flow the added frictional drag demanded more torque from the motor; the constant power motor responded by losing speed. The knee of curve A in Fig.

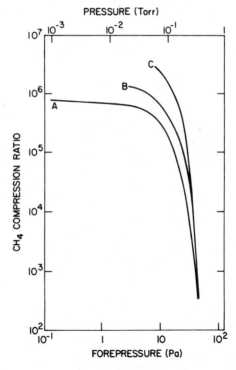

Fig. 21.4 Methane compression ratio as a function of argon gas flow in a Balzers 250-L/s turbomolecular pump: (a) $Q(Ar) = 0$, (b) $Q(Ar) = 2$ Pa-L/s, (c) $Q(Ar) = 8$ Pa-L/s. Adapted with permission from *Trans. Conference and School on Elements and Techniques and Applications of Sputtering*, Brighton, p. 105. Copyright 1971, Materials Research Corp., Orangeburg, NY 10962.

21.5 will move slightly to the right for a forepump larger than the 35-m²/h pump used here, and to the left for a smaller forepump. When using the pump described here, it is important to keep the pump running at full rotational velocity to maintain adequate hydrogen pumping speed. The blades will run at full velocity as long as the inlet pressure is suitably throttled; for example, the pump used to take the data in Fig. 21.5 should be throttled to an inlet pressure of 0.9 Pa or less when backed by a 35 m³/h mechanical pump. This pump has an inlet argon speed of 225 L/s at an inlet pressure of 0.9 Pa of argon when backed by the 35 m³/h mechanical pump; it yields a maximum argon throughput of 200 Pa-L/s, the same value that can be obtained with a 6-in. diffusion pump. The numerical values recorded for this pump are not directly applicable to any other. Some turbomolecular pumps are driven by constant-speed rather than constant-power motors. Even so, the pressure at which the argon speed decreases is dependent in a similar manner on the size of the forepump.

The maximum inlet pressure is a function of the size of the forepump and the design of the turbomolecular pump. To obtain the maximum suitable throughput from the pump the staging ratio, or ratio of the turbomolecular pump speed to the forepump speed should be small, perhaps as low as 20:1 or 30:1.

Figure 21.6 shows a turbomolecular pump configuration suitable for a sputtering application. A high pressure ionization gauge of the Schulz-Phelps type is located at the pump inlet, below the throttle valve, to monitor the inlet pressure. The gauge tube should be mounted off-axis and its entrance protected by a wire mesh. The gas flow can be increased and the throttle adjusted to maintain the chamber at the desired pressure and the pump inlet pressure should be low enough to allow the

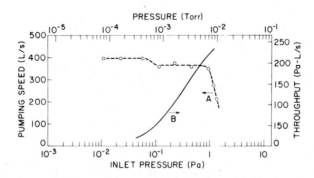

Fig. 21.5 Hydrogen pumping speed (*a*), and argon throughput (*b*) as a function of argon inlet pressure in a Balzers 400-L/s turbomolecular pump backed by a 35-m³/h rotary vane pump. Reprinted with permission from *J. Vac. Sci. Technol.*, **16**, p. 724, J. F. O'Hanlon. Copyright 1979, The American Vacuum Society.

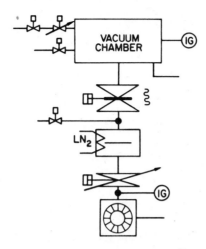

Fig. 21.6 Component location in a turbomolecular pumped sputtering system.

pump to run at full rotational velocity. A liquid nitrogen trap for pumping condensables is shown below the high vacuum valve. Alternatively, it
may be placed in the sputtering chamber as a Meissner trap, where it will
have increased pumping speed but will require cyclical venting.

The turbomolecular pump is well suited for use as a high flow, medium
vacuum pump, provided that it is throttled to keep the blades running at
full velocity and is backed by an adequately large mechanical pump. For
large turbomolecular pumps (1000 L/s and larger) this will usually mean
use of a Roots pump and a suitably large foreline. When operated in this
manner, it will have a high pumping speed and a high compression ratio
for light gases. The turbomolecular pump thus serves as a one-way baffle
for light gases and hydrocarbons, whereas most of the pumping is done
by the forepump. The turbomolecular pump is not well suited for pumping on high-pressure plasma polymer deposition systems because material
may deposit on the rotors, unbalance them, and destroy the pump [14].

21.1.3 Cryogenic Pump

Like turbomolecular and diffusion pumps, the cryogenic pump should
be capable of evacuating the chamber to an adequate base pressure and
pumping a large gas flow. The ultimate, or base, pressure is determined
by chamber outgassing and the temperature and history of the cold stage.
In Chapter 15 we discussed how the temperature was determined by a
balance between the refrigeration capacity and the heat loads. The
temperature alone was not the only factor that determined the pumping
speed for a gas. It was found to be a function of the nature and quantity
of gases previously sorbed on the cryosurface.

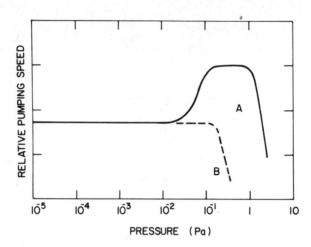

Fig. 21.7 Pressure dependence of cryogenic pumping speed: (*a*) free surface, (*b*) practical baffled pump.

It was also observed that the heat load carried to the pumping surfaces by the incoming gases under high vacuum conditions (low gas through-put) was insignificant in comparison to the radiant flux. If nitrogen was pumped with a typical cryogenic pump consisting of two cooled stages, one at 80 K, and the other at, say, 20 K, then the time to build up a condensed layer of solid nitrogen 1 mm thick would be about 10^4 h at a pressure of 10^{-5} Pa [15]. Therefore neither the heat load of the incoming gas nor the resulting solid deposit is a major concern in the high vacuum region. This is not so at high gas flows. As the gas flow to a cryogenic pump is increased, the pumping speed changes. Figure 21.7 sketches the pressure dependence of the pumping speed over several flow regions. In the free molecular flow region the pumping speed is constant. At some-what higher pressures the speed increases due to the increased conduc-tance as the gas enters the transition flow region. Under some circum-stances this flow will reach a maximum value (choked or critical flow) that is characteristic of the sonic velocity of the gas. At higher pressures the heat conductivity of the gas becomes large and heat from the walls of the chamber flows to the cooled surfaces by gas collisional energy trans-fer. As these surfaces warm the sticking coefficients decrease and pump-ing ceases. This behavior has been observed by Dawson and Haygood [16] for CO_2 and by Bland [17] for water vapor. Loss of pumping speed in a practical pump usually occurs at about 0.2 to 0.4 Pa, where the heat loads exceed the capacity of the refrigerator. The pumping speed of a typical pump in the high pressure region is sketched curve B in Fig. 21.

At high gas throughputs the major heat flow to the cryogenic surface is carried by the gas molecules; for example an argon flow of 180 Pa-L/s

Table 21.1   Enthalpy of Gases Frequently Pumped
at High Flow Rates[a]

Gas or Vapor	Total Enthalpy in Vacuum kJ/(kg-mole)		
	300 K	80 K	20 K
Ar	13,950	9,370	**88**
N_2	15,580	9,190	**134**
CCl_4	49,750	**2,950**	738
CF_4	26,000	16,000	**670**
CF_3H	30,750	**2,678**	670
CF_3Cl	31,300	**20,230**	678

[a]Approximate total enthalpy at ambient temperature and at the nominal temperatures of the first and second stages of a cryogenic pump for several gases and vapors used in high gas flow applications. The enthalpy is shown in bold for the surface on which the gas or vapor solidifies.

corresponds to an incident heat flux of 1 W on a 20-K surface. For gases such as nitrogen which have heats of condensation greater than argon the heat flux will be proportionately larger. Most of this heat flux for nitrogen or argon will be absorbed by the cold stage. As an example consider a two-stage cryogenic pump with surfaces at 20 and 80 K, respectively, in which the argon collides with the 80-K baffle, passes through the baffle, and finally collides with the 20-K surface, where it is pumped. For each mole of argon that flows into the pump a total of 13,862 kJ must be removed by the expander. This value is obtained by taking the difference in enthalpy between 300 and 20 K. See Table 21.1. If all of the argon were to be cooled to 80 K on impact with the 80-K baffle, a total of 4580 kJ/(kg-mole) would be removed. This value corresponds to 33% of the total heat that is removed during the pumping process. In practice the gas is not cooled to 80 K on impact with the warm stage because the accommodation coefficient is not unity. Adequate cooling does take place, however. If, for example, the warm stage had an efficiency of 0.5 for cooling the argon, then 85% of the total heat of the gas would remain to be removed by the 20-K stage. The cold stage removes at least 67% of the heat from argon because its vapor pressure is so low and its heat of condensation is so large. Table 21.1 gives the approximate total enthalpy at ambient temperature at the nominal temperatures of the first and second stages of a cryogenic pump for argon and nitrogen, and for some gases frequently used in reactive ion etching. The maximum throughput for each gas is determined by the maximum power that can be removed by the stage on which the gas or vapor condenses. Cryogenic pumps do not show the same maximum throughput limits for all gases. CF_4 has a maximum throughput of half the value quoted by the manufacturer for argon. Several other gases used for reactive-ion etching deposit on the 80-K stage and can completely close the baffles and quickly render the pump useless for high vacuum pumping unless it is first regenerated.

At high flow the heat absorbed by the expander will be proportional to the total gas throughput until the pressure is large enough for heat conduction from the walls to be appreciable. Heat conduction from the walls begins at a Knudsen number of 1 and reaches its maximum or high pressure value at Kn = 0.01. This heat flow is added to the heat flow due to heat of condensation but not in a linear manner. Most of the heat delivered by gas conduction flows to the 80-K stage. This added load on the warm stage reduces the refrigeration capacity of the cold stage. For high gas flow applications a cryogenic pump should be designed to minimize the heat load from the walls so that the expander may be most efficiently used to remove the heat of condensation of the incoming gas. For a pump of typical dimensions the Knudsen number for air at 0.2 Pa is about 0.3. At this value the heat capacity is a substantial fraction of the high pressure value which allows several watts to flow from the chamber walls to the warm stage. To reduce this heat load the warm stage is completely insulated except for the baffled entrance to the cold stage or surrounded with an added liquid nitrogen shroud. See Fig. 21.8. In either case the heat flow to the warm stage is reduced. The best insulation material is one whose surface has a low emissivity. A liquid nitrogen shroud provides an alternative sink for the conductive heat flow and substantially eliminates this load on the warm stage.

In some high gas flow situations continued pumping of hydrogen is important but does not always take place. Pumps are designed so that

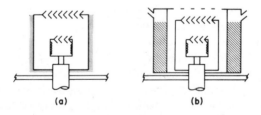

Fig. 21.8 Sectional view of (a) gas refrigerator cryopump with insulation around the warm stage, and (b) gas refrigerator cryopump with a liquid nitrogen shroud surrounding the warm stage.

the inner surface of the cold stage is baffled from the warm stage and covered with a layer of activated charcoal. This surface will pump hydrogen, neon, and helium effectively if it is adequately cooled. At 10 K there is adequate sorption capacity for hydrogen on activated charcoal or molecular sieve. Argon will also be pumped on this surface. To avoid or reduce the argon pumping on the inner surface it is baffled from the remainder of the system. If the baffle is optically dense to keep argon reaching the inner surface, the hydrogen pumping speed will be reduced, and if the baffle is open the argon will readily pass through, condense on the inner surface, and cover the sorbent with solid argon. All pump designs are a compromise between these two concerns. In a high flow application, however, the sorbent surface will become coated with argon rather quickly, regardless of the nature of the baffling. Once the sorbent becomes coated, cryotrapping is the only mechanism by which hydrogen can be pumped. Hengevoss [18] has shown that the cryotrapping of hydrogen in argon is strongly temperature dependent and becomes nil for solid argon temperatures greater than 20 K. The value of argon throughput which will keep the second stage temperature below 20 K may be below the stated maximum value for a particular pump.

The constraints placed on a cryopump system for high gas flow are considerably different than those placed on pumps used for high vacuum. The system requires a throttle valve to keep the pressure in the cryopump below a value of 0.2 and 0.4 Pa and the pump must be able to handle the conductive heat load from the walls. One way of reducing this heat contribution is to use a two-stage gas refrigerator of moderate capacity surrounded by a liquid nitrogen baffle, while another design involves only the use of a large capacity gas refrigerator and insulation. Regeneration of a high gas flow system will obviously be more frequent in a pump used for high gas flow than in a pump used for high vacuum, but automatic controllers are available for performing this function on idle time. If a cryopump is used at the design-limit argon throughput, the cold stage will be heated to a temperature at which it will not pump hydrogen or helium; either supplemental pumping or throughput reduction will be necessary to pump these gases. Furthermore, the pump's maximum throughput is dependent on the vapor pressure, specific heat, and heat of sublimation of the gas or vapor being pumped. Cryogenic pumps also suffer from the phenomenon of overloading. Irreversible warming of a pump can be triggered by a gas burst entering a pump that is running near its throughput limit. It will then cease to pump and require regeneration. Cryogenic pumping of toxic or explosive gases presents serious safety concerns. If the pump were to suddenly warm, a large quantity of gas could be emptied into the exhaust system. If the pump is condensing an explosive gas, operation of an ionization gauge in the pump body during release of the gas presents a serious problem. The ion gauge tube should be located outside the pump body and interlocked so that its filament cannot be operated when the gas is being released. A prudent operator would not choose cryogenic pumping for certain gases. These concerns should be understood before cryopumps are chosen for high gas flow applications.

REFERENCES

1. J. F. O'Hanlon and D. B. Fraser, *J. Vac. Sci. Technol. A*, **6**(3), 1226 (1988).
2. E. Stern and H. L. Caswell, *J. Vac. Sci. Technol.*, **4**, 128 (1967).
3. J. J. Cuomo, P. A. Leary, D. Yu, W. Reuter, and M. Frisch, *J. Vac. Sci. Technol.*, **16**, 299, (1979).
4. L. T. Lamont, *J. Vac. Sci. Technol.*, **10**, 251 (1973).
5. J. Visser, *Trans. Conference and School on Elements and Techniques and Applications of Sputtering*, Brighton, November 7–9, 1971, p. 105.
6. L. I. Maissel, *Physics of Thin Films*, Vol. 3, G. Hass and R. E. Thun, Eds., Academic, New York, 1966, p. 106.
7. G. A. Shirn and W. L. Patterson, *J. Vac. Sci. Technol.*, **7**, 453 (1970).
8. H. C. Theuerer and J. J. Hauser, *Appl. Phys.*, **35**, 554 (1964).
9. F. M. d'Heurle, *Metall. Trans.*, **1**, 625 (1970).
10. A. G. Blachman, *Metall. Trans.*, **2**, 699 (1971).
11. The author is indebted to Dr. G. Rettinghaus of Balzers High Vacuum for this discussion.
12. G. Rettinghaus and W. Huber, *Vacuum*, **24**, 249 (1974).
13. J. F. O'Hanlon, *J. Vac. Sci. Technol.*, **16**, 724 (1979).
14. J. Vossen, Dry Etching Seminar, New England Combined Chapter and National Thin Film Division of AVS., Oct. 10-11, 1978, Danvers, MA.
15. G. Davey, *Vacuum*, **26**, 17 (1976).
16. J. P. Dawson and J. D. Haygood, *Cryogenics*, **5**, 57 (1965).
17. M. E. Bland, *Cryogenics*, **15**, 639 (1975).
18. J. Hengevoss, *J. Vac. Sci. Technol.*, **6**, 58 (1969).

PROBLEMS

21.1 †Figure 21.3*a* shows a system with a leak valve between the gas tank and the chamber, and a throttle valve between the chamber and the pump. Which valve can independently control the chamber pressure, and which valve can independently control the gas flow?

21.2 †Would you use plastic tubing to connect a gas tank to a leak valve on a sputtering system?

21.3 The 1500-L/s high vacuum pump on a sputtering system has been replaced with a 3000-L/s cryogenic pump. The operator pumps the system to a predetermined base pressure and initiates the sputtering discharge. Since the pump size has been increased, the system pumps to the base pressure in less time. What problems will this cause in the performance of the sputtering system?

21.4 Given the following compatible equipment: gas sources, pressure gauges with electrical output, valves with servo operated controllers, and thermal mass flow sensing elements with electrical outputs. (a) Connect the components to control flows Q_1 and Q_2 of two gases into a vacuum chamber such that the total pressure is controlled by the majority gas Q_1, while the minority constituent remains constant. (b) Connect the equipment in such a way that the ratio of the major to minor gas flow remains at a predetermined value, and the system pressure is recorded.

21.5 In problem 21.5 above, a large burst of gas entered the chamber from wall desorption. Describe qualitatively the transient behavior of (a) the chamber pressure, (b) the flow meter and (c) the throttle valve.

21.6 How would you optimize the blade design of a turbopump for pumping argon? Would such a pump have a high speed for hydrogen?

21.7 A gas line contains gas flowing at the rate of 10,000 cc/min and is leak tight to a level of 10^{-5} T-L/s. (a) What is the contamination of the gas due to the leak? (b) What leak tightness would be required to keep this same contaminant level with only 1 cc/min of gas flow?

21.8 Assume we have a sputtering chamber of 100 L volume and internal surface area of 1 m^2. Assume a normal outgassing rate for clean stainless steel and determine what flow of argon is necessary to keep the outgassing contaminant level below 1 ppm.

21.9 Which pump will produce the cleanest vacuum for a sputtering system, a diffusion pump, a turbomolecular pump or a cryogenic pump?

21.10 What are the relative pumping speeds and throughputs of the pumps shown in Fig. 21.9? Assume the drawings are to scale and the power inputs to each boilers are equal.

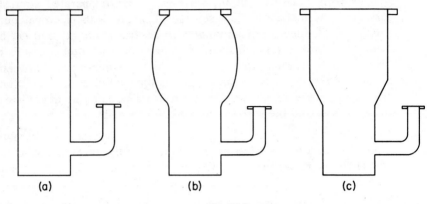

Fig. 21.9

CHAPTER 22

Leak Detection

At some time we will be confronted with a system that does not behave normally—behavior that could be the result of a component malfunction, initial outgassing, or leak. At another time we may have to check out a new system. When and how to hunt for leaks are two useful skills discussed in this section. This leak detection discussion is based largely on the response of a mass spectrometer to a tracer gas flow through a leak in a vacuum wall.

The decision to search for leaks is as important as the method chosen for their detection. Each new component or subassembly should be routinely leak tested on a helium mass spectrometer leak detector after welding or brazing. It is premature, however, to leak-test a new system a few hours after placing it in operation simply because its performance does not meet the user's expectations. New systems often pump slowly because of outgassing of the fixturing, seals, or fresh pump fluids. The patient operator will usually wait a few days before criticizing the base pressure. At that time it is useful to take an RGA scan of the system background or to leak-test the system.

Searching for a leak in an established system is straightforward when a history of the system is known. A well-documented log book assists the operator in determining the cause of the poor performance. A log book should contain information such as the system pressure versus pumping time, base pressure, rate of rise, maintenance history, and perhaps a background RGA scan.

The sensitive step in the leak checking procedure is done with a mass spectrometer leak detector (MSLD) or residual gas analyzer. The mass spectrometer leak detector is a mass spectrometer permanently tuned to helium with a separate, self-contained pumping system. The operation of the mass filter was described in Chapter 8. The two most important parameters in leak detection with both instruments are sensitivity and response time.

22.1 LEAK DETECTOR SENSITIVITY

Either the RGA or the MSLD is sensitive to a threshold partial pressure of the tracer gas used to probe the leak. In the best case the minimum detectable partial pressure is the absolute sensitivity of the instrument above the background noise. In a typical operating system a residual background pressure of the tracer gas exists because it is regurgitated from a pump, back diffuses, or is released from a trap surface. This residual tracer gas pressure may be considerably greater than the ultimate detectability of the instrument and significantly increases the minimum detectable tracer pressure. The minimum leak flux Q_{min} which can be detected is given by $Q_{min} = P_{min} S$. The leak detector is capable of measuring all leaks that produce a tracer gas partial pressure greater than P_{min} in a system in which the total pumping speed is given by S.

The minimum leak flux Q_{min} is distinct from the size of the leak. A large tracer gas pressure drop across a small leak conductance can give the same flux as a small tracer gas pressure drop across a large leak conductance. The minimum detectable leak conductance thus depends on the sensitivity of the instrument, the background tracer gas pressure in the detector, the external pressure of the tracer gas, and the speed of the pumps attached to the chamber. All of these parameters can be optimized to increase leak detection sensitivity. In practice external tracer gas pressures are usually atmospheric, the instrument sensitivity is fixed, and the background tracer gas pressure is dependent on the tracer gas, the type of pump, and the system. Helium is the most frequently used tracer gas. In the molecular flow regime the ratio of helium flow to air flow through a leak [(3.19)] is $Q(He)/Q(air) = (M_{air}/M_{He})^{1/2} = 2.74$. For laminar viscous flow this advantage no longer holds true. From (3.12) we see that $Q(He)/Q(air) = (\eta_{air}/\eta_{He}) = 0.92$. Signal strength is not essential in the detection of large viscous leaks, however. If the leak detecting is done with an RGA, a tracer gas other than helium may reduce the background. Increasing the external tracer gas pressure is not an effective technique because the molecular leak flux is only linearly proportional to pressure. With a commercial helium MSLD, the only variable that can be changed easily to improve the sensitivity is the pumping speed.

The maximum detectable sensitivity is reached at zero pumping speed. Closing the valve between the pump and chamber allows the leaking tracer gas to accumulate in the chamber. This is easily accomplished when leak detecting with an RGA and with MSLD units that are equipped with a valve between the ionizer and self-contained pump. For molecular gas flow the partial pressure of the tracer gas will increase linearly with time at zero pumping speed. After time t_1 a detector will measure partial pressure P_1; the leak flux will be given by $Q = P_1 V/t_1$. This technique for increasing the basic sensitivity is called the accumulation technique. Small volumes may also be tested effectively by throttling the MSLD to a low but nonzero value. In this technique known as foreline sampling, S is reduced to increase the sensitivity but some pumping speed is retained to reduce the system background pressure. No increase in sensitivity is realized by this technique for tracer gases such as oxygen which may also desorb from the walls. Desorption increases the background pressure and negates any increase in sensitivity.

The minimum leak flux detectable by an RGA or MSLD is approximately 10^{-9} Pa-L/s [1]. Under clean conditions the accumulation technique is sensitive to helium leaks as small as 10^{-11} Pa-L/s [2]. Leaks as small as these, which were measured across layers of glass and thin-film interconnection metallurgy were soon hydrolized shut by the moisture in the air. In normal vacuum system operation leaks less that 10^{-8} Pa-L/s are rarely found [3].

22.2 LEAK DETECTOR RESPONSE TIME

The maximum sensitivity of the leak detector can be realized only if the tracer gas has had time to reach the steady-state value. For a system of volume V evacuated by a pump of speed S, the pressure change due to a sudden application of a tracer gas to a leak is given by

$$P_t = P_{to} + \frac{Q_t}{S}(1 - e^{-St/V}) \tag{22.1}$$

where P_{to} is the background pressure of the tracer gas. At time zero, the pressure of the tracer gas in the system is P_{to}. The pressure slowly builds to a steady-state value. 63% of the steady-state pressure is reached in a time equal to the system time constant V/S, and five time constants are required to reach 99% of the response. This means that a 100-L system pumped by a leak detector with a speed of 1 L/s will require application of the tracer gas for at least 100 s to realize the maximum sensitivity of the instrument. This is the case when an MSLD is connected to a chamber that has been valved from the high vacuum pump (see Fig. 22.1a). The time constant can be reduced by placing the pumps in parallel (Fig. 22.1b) but the sensitivity is also reduced. If the leak detector can handle the gas load, the fast time constant can be retained in turbomolecular and diffusion pumps without loss of sensitivity by placing the leak detector in the foreline and valving the mechanical pump from the system (see Fig. 22.1c). This technique may be used if the gas flow is too large for the

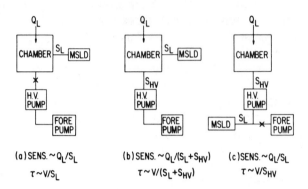

Fig. 22.1 Methods for mass spectrometer leak detecting a vacuum chamber pumped by a diffusion or turbomolecular pump. Methods a and c have a high sensitivity and methods b and c have a short response time to the influx of a tracer gas.

leak detector by allowing the leak detector to pump at its maximum flow rate while pumping the remainder of the gas with the forepump. In this manner the fast time constant is retained and the sensitivity is reduced only by the ratio of gas flows to the forepump and leak detector.

Leak detection in an ion- or cryogenic-pumped system is different than in a turbomolecular- or diffusion-pumped system because the first two pumps have no forepumps. In the latter two systems the MSLD must be appended directly to the chamber, where the high speed of the cryo or turbopump causes a loss of sensitivity. An ion pump may be helium leak checked by momentarily removing power to the pump. A cryogenic pump cannot be shut down for leak checking because the evolved gas load will overload the pump in the MSLD. If the temperature in the cold stage is increased to 20 K, the helium pumping of the sorbent bed will drop to zero and the leak detector can be operated at its maximum sensitivity. The pump must be equipped with a heating element on the cold stage for this purpose. Moraw and Prasol [4] have discussed the optimum conditions for leak detecting large space chambers. If the cold stage is not heated during leak detection helium will accumulate. The next time the pump is exposed to an air load the helium will desorb. A sufficient quantity of desorbed helium will thermally short the pump. As we discussed in Section 19.4.2, no amount of pumping will help; complete regeneration is required.

22.3 PROCEDURES

It is not possible to give a definitive or complete procedure for leak detecting a vacuum system but it is helpful to review some methods. A thorough understanding of the system under test is invaluable. The system volume and pumping speed must be known in order to calculate the system time constant.

A system that cannot be pumped below the operating range of the roughing pump has a gross leak or a malfunctioning pump. First close all the valves in the system and observe the pressure in the mechanical pump with a thermal conductivity gauge. In an unvalved system the mechanical pump should be disconnected and blanked with a flange containing a thermocouple gauge. Poor pressure at the inlet to the blanked-off roughing pump may be a result of low oil level, oil contamination, or an internal difficulty such as a sticking vane or exhaust valve. A pump used on corrosive gases may be badly etched or contaminated. If the pump is operating properly, sections of the foreline and roughing line can be pumped sequentially until the leaky section is isolated. Helium may be sprayed around suspected seals and welds while listening for a change in the sound of the motor. Alternatively, alcohol sprayed on a leak causes a large upward deflection of the thermal conductivity gauge. Helium sprayed on a leak in a sorption pump roughing line causes a large increase in pressure.

If the system pumps to the high vacuum range but cannot reach its usual base pressure, there may be a leak. A leak is not the only reason for poor performance. There may also be a faulty high vacuum pump, a leak on the pump side of the gate valve, a contaminated gauge, or considerable outgassing. Begin by closing the high vacuum gate valve and observing the downstream pressure. A low pressure on the pump side of the gate valve tells us the pump is operating properly. If the blank-off pressure is not acceptable, there could be a leak in the pump or a faulty pump. A leak detector can be attached to the foreline of a diffusion or turbomolecular pump or to a flange adjacent to a cryogenic or ion pump. A valve in the foreline (Fig. 22.1c) allows the leak detector to be attached without removing power from the pumps.

When the system has been shown to be leak free from the mechanical pump to the top of the high vacuum pump, the gate valve to the chamber may be opened. Poor chamber pressure may result from outgassing, external leaks, or internal leaks. One simple way to distinguish between leaks and outgassing is to examine a plot of the system pressure versus time after closing the high vacuum valve. See Fig. 22.2. A leak causes a continual increase in pressure that is linear in time for a molecular leak. Outgassing causes the pressure to rise to a steady-state value that is determined by the vapor pressures of the desorbing species. If the system contains an RGA, a quick scan tells us whether the poor performance is due to an atmospheric leak or outgassing, although outgassing of water vapor may be difficult to distinguish from a leak in a water line.

External leak checking with helium should begin at the top of the chamber; only a small flow rate is necessary. In some cases it may be

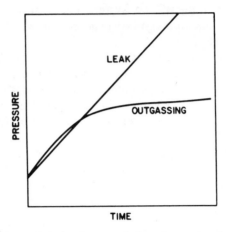

Fig. 22.2 Response of a sealed chamber to a leak and to outgassing from interior walls. Adapted with permission from *Handbook of Vacuum Leak Detection*, N. G. Wilson and L. C. Beavis, Eds. Copyright 1976, The American Vacuum Society.

necessary to wrap plastic around an area and flush it with nitrogen to prevent helium from entering more than one potential leak site. Alcohol freezes in a small leak and allows adjacent areas to be checked without confusion. The alcohol can be removed with a heat gun. The most obvious places, such as welds and seals should be checked first. Welds and seals are the most common leak sites. However leaks have been observed in plates and flanges. See Chapter 17. These leaks through voids are usually masked with grease and appear only after baking. The helium background pressure may increase slowly with time when a search is made for small leaks with the MSLD in systems sealed with elastomer O-rings. This pressure rise is brought about by helium permeation [5]. The permeation time is about 1/2 to 1 h for most O-rings. The helium background will not decrease until it has been pumped from the gaskets.

Interior water lines are difficult to check because water may slow the diffusion of helium through the leak. They are best checked by draining, connecting directly to the leak detector, and warming with a heat gun before spraying with helium from the interior of the chamber.

Interior leaks in liquid nitrogen traps are difficult to locate. Leaks have been observed when LN_2 has been added to the trap, or suspected when a large mass 28 (and 14) spectrum is seen on an RGA. Trap leaks may be verified by bubbling helium through the liquid nitrogen and looking for its appearance on an RGA or MSLD. The trap will have to be removed from the system and leak checked with an MSLD to pinpoint the leak. Sometimes the source of a large mass 28 peak is not obvious. Nitrogen is often used to release chambers to atmosphere. A through leak in the bleed valve connecting the dry nitrogen line to the chamber cannot be detected by spraying the external walls with helium. Internal valve seat leaks in this line can be located by pumping the nitrogen from a section of this line and replacing it with helium.

The RGA allows the user to determine the state of the system with little additional data. Air leaks and water outgassing are easily differentiated. Air leaks are discerned by the presence of oxygen at mass 32, except in large TSP systems in which the oxygen pumping speed is so large that the oxygen signal will not be seen, and in baked UHV systems where it is adsorbed on the stainless steel walls. Outgassing and water-line leaks each can produce a large mass 18 peak, but they can be distinguished by the rate of rise. The RGA has the added advantage of functioning with tracer gases other than helium. Oxygen is often used for leak checking sputter-ion pumped systems and argon for TSPs. Additional operational hints are given in the AVS leak detection handbook [4].

The helium mass spectrometer leak detector has been universally accepted for sensitive leak detection of vacuum components and systems. The instrument is most valuable if it has an internal calibration standard and valves for throttling the internal pumping speed (accumulation technique) and the inlet flow (foreline sampling technique). An important consideration in selecting a commercial instrument is ease of operation. Operators are reluctant to use a machine whose operation is difficult. Most instruments can be disconnected for a few minutes while being transported to the test site without going through a complete diffusion pump shut-down procedure, and completely portable instruments that do not use liquid nitrogen are commercially available. Although it is the most common and sensitive leak detection method, mass spectrometric leak detection is not the only technique. Guthrie [6], Santeler [7], and McKinney [8] discuss leak detection with the ion gauge, the halogen leak detector, and other techniques.

REFERENCES

1. L. C. Beavis, *Vacuum*, **20**, 233 (1970).

2. J. F. O'Hanlon, K. C. Park, A. Reisman, R. Havreluk, and J. G. Cahill, *IBM J. Res. Dev.*, **22**, 613 (1978).

3. N. G. Wilson and L. C. Beavis, *Handbook of Leak Detection*, W. R. Bottoms, Ed., American Vacuum Society, New York, 1976.

4. M. Moraw and H. Prasol, *Vacuum*, **28**, 63 (1978).

5. J. R. Young, *J. Vac. Sci. Technol.*, **7**, 210 (1970).

6. A. Guthrie, *Vacuum Technology*, Wiley, New York, 1963, p. 456.

7. D. J. Santeler et al., *Vacuum Technology and Space Simulation*, NASA SP-105, National Aeronautics and Space Administration, Washington, D.C., p. 275.

8. H. F. McKinney, *J. Vac. Sci. Technol.*, **6**, 958 (1969).

PROBLEMS

22.1 Describe a procedure for calibrating a helium leak detector.

22.2 †Why should connecting lines to the leak detector be as short and as large in diameter as possible?

22.3 Why should you turn off the high voltage, e.g., on a sputtering system or ion-pumped system, before spraying a feedthrough with helium? detector and the part or system to be tested? Explain your choice.

22.4 †When using a MSLD to locate a leak, the helium should be allowed to remain at the leak site until the response: (a) increases to its maximum amount, or (b) it reaches about 4 or 5 times the background signal level?

22.5 †A mechanically pumped CVD reactor is "leak checked" by pumping overnight and monitoring the base pressure the next morning. Is this a good procedure?

22.6 A combination of a narrow, shallow O-ring grove and a large O-ring diameter, all within tolerances, can lead to trapped gas pockets on the inner and outer diameters at the base of the groove. A measurement of base pressure and/or rate of rise will suggest a leak. Helium leak detecting from the vessel exterior will show no leak even if there is a leak path across the base of the O-ring. How would you modify the flange to eliminate this problem?

22.7 Calculate the leak rate of a human hair lying across an elastomer O-ring gasket whose nominal diameter is 6 mm. Use the method of Santeler described in Section 3.3.3.

22.8 Why don't we see oxygen when we have an air leak in a TSP-ion-pumped system?

22.9 Why can gas ballasting in a mechanical pump be used to differentiate between a leak and oil contaminated with a high vapor pressure impurity?

22.10 A cryopumped chamber has a suspiciously high base pressure. All of the gaskets are helium leak checked. After considerable searching, the leaking flange is located. It is tightened, the leak disappears, and the pressure drops to an acceptable level. Next an adjoining load lock is evacuated to 1 Pa with a separate pump. The connecting valve to the main cryopump is then opened. The pressure in the main chamber suddenly rises, and the cryopump ceases to operate. Describe what happened. Give a procedure for correcting the situation.

Symbols

Symbol	Quantity	Units
A	Area	m²
B	Magnetic field strength	T (tesla)
C	Conductance (gas)	L/s
D	Diffusion constant	m²/s
E_o	Heat transfer	J-s⁻¹-m⁻²
F	Force	N (newton)
G	Electron multiplier gain	
H	Heat flow	J/s
K	Compression ratio (gas)	
K_p	Permeability constant	m²/s
Kn	Knudsen's number	
K_R	Radiant heat conductivity	J-s⁻¹-m⁻¹-K⁻¹
K_T	Thermal conductivity	J-s⁻¹-m⁻¹-K⁻¹
M	Molecular weight	
N	Number of molecules	
P	Pressure	Pa (pascal)
Q	Gas flow	Pa-m³/s
R	Reynolds' number	
S	Pumping speed	L/s
S'	Gauge sensitivity	Pa⁻¹
T	Absolute temperature	K (kelvin)
U	Average gas stream velocity	m/s
U	Mach number	
V	Volume	m³
V_a	Acceleration potential	V
V_b	Linear blade velocity	m/s
W	Ho coefficient	
a	Transmission probability	

b	Turbomolecular pump blade chord length	m
c	Condensation coefficient	
c_p	Specific heat at constant pressure	J-(kg-mole)$^{-1}$-K^{-1}
c_v	Specific heat at constant volume	J-(kg-mole)$^{-1}$-K^{-1}
d	Diameter dimension	m
d_o	Molecular diameter	m
d^1	Average molecular spacing	m
i_e	Emission current	A
i_p	Plate current	A
l	Length dimension	m
m	Mass of molecule	kg
n	Gas density	m^{-3}
q	Outgassing rate	W/m^2
q_k	Permeation rate	W/m^2
r	Radius	m
s	Turbomolecular pump blade spacing	m
s_r	Turbomolecular pump blade speed ratio	
u	Local gas stream velocity	m/s
v	Average particle velocity	m/s
Γ	Particle flux	m^{-2}-s^{-1}
Λ	Free molecular heat conductivity	J-s^{-1}-m^{-2}-K^{-2}-Pa^{-1}
α	Accommodation coefficient	
β	Molecular slip constant	
γ	Ratio c_p/c_v	
ε	Emissivity	
λ	Mean free path	m
η	Dynamic viscosity	Pa-s
ρ	Mass density	kg/m^3
ω	Angular frequency	rad/s
ϕ	Turbomolecular pump blade angle	degrees

Appendixes

APPENDIX A

Units and Constants

Appendix A.1   Physical Constants

k	Boltzmann's constant	1.3804×10^{-23} J/K
m_e	Rest mass of electron	9.108×10^{-31} kg
m_p	Rest mass of proton	1.672×10^{-27} kg
N_o	Avogadro's number	6.02252×10^{26}/(kg-mole)
R	Gas constant	8314.3 J-(kg-mole)$^{-1}$-K^{-1}
V_o	Normal specific volume of an ideal gas	22.4136 m^3/(kg-mole)
σ	Stefan-Boltzmann constant	5.67×10^{-8} J-s^{-1}-m^{-1}-K^{-4}

Appendix A.2   SI Base Units

Length	meter	m
Mass	kilogram	kg
Time	second	s
Electric current	ampere	A
Thermodynamic temperature	kelvin	K
Amount of substance	kg-mole	kg-mole

Appendix A.3   Conversion Factors

Conventional unit	→ multiply by →	to get SI unit
Mass		
lb	0.45359	kg
Length		
micrometer	0.000001	m
mil	0.00254	cm
inch	0.0254	m
foot	0.3048	m
angstrom	1.0×10^{-10}	m
Area		
ft^2	0.0929	m^2
in.^{2}	6.452	cm^2
ft^2	929.03	cm^2
Volume		
cm^3	0.001	L
in.^{3}	0.0164	L
gal (US)	3.7879	L
ft^3	28.3	L
L	1000.	cm^3
Pressure		
micrometer (Hg)	0.13332	Pa
N/m^2	1.0	Pa
millibar	100.	Pa
Torr	133.32	Pa
in. (Hg)	3386.33	Pa
lb/in.^{2}	6895.3	Pa
atmosphere	101,323.2	Pa
Conductance or pumping speed		
L/h	0.000277	L/s
L/s	0.001	m^3/s
L/min	0.0166	L/s
m^3/h	0.2778	L/s
ft^3/min	0.4719	L/s
ft^3/min	1.6987	m^3/h
to get conventional unit ←	divide by ←	SI unit

Conventional unit →	multiply by →	to get Si unit
	Gas flow	
micron-L/s	0.13332	Pa-L/s
Pa-L/s	3.6	Pa-m^3/h
atm-cc/s	101.323	Pa-L/s
Torr-L/s	133.32	Pa-L/s
Torr-L/s	0.133	J/s
watt	1000	Pa-L/s
kg-mole/s (at 0°C)	2.48 × 10^9	Pa-L/s
molecules/s (at 0°C)	4 × 10^{-18}	Pa-L/s
	Outgassing rate	
Pa-L/(m^2-s)	0.001	W/m^2
Pa-m^3/(m^2-s)	1.0	W/m^2
μL/(cm^2-s)	1.33	W/m^2
Torr-L/(cm^2-s)	1333.2	W/m^2
	Dynamic viscosity	
poise	10	Pa-s
Newton-s/m^2	1	Pa-s
	Kinematic viscosity	
centistoke	1	mm^2/s
	Diffusion constant	
cm^2/s	0.0001	m^2/s
	Heat conductivity	
watt-cm^{-1}-K^{-1}	100	J-s^{-1}-m^{-1}-K^{-1}
	Specific Heat	
cal-(g-mole)$^{-1}$-K^{-1}	4184.	J-(kg-mole)$^{-1}$-K^{-1}
J-kg^{-1}-K^{-1}	M	J-(kg-mole)$^{-1}$-K^{-1}
BTU-lb^{-1}-°F^{-1}	4186M	J-(kg-mole)$^{-1}$-K^{-1}
	Heat capacity	
cal-(g-mole)$^{-1}$	4184	J-(kg-mole)$^{-1}$
J/kg	M	J-(kg-mole)$^{-1}$
BTU/lb	2325.9M	J-(kg-mole)$^{-1}$
	Energy, work, or quantity of heat	
kW-h	3.6	MJ
kcal	4184	J
BTU	1055	J
ft-lb	1.356	J
to get Conventional unit ←	divide by ←	SI unit

APPENDIX B

Gas Properties

Appendix B.1   Mean Free Paths of Gases as a Function of Pressure[a]

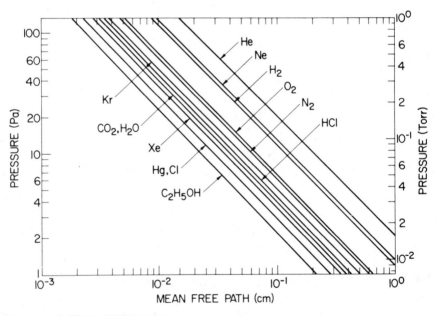

Source. Reprinted with permission from *Vacuum Technology*, p. 505, A. Guthrie. Copyright 1963, John Wiley & Sons, New York, 1963.
[a] $T = 20°C$.

Appendix B.2   Physical Properties of Gases and Vapors at $T = 0\,^{\circ}C$

Gas	Symbol	MW[a]	Molecular Diameter[b] (nm)	Average Velocity[c] (m-s^{-1})	Thermal Cond.[a,d] (mJ-s^{-1}-K^{-1})	Dynamic Viscosity[a,d] (μPa-s)	Diffusion in Air[d,e] (10^{-6}m^2-s^{-1})
Helium	He	4.003	0.218	1197.0	142.0	18.6	58.12
Neon	Ne	20.183	0.259	533.0	45.5	29.73	27.63
Argon	Ar	39.948	0.364	379.0	16.6	20.96	17.09
Krypton	Kr	83.8	0.416	262.0	6.81[f]	23.27	13.17
Xenon	Xe	131.3	0.485	209.0	4.50[g]	21.9	10.60
Hydrogen	H$_2$	2.016	0.274	1687.0	173.0	8.35	63.4[a]
Nitrogen	N$_2$	28.0134	0.375	453.0	24.0	16.58	18.02
Air		28.966	0.372	445.0	24.0	17.08	18.01
Oxygen	O$_2$	31.998	0.361	424.0	24.5	18.9	17.8[a]
Hydrogen chloride	HCl	36.46	0.446	397.0	12.76	14.25[h]	14.11
Water vapor	H$_2$O	18.0153	0.46	564.0	24.1[i]	12.55[i]	23.9[a,j]
Hydrogen sulfide	H$_2$S	34.08	0.47[k]	412.0	12.9	11.66	14.62[k]
Nitric oxide	NO	30.01	0.372[k]	437.0	23.8	17.8	19.3[k]
Nitrous oxide	N$_2$O	44.01	0.47[k]	361.0	15.2	13.5	13.84[k]
Ammonia	NH$_3$	17.03	0.443	581.0	21.9	9.18	17.44
Carbon monoxide	CO	28.01	0.312[a]	453.0	23.0	16.6	21.49
Carbon dioxide	CO$_2$	44.01	0.459	361.0	14.58	13.9	13.9[a]
Methane	CH$_4$	16.4	0.414	592.0	30.6	10.26	18.98
Ethylene	C$_2$H$_4$	28.05	0.495	452.0	17.7	9.07	13.37
Ethane	C$_2$H$_6$	30.07	0.53	437.0	16.8	8.48	12.14

[a] Reprinted with permission from *Handbook of Chemistry and Physics*, 58 ed., R. C. Weast, Ed. Copyright 1977, Chemical Rubber Co., CRC Press, West Palm Beach, FL.
[b] Reprinted with permission from *Kinetic Theory of Gases*, E. H. Kennard, p. 149. Copyright 1938, McGraw-Hill, New York.
[c] Calculated from Eq. (2.2).
[d] At atmospheric pressure.
[e] Calculated from Eq. (2.30).
[f] $T = 210$ K. Reprinted with permission from *Cryogenic and Industrial Gases*, May/June 1975, p. 62. Copyright 1975, Thomas Publishing Co., Cleveland, OH.
[g] Footnote e, $T = 240$ K;
[h] $T = 18\,^{\circ}C$;
[i] $T = 100\,^{\circ}C$;
[j] $T = 8\,^{\circ}C$;
[k] Calculated from viscosity data.

Appendix B.3   Cryogenic Properties of Gases

Property	Units	He	H_2	Ne	N_2	Ar	O_2	Xe	CF_4
nbp liq.[a]	K	4.125	20.27	27.22	77.35	87.29	90.16	164.83	145.16
mp (1 atm)[b]	K		14.01	24.49	63.29	83.95	54.75	161.25	123.16
den liq nbp[a]	kg/m^3	124.8	70.87	1208.	810.0	1410.	1140.	3058.0	1962.0
vol liq nbp	$(m^3/kg) \times 10^{-3}$	8.01	14.1	0.83	1.24	0.709	0.877	0.327	0.597
vol gas at 273 K[a]	m^3/kg	5.602	11.12	1.11	0.79	0.554	0.698	0.169	0.274
Ratio V^g_{273}/V^l_{nbp}		699.4	788.7	1337.	637.5	781.5	796.3	516.8	458.3
h of vap nbp[b]	kJ/(kg-mole)	95.8	911.0	1740.	5580.	6502.	6812.	12640.	12,000.
h of fus mp[b]	kJ/(kg-mole)	16.75	118.0	338.0	714.0	1120.0	438.0	1812.0	699.0
sp h, c_p^{vap}, 300 K[b]	$kJ\text{-}(kg\text{-}mole)^{-1}\text{-}K^{-1}$	20.94[c]	28.63	20.85	29.08	20.89	29.45	20.85	62.23

[a] Reprinted with permission from *Cryogenic and Industrial Gases*, May/June 1975. Copyright 1975, Thomas Publishing Co., Cleveland, OH.
[b] Reprinted with premission from *Handbook of Chemistry and Physics*, 58th ed., R. C. Weast, Ed. Copyright 1977, Chemical Rubber Co., CRC Press, West Palm Beach, FL.
[c] at -180°C.

Note: In these formulas pressure is in pascal, volume in m³, length in m, pumping speed in m³/s, gas flow in Pa-m³/s, velocity in m/s, particle density in m⁻³ and temperature in kelvins unless otherwise stated.

Characteristic Numbers

Knudsen's number

$$Kn = \frac{\lambda}{d}$$

$$Kn = \frac{6.6}{P(Pa)d(mm)} \quad (air, 22°C)$$

Reynolds' number

$$R = \frac{4m}{\pi kT\eta} \frac{Q}{d}$$

$$R = 8.41 \times 10^{-4} \frac{Q(Pa - L/s)}{d(m)} \quad (air, 22°C)$$

Mach number

$$U = \frac{4Q}{\pi d^2 P U_{sound}}$$

Langhaar's number

$$l_e = 0.0568d R$$

Quantities from Kinetic Theory

Most probable velocity

$$v_p = (2kT/m)^{1/2}$$

Average velocity

$$v = \left(\frac{8kT}{\pi m} \right)^{1/2} = 1.128 v_p$$

$$v = 463 \text{ m/s} \quad (air, 22°C)$$

RMS velocity

$$v_{rms} = 3kT/m = 1.225 v_p$$

Mean free path, one-component gas

$$\lambda = \frac{1}{2^{1/2}d_o^2 n}$$

$$\lambda(mm) = \frac{6.6}{P} \quad (air, 22°C)$$

Mean free path, gas a in gas b

$$\lambda_a = \frac{1}{\left[2^{1/2} \pi n_a d_a^2 + \left(1 + \frac{v_b^2}{v_a^2} \right)^{1/2} n_b \frac{\pi}{4} (d_a + d_b)^2 \right]}$$

Particle flux

$$\Gamma = n \left(\frac{kT}{2\pi m} \right)^{1/2} = \frac{nv}{4}$$

Monolayer formation time

$$t_{ml} = \frac{1}{\Gamma d_o^2} = \frac{4}{nvd_o^2}$$

Ideal gas law

$$P = nkT$$

$$\frac{P_1 V_1}{T_1} = \frac{P_2 V_2}{T_2}$$

Specific heat ratio

$$\gamma \sim 1.4 \quad \text{diatomic gas}$$

$$\gamma \sim 1.667 \quad \text{monatomic gas}$$

$$\gamma \sim 1.333 \quad \text{triatomic gas}$$

Viscosity at normal pressures

$$\eta(Pa - s) = \frac{0.499(4mkT)^{1/2}}{\pi^{3/2} d_o^2}$$

Viscosity at reduced pressures (free molecular)

$$\eta_{fm}(Pa - s) = \left(\frac{Pmv}{4kT} \right)$$

Heat conductivity at normal pressures

$$K = \frac{1}{4}(9\gamma - 5)\eta c_v$$

Diffusion constant, gas 1 in gas 2

$$D_{12} = \frac{8\left(\frac{2kT}{\pi} \right)^{1/2} \left(\frac{1}{m_1} + \frac{1}{m_2} \right)^{1/2}}{3\pi(n_1 + n_2)(d_{01} + d_{02})^2}$$

Diffusion constant, self-diffusion

$$D_{11} = \frac{4}{3\pi n d_o^2}\left(\frac{kT}{\pi m}\right)^{1/2}$$

Diffusion constant, molecular (Knudsen), pipe of radius r

$$D = \frac{2}{3}rv$$

Speed of sound in a gas

$$U(m/s) = v\left(\frac{\pi\gamma}{8}\right)^{1/2}$$

$$U = 343.1 \ \ m/s \ \ \ (air, \ 22°C)$$

Conductance

$$C = \frac{Q}{P_2 - P_1}$$

Pumping speed

$$S = \frac{Q}{P}$$

Flow Regimes

Turbulent flow	$R > 2200$
Choked flow	$U = 1$
Viscous flow	$R < 1200$ and $Kn < 0.01$
Poiseuille flow	$U < 1/3, R < 1200, Kn < 0.01$ and $l_e \ll l.$
Molecular flow	$Kn > 1$

Conductance Formulas

Continuum flow conductance, thin aperture, any gas

$$C(m^3/s) = \frac{A}{(1-P_2/P_1)}\left(\frac{2\gamma}{\gamma-1}\frac{kT}{m}\right)^{1/2}\left(\frac{P_2}{P_1}\right)^{1/\gamma}\left[1-\left(\frac{P_2}{P_1}\right)^{(\gamma-1)/\gamma}\right]^{1/2}$$

$$for \ \ 1 > P_2/P_1 \geq 0.52$$

Choked flow limiting conductance, thin aperture, any gas

$$C(m^3/s) = \frac{A}{(1-P_2/P_1)}\left(\frac{kT}{m}\frac{2\gamma}{\gamma+1}\right)^{1/2}\left(\frac{2}{\gamma+1}\right)^{1/(\gamma-1)}$$

$$for \ \ P_2/P_1 \leq 0.52$$

Continuum flow conductance, thin aperture, air 22°C

$$C(L/s) = \frac{7.66 \times 10^5 A(m^2)}{(1-P_2/P_1)} \left(\frac{P_2}{P_1}\right)^{0.712} \left[1-\left(\frac{P_2}{P_1}\right)^{0.288}\right]^{1/2}$$

for $1 > P_2/P_1 \geq 0.52$

Choked flow limiting conductance, thin aperture, air 22°C.

$$C(L/s) \sim 2 \times 10^5 \frac{A(m^2)}{(1-P_2/P_1)}$$

for $P_2/P_1 \leq 0.52$

Viscous flow conductance, long circular tube (Poiseuille)

$$C(m^3/s) = \frac{\pi d^4}{128\eta\ell} \frac{(P_1+P_2)}{2}$$

$$C(L/s) = 1.38 \times 10^6 \frac{d^4}{\ell} \frac{(P_1+P_2)}{2} \qquad (air,\ 22°C)$$

Molecular flow conductance

$$C(m^3/s) = \frac{a'v}{4} A$$

$$C(L/s) = 1.16 \times 10^5 a' A(m^2) = 11.6 a' A(cm^2) \quad (air,\ 22°C)$$

where a' is the transmission coefficient and A is the entrance area.

Useful transmission coefficients

Five transmission coefficients for geometries of common interest are given here. Others are found in Figs. 3.4 to 3.11.

1. Very thin aperture, length l << diameter d

$a' = 1$ *2. Round pipe; any length l, radius r*

Use a' from Table 3.1 or Fig. 3.3, or use the following formula given by A. S. Berman, *J. Appl. Phys.*, **36**, 3356 (1965). $a' = (K_1 - K_2)$. K_1 and K_2 are given by the following formulas with $L = l/r$.

$$K_1 = 1 + L^2/4 - (L/4)[L^2 + 4]^{1/2}$$

$$K_2 = \frac{[(8-L^2)(L^2+4)^{1/2} + L^3 - 16]^2}{72L(L^2+4)^{1/2}288 \ln[L + (L^2+4)^{1/2}] + 288 \ln 2}$$

3. Rectangular pipe, any width-to-thickness ratio (a/b), length l

Use a' from Fig. 3.6, or use the following formula given by I. G. Neuda-chin, et al., *Soviet Physics, Technical Physics,* **17,** 1036 (1972), where $a' = (1/2-J_1/K_1)$ and J_1 and K_1 are given by the following formulas.

$$J_1 = \frac{l^2}{4} - \frac{l}{2ab}\left(ba^2 \text{Arsh}\frac{a}{b} + a^2 b \text{Arsh}\frac{b}{a} + \frac{(a^3+b^3)}{3} - \frac{(a^2+b^2)^{3/2}}{3} \right)$$

$$K_1 = \frac{l^2}{2}\left(1 + \frac{2}{\pi ab}\frac{\ln l^2}{l^2} \right)$$

4. Rectangular pipe; thin, slit-like (thickness b << width a), any length l

Use a' from Table 3.2 or use the following formula given by A. S. Berman, *J. Appl. Phys.,* **36,** 3356 (1965), and erratum **37,** 4509 (1966). $a' = (K_1-K_2)$. K_1 and K_2 are given by the following formulas with $L = l/b$.

$$K_1 = (1/2)[1 + (1 + L)^{1/2} - L]$$

$$K_2 = \frac{3/2[L - \ln (L + (L^2 + 1)^{1/2})]^2}{L^3 + 3L^2 + 4 - (L^2 + 4)(L^2 + 1)^{1/2}}$$

5. Annular cylindrical pipe, length l, inner radius r_i and outer radius r_o

Use a' from Fig. 3.5 or use the following formula given by A. S. Berman, *J. Appl. Phys.,* **40,** 4991 (1969).

$$a' = [1 + L(1/2 - A \tan^{-1}\frac{L}{B})]^{-1}$$

$L = l/(r_o - r_i)$. A and B are given in the formulas below. In these formulas $\sigma = r_i/r_o$ and has a range $0 < \sigma < 0.9$, and L has a range $0 \le L \le 100$.

$$A = \frac{(0.0741 - 0.014\sigma - 0.037\sigma^2)}{(1 - 0.918\sigma + 0.050\sigma^2)}$$

and

$$B = \frac{(5.825 - 2.86\sigma - 1.45\sigma^2)}{(1 + 0.56\sigma - 1.28\sigma^2)}$$

Combining conductances in molecular flow:

1. Parallel

$$C_T = C_1 + C_2 + C_3 + \ldots$$

2. Series, isolated

$$\frac{1}{C_T} = \frac{1}{C_1} + \frac{1}{C_2} + \frac{1}{C_3} + \ldots$$

3. Series, not isolated, equal entrance and exit areas (Oatley)

$$\frac{1-a'}{a'} = \frac{1-a_1}{a_1} + \frac{1-a_2}{a_2} + \ldots$$

4. Series, not isolated, unequal entrance and exit areas (Haefer)

$$\frac{1}{A_1}\left(\frac{1-a_t}{a_t}\right) = \sum_1^n \frac{1}{A_i}\left(\frac{1-a_i}{a_i}\right) + \sum_1^{n-1}\left(\frac{1}{A_{i+1}} - \frac{1}{A_i}\right)\delta_{i,i+1}$$

where $\delta_{i,i+1} = 1$ for $A_{i+1} < A_i$, and $\delta_{i,i+1} = 0$ for $A_{i,i+1} \geq A_i$.

Transition conductance, Knudsen's method

$$C = \frac{Q}{(P_2 - P_1)}$$

$$Q = Q_{\text{viscous}} + Z'Q_{\text{molecular}}$$

where

$$Z' = \frac{1 + 2.507\left(\dfrac{d}{2\lambda}\right)}{1 + 3.095\left(\dfrac{d}{2\lambda}\right)}$$

Appendix B.5 Vapor Pressure Curves of Common Gases

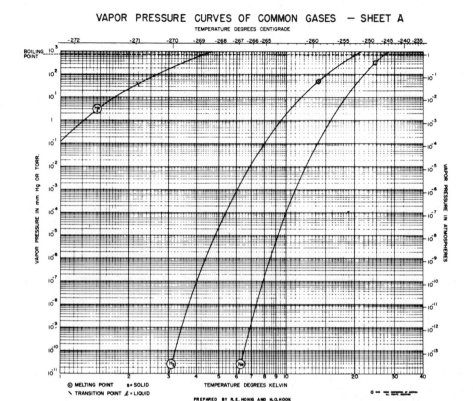

VAPOR PRESSURE CURVES OF COMMON GASES — SHEET A

PREPARED BY R.E. HONIG AND H.O. HOOK
RADIO CORPORATION OF AMERICA   PRINCETON, N. J.

VAPOR PRESSURE CURVES OF COMMON GASES — SHEET B

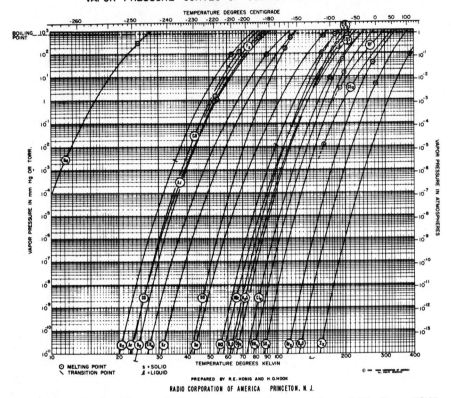

Source. Reprinted with permission from *RCA Review*, **21**, p. 360, Sept. 1960, *Vapor Pressure Data for Some Common Gases*, by R. E. Honig and H. O. Hook. Copyright 1960, RCA Corp.

Appendix B.6 Appearance of Discharges in Gases and Vapors at Low Pressures

Gas	Negative Glow	Positive Column
Argon	Blue	Violet
Carbon tetrachloride	Light green	Whitish green
Carbon monoxide	Greenish white	White
Carbon dioxide	Blue	White
C_2H_5OH	-	Whitish
Cadmium	Red	Greenish blue
Hydrogen	Light blue	Pink
Mercury	Whitish yellow	Blue green
Potassium	Green	Green
Krypton	Violet	Yellow pink
Air	Blue	Reddish
Nitrogen	Blue	Red-yellow
Sodium	Whitish	yellow
Oxygen	Yellowish white	Lemon yellow with pink core
Thallium	Green	Green
Xenon	Pale blue	Blue violet

Source. Reprinted with permission from *Materials for High Vacuum Technology*, **3**, p. 393, W. Espe. Copyright 1968, Pergamon Press.

APPENDIX C

Material Properties

Appendix C.1   Outgassing Rates of Vacuum Baked Metals

Material	Treatment	q $(10^{-11}$ W/m$^2)$
Aluminum [a]	15 h at 250°C	53.0
Aluminum [b]	20 h at 100°C	5.3
6061 Aluminum [c]	glow disch. + 200°C bake	1.3
Copper [b]	20 h at 100°C	146.0
304 Stainless Steel [a]	30 h at 250°C	400.0
Stainless Steel [d]	2 h at 850/900°C vac. furnace	27.0
316L Stainless Steel [e]	2 h at 800°C vac. furnace	46.0
U15C Stainless Steel [f]	3 h vac. furn. 1000°C + 25-h *in situ* vac. bake at 360°C	2.1

Source. Adapted with permission from *Vacuum*, **25**, p. 347, R. J. Elsey. Copyright 1975, Pergamon Press.
[a] J. R. Young, *J. Vac. Sci. Technol.*, **6**, 398 (1969);
[b] G. Moraw, *Vacuum*, **24**, 125 (1974);
[c] H. J. Halama and J. C. Herrera, *J. Vac. Sci. Technol.*, **13**, 463 (1976);
[d] R. L. Samuel, *Vacuum*, **20**, 295 (1970);
[e] R. Nuvolone, *J. Vac. Sci. Technol.*, **14**, 1210 (1977);
[f] R. Calder and G. Lewin, *Brit. J. Appl. Phys.*, **18**, 1459 (1967).

Appendix C.2 Outgassing Rates of Unbaked Metals[1]

Material	q_1 $(10^{-7}$ W/m$^2)$	α_1	q_{10} $(10^{-7}$ W/m$^2)$	α_{10}
Aluminum (fresh)[a]	84.0	1.0	8.0	1.0
Aluminum (degassed 24-h)[a]	55.2	3.2	4.08	0.9
Aluminum (3-h in air)[a]	88.6	1.9	6.33	0.9
Aluminum (fresh)[a]	82.6	1.0	4.33	0.9
Aluminum (anodized 2 μm pores)[a]	3679.0	0.9	429.0	0.9
Aluminum (bright rolled)[b]	-	-	100.0	1.0
Duraluminum[b]	2266.0	0.75	467.0	0.75
Brass (wave guide)[b]	5332.0	2.0	133.0	1.2
Copper (fresh)[a]	533.0	1.0	55.3	1.0
Copper (mech. polished)[a]	46.7	1.0	4.75	1.0
OHFC copper (fresh)[a]	251.0	1.3	16.8	1.3
OHFC copper (mech. polished)[a]	25.0	1.1	2.17	1.1
Gold (wire fresh)[a]	2105.0	2.1	6.8	1.0
Mild steel[b]	7200.0	1.0	667.0	1.0
Mild steel (slightly rusty)[b]	8000.0	3.1	173.0	1.0
Mild steel (chromium plated polished)[b]	133.0	1.0	12.0	-
Mild steel (aluminum spray coated)[b]	800.0	0.75	133.0	0.75
Steel (chromium plated fresh)[a]	94.0	1.0	7.7	1.0
Steel (chromium plated polished)[a]	121.0	1.0	10.7	1.0
Steel (nickel plated fresh)[a]	56.5	0.9	6.6	0.9
Steel (nickel plated)[a]	368.0	1.1	3.11	1.1
Steel (chemically nickel plated fresh)[a]	111.0	1.0	9.4	1.0
Steel (chemically nickel plated polished)[a]	69.6	1.0	6.13	1.0
Steel (descaled)[a]	4093.0	0.6	3933.0	0.7
Molybdenum[a]	69.0	1.0	4.89	1.0
Stainless steel EN58B (AISI 321)[b]	-	-	19.0	1.6
Stainless steel 19/9/1-electropolished[c]	-	-	2.7	-
-vapor degreased[c]	-	-	1.3	-
-Diversey cleaned[c]	-	-	4.0	-
Stainless steel[b]	2333.0	1.1	280.0	0.75
Stainless steel[b]	1200.0	0.7	267.0	0.75
Stainless steel ICN 472 (fresh)[a]	180.0	0.9	19.6	0.9
Stainless steel ICN 472 (sanded)[a]	110.0	1.2	13.9	0.8
Stainless steel NS22S (mech. polished)[a]	22.8	0.5	6.1	0.7
Stainless steel NS22S (electropolished)[a]	57.0	1.0	5.7	1.0
Stainless steel[a]	192.0	1.3	18.0	1.9
Zinc[a]	2946.0	1.4	429.0	0.8
Titanium[a]	150.0	0.6	24.5	1.1
Titanium[a]	53.0	1.0	4.91	1.0

Source. Reprinted with permission from *Vacuum*, **25**, p 347, R. J. Elsey. Copyright 1975, Pergamon Press.

[1] $q_n = qt^{-\alpha_n}$, where n is in hours.

[a] A. Schram, *Le Vide*, No. 103, 55 (1963),

[b] B. B. Dayton, *Trans. 6th Nat. Vac. Symp. (1959)*, Pergamon Press, New York, 1960, p. 101,

[c] R. S. Barton and R. P. Govier, *Proc. 4th Int. Vac. Congr. (1968)*, Institute of Physics and the Physical Society, London, 1969, p. 775, and *Vacuum*, **20**, 1 (1970).

Appendix C.3 Outgassing Rates of Ceramics and Glasses[1]

Material	q_1 (10^{-7} W/m^2)	α_1	q_{10} (10^{-7} W/m^2)	α_{10}
Steatite[a]	1200.0	1.0	127.0	-
Pyrophyllite[b]	2667.0	1.0	267.0	-
Pyrex (fresh)[c]	98.0	1.1	7.3	-
Pyrex (1 month in air)[c]	15.5	0.9	2.1	-

Source. Reprinted with permission from *Vacuum*, **25**, p. 347, R. J. Elsey. Copyright 1975, Pergamon Press.
[1] $q_n = qt^{-\alpha_n}$, where n is in hours.
[a] R. Geller, *Le Vide*, No. 13, 71 (1958);
[b] R. Jaeckel and F. Schittko, quoted by Elsey;
[c] B. B. Dayton, *Trans. 6th Nat. Symp. Vac. Technol. (1959)*, Pergamon Press, New York, 1960, p. 101.

Appendix C.4 Outgassing Rates of Elastomers[1]

Material	q_1 (10^{-5} W/m^2)	α_1	q_4 (10^{-5} W/m^2)	α_4
Butyl DR41[a]	200.0	0.68	53.0	0.64
Neoprene[a]	4000.0	0.4	2400.0	0.4
Perbunan[a]	467.0	0.3	293.0	0.5
Silicone[b]	930.0	-	267.0	-
Viton A (fresh)[c]	152.0	0.8	-	-
Viton A (bake 12 h at 200°C)[d]	-	-	0.027[e]	-
Polyimide (bake 12 h at 300°C)[d]	-	-	0.005[e]	-

Source. Adapted with permission from *Vacuum*, **25**, p. 347, R. J. Elsey. Copyright 1975, Pergamon Press.
[1] $q_n = qt^{-\alpha_n}$, where n is in hours.
[a] J. Blears, E. J. Greer and J. Nightengale, *Adv. Vac. Sci.Technol.*, **2**, E. Thomas, Ed., Pergamon Press, 1960, p. 473;
[b] D. J. Santeler, et al., *Vacuum Technology and Space Simulation*, NASA SP-105, National Aeronautics and Space Administration, Washington, DC, 1966, p. 219;
[c] A. Schram, *Le Vide*, No. 103, 55 (1963);
[d] P. Hait, *Vacuum*, **17**, 547 (1967);
[e] Pumping time is 12 h.

Appendix C.5. Permeability of Polymeric Materials[a]

Material	Permeability (10^{-12} m²/s)					
	Nitrogen	Oxygen	Hydrogen	Helium	Water Vapor	Carbon Dioxide
PTFE[b]	2.5	8.2	20.0	570.0	–	–
Perspex[b]	–	–	2.7	5.7	–	–
Nylon 31[b]	–	–	0.13	0.3	–	–
Neoprene CS2368B[b]	0.21	1.5	8.2	7.9	–	–
Viton-A[c]	0.05	1.1	2.2	8.9	–	5.9
Kapton[c]	0.03	0.1	1.1	1.9	–	0.2
Buna-S[d]	4.8 (30)	–	–	–	–	940.0 (30)
Perbunan[d]	0.8	–	–	–	–	23.0 (30)
Delrin[d]	–	48.0	–	–	17.0	93.0
Kel-F[d]	0.99 (30)	0.46 (30)	–	–	0.22 (25)	–

[a] Measurements made at 23°C unless noted in parenthesis after the value.
[b] Reprinted with permission from *Vacuum*, **25**, p. 469, G. F. Weston. Copyright 1975, Pergamon Press. Data derived by Weston from measurements made by Barton reported by J. R. Bailey in *Handbook of Vacuum Physics*, **3**, Part 4, Pergamon Press, Oxford, 1964;
[c] Reprinted with permission from *J. Vac. Sci. Technol.*, **10**, p. 543, W. G. Perkins. Copyright 1973, The American Vacuum Society;
[d] Reprinted with permission from *Vacuum Science and Space Simulation*, D. J. Santeler et al., NASA SP-105, National Aeronautics and Space Administration, Washington, DC, 1966, p. 216.

Appendix C.6 Vapor Pressure Curves of the Solid and Liquid Elements

VAPOR PRESSURE CURVES OF THE ELEMENTS

SHEET A

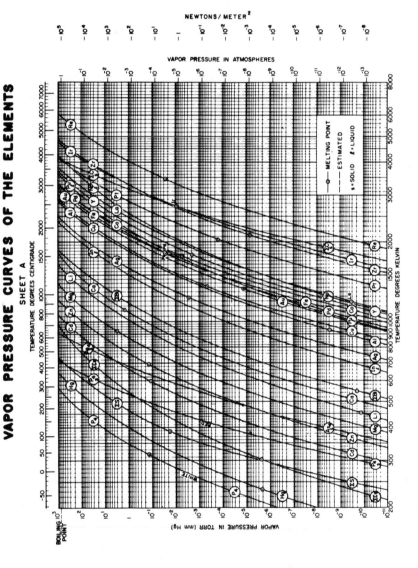

Appendix C.6 (Continued)

VAPOR PRESSURE CURVES OF THE ELEMENTS
SHEET B

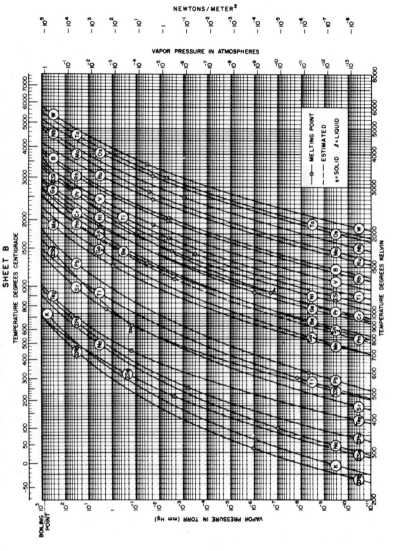

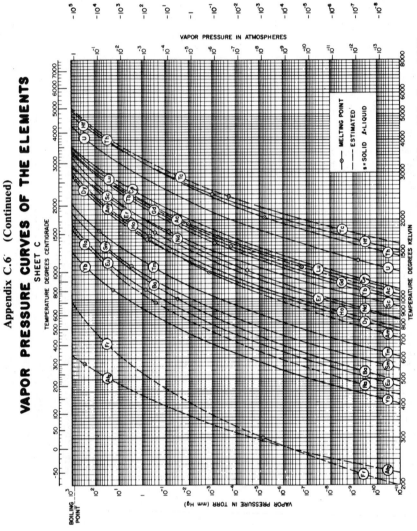

Appendix C.6' (Continued)

VAPOR PRESSURE CURVES OF THE ELEMENTS

SHEET C

Source. Reprinted with permission from *RCA Review*, **30**, p.285, June 1969, *Vapor Pressure Data for the Solid and Liquid Elements*, by R. E. Honig and D. A. Kramer. Copyright 1969, RCA Corp.

Appendix C.7   Outgassing Rates of Polymers[1]

Material	q_1 (10^{-5} W/m^2)	α_1	q_{10} (10^{-5} W/m^2)	α_{10}
Araldite (molded)[a]	155.0	0.8	47.0	0.8
Araldite D[b]	253.0	0.3	167.0	0.5
Araldite F[b]	200.0	0.5	97.0	0.5
Kel-F[c]	5.0	0.57	2.3	0.53
Methyl Methacrylate[d]	560.0	0.9	187.0	0.57
Mylar (24-h at 95 RH)[e]	307.0	0.75	53.0	-
Nylon[f]	1600.0	0.5	800.0	0.5
Plexiglas[g]	961.0	0.44	36.0	0.44
Plexiglas[b]	413.0	0.4	240.0	0.4
Polyester-glass Laminate[c]	333.0	0.84	107.0	0.81
Polystyrene[c]	2667.0	1.6	267.0	1.6
PTFE[h]	40.0	0.45	26.0	0.56
PVC (24-h at 95 RH)[e]	113.0	1.0	2.7	-
Teflon[g]	8.7	0.5	3.3	0.2

Source. Reprinted with permission from *Vacuum*, **25**, p. 347, R. J. Elsey, Copyright 1975, Pergamon Press.

[1] $q_n = qt^{-\alpha_n}$, where n is in hours.
[a] A. Schram, *Le Vide*, No. 103, 55 (1963);
[b] R. Geller, *Le Vide*, No.13, 71 (1958);
[c] B. B. Dayton, CVC Technical Report;
[d] J. Blears, E. J. Greer and J. Nightengale, *Adv. Vac. Sci. Technol.*, **2**, E. Thomas, Ed., Pergamon Press, 1960, p. 473;
[e] D. J. Santeler, *Trans. 5th Symp. Vac. Tech. (1958)*, Pergamon Press, New York, 1959, p. 1;
[f] B. D. Power and D. J. Crawley, *Adv. Vac. Sci. Technol.*, **1**, E. Thomas, Ed., Pergamon Press, New York, 1960, p. 207;
[g] G. Thieme, *Vacuum*, **13**, 55 (1963);
[h] B. B. Dayton, *Trans. 6th Nat. Vac. Symp. Vac. Technol. (1959)*, Pergamon Press, New York, 1960, p.101.

AISI type	General Forming	High temperature use	Cryogenic use	Resist carbide precipitation	High yield strength	Non-magnetic at cryogenic temperatures	Free machining	Weldability	0.2% proof (yield) kgf mm^2	Ultimate TS kgf mm^2	Charpy-V J	Price index
	Performance - recommend for								Tensile properties Typical, at R.T. Annealed		Impact strength at -196°C	
302	x							E	22	60	70	100
303S, Se							x	P	22	60		105
304	x		x					E	20	50	70	100
304L	x		x	x				E	18	48	80	115
304N	x		x		x			E	30	55-75	min 63	110
304LN	x		x	x	x	x		E	28	55-75	min 55	125
310	x	x			x	x		G	23	65		210
316	x		x				~	E	22	50		130
316L	x		x	x			~	E	20	45	120	150
316N	x		x		x	x		E	30	60		140
316LN	x		x	x	x	x		E	30	60-80	min 55	160
317	x	x					~	E	24	60		200
321	x	x		x				E	21	50	65	130
347	x	x		x				G	22	50	65	150

Source. Reprinted with permission from *Vacuum*, **26**, p. 287, C. Geyari. Copy-
[a] Selection guide for vacuum and cryogenic equipment. X = recommended,
P = poor.

Austenitic Stainless Steels[a]

Typical composition, %						
C max	Cr	Ni	Mo	N	Others	Description
0.12	17-19	8-10				General purpose. Good resistance to atmosphere corrosion. Good mechanical properties.
0.15	17-19	8-10			S, Se≥0.15	Free machining type. Good corrosion resistance
0.08	18-20	8-12				Low C variation of 302. Improved corrosion resistance after welding.
0.03	18-20	8-12				Extra low C prevents carbide precipitation.
0.06	18.5	9.5		0.25	Mn 2	Improved mechanical properties.
0.03	18.5	9.5		0.18	Mn 2	Improved mechanical properties. Extra low C prevents carbide precipitation.
0.25	24-26	19-22				High scale resistance. Superior corrosion resistance.
0.08	16-18	10-14	2-3			Very good corrosion resistance in most media.
0.03	16-18	10-14	2-3			Extra low C variation of 316.
0.07	17-18	10-13	2.5-3	0.2	Mn 0.5-2	Improved mechanical properties.
0.03	17.5	13	2.8	0.18	Mn 2	Improved mechanical properties. Extra low C prevents carbide precipitation.
0.08	18-20	11-15	3-4			Higher alloy content improves creep and corrosion resistance of 316.
0.08	17-19	9-12			Ti≥5xC	Stabilized--Ti prevents carbide precipitation. Improved corrosion resistance after welding.
0.08	17-19	9-13			Nb, Ta≥10xC	Stabilized--Nb, Ta prevent carbide precipitation. Improved corrosion resistance after welding.

right 1976, Pergamon Press.
~ = probable–should be tested, E = excellent, G = good with precautions,

Isotopic Abundances

Element	AMU	Relative Abundance	Element	AMU	Relative Abundance
H	1	99.985	S	32	95.06
	2	0.015		33	0.74
He	3	0.00013		34	4.18
	4	~100.0		36	0.016
Li	6	7.42	Cl	35	75.4
	7	92.58		37	24.6
Be	9	100.0	Ar	36	0.337
B	10	19.78		38	0.063
	11	80.22		40	99.600
C	12	98.892	K	39	93.08
	13	1.108		40	0.0119
N	14	99.63		41	6.91
	15	0.37	Ca	40	96.97
O	16	99.759		42	0.64
	17	0.0374		43	0.145
	18	0.2039		44	2.06
F	19	100.0		46	0.0033
Ne	20	90.92		48	0.185
	21	0.257	Sc	45	100.0
	22	8.82	Ti	46	7.95
Na	23	100.0		47	7.75
Mg	24	78.60		48	73.45
	25	10.11		49	5.51
	26	11.29		50	5.34
Al	27	100.0	V	50	0.24
Si	28	92.27		51	99.76
	29	4.68	Cr	50	4.31
	30	3.05		52	83.76
P	31	100.0		53	9.55
				54	2.38

Element	AMU	Relative Abundance	Element	AMU	Relative Abundance
Mn	55	100.0	Zr	90	51.46
Fe	54	5.82		91	11.23
	56	91.66		92	17.11
	57	2.19		94	17.4
	58	0.33		96	2.8
Co	59	100.0	Nb	93	100.0
Ni	58	67.76	Mo	92	15.86
	60	26.16		94	9.12
	61	1.25		95	15.70
	62	3.66		96	16.50
	64	1.16		97	9.45
Cu	63	69.1		98	23.75
	65	30.9		100	9.62
Zn	64	48.89	Ru	96	5.47
	66	27.82		98	1.84
	67	4.14		99	12.77
	68	18.54		100	12.56
	70	0.617		101	17.10
Ga	69	60.2		102	31.70
	71	39.8		104	18.56
Ge	70	20.55	Rh	103	100.0
	72	27.37	Pd	102	0.96
	73	7.67		104	10.97
	74	36.74		105	22.23
	76	7.67		106	27.33
As	75	100.0		108	26.71
Se	74	0.87		110	11.81
	76	9.02	Ag	107	51.82
	77	7.58		109	48.18
	78	23.52	Cd	106	1.22
	80	49.82		108	0.87
	82	9.19		110	12.39
Br	79	50.52		111	12.75
	81	49.48		112	24.07
Kr	78	0.354		113	12.26
	80	2.27		114	28.86
	82	11.56		116	7.85
	83	11.55	In	113	4.23
	84	56.90		115	95.77
	86	17.37	Sn	112	0.95
Rb	85	72.15		114	0.65
	87	27.85		115	0.34
Sr	84	0.56		116	14.24
	86	9.86		117	7.57
	87	7.02		118	24.01
	88	82.56		119	8.58
Y	98	100.0		120	32.97
				122	4.71
				124	5.98

Element	AMU	Relative Abundance	Element	AMU	Relative Abundance
Sb	121	57.25	Eu	151	47.77
	123	42.75		153	52.23
Te	120	0.089	Gd	152	0.20
	122	2.46		154	2.15
	123	0.87		155	14.73
	124	4.61		156	20.47
	125	6.99		157	15.68
	126	18.71		158	24.87
	128	31.79		160	21.90
	130	34.49	Tb	159	100.0
I	127	100.0	Dy	156	0.052
Xe	124	0.096		158	0.090
	126	0.090		160	2.294
	128	1.92		161	18.88
	129	26.44		162	25.53
	130	4.08		163	24.97
	131	21.18		164	28.18
	132	26.89	Ho	165	100.0
	134	10.44	Er	162	0.136
	136	8.87		164	1.56
Cs	131	100.0		166	33.41
Ba	130	0.101		167	22.94
	132	0.097		168	27.07
	134	2.42		170	14.88
	135	6.59	Tm	169	100.0
	136	7.81	Yb	168	0.140
	137	11.32		170	3.03
	138	71.66		171	14.31
La	138	0.089		172	21.82
	139	99.911		173	16.13
Ce	136	0.193		174	31.84
	138	0.250		176	12.73
	140	88.48	Lu	175	97.40
	142	11.07		176	2.60
Pr	141	100.0	Hf	174	0.18
Nd	142	27.13		176	5.15
	143	12.20		177	18.39
	144	23.87		178	27.08
	145	8.30		179	13.78
	146	17.18		180	35.44
	148	5.72	Ta	180	0.012
	150	5.62		181	99.988
Sm	144	3.16	W	180	0.135
	147	15.07		182	26.4
	148	11.27		183	14.4
	149	13.84		184	30.6
	150	7.47		186	28.4
	152	26.63	Re	185	37.07
	154	22.53		187	62.93

Element	AMU	Relative Abundance	Element	AMU	Relative Abundance
Os	184	0.018		199	16.84
	186	1.59		200	23.13
	187	1.64		201	13.22
	188	13.3		202	29.80
	189	16.1		204	6.85
	190	26.4			
	192	41.0	Tl	203	29.50
Ir	191	37.3		205	70.50
	193	62.7	Pb	204	1.48
Pt	190	0.012		206	23.6
	192	0.78		207	22.6
	194	32.8		208	52.3
	195	33.7			
	196	25.4	Bi	209	100.0
	198	7.21	Th	232	100.0
Au	197	100.0	U	234	0.0057
Hg	196	0.15		235	0.72
	198	10.02		238	99.27

Source. Reprinted with permission from *Mass Spectroscopy for Science and Technology*, F. A. White, p. 339. Copyright 1968, John Wiley & Sons.

APPENDIX E

Cracking Patterns

			Appendix E.1	Cracking Patterns of Pump Fluids		
AMU	Welch 1407[a]	Fomblin Y-25[b]	DC-704[c]	DC-705[c]	Octoil-S[c]	Convalex-10[c]
18					1.86	
27					14.40	23.20
28			17.19	67.18	8.37	
29				2.56	28.53	23.20
30				3.52	1.10	20.20
31		31.49			1.27	6.00
32			2.15	10.08		
35						
36						6.50
37						
38					0.65	4.20
39				2.22	8.82	14.90
40				4.44	2.21	15.50
41	40			3.59	54.03	31.00
42					17.12	8.30
43	74		2.40	7.69	66.58	29.20
44			1.98	12.47	3.17	50.00
45					1.98	
47		20.67				
50		15.30				5.40
51		27.66	2.61		0.84	11.30
52		2.92				
53	3				3.41	
54					4.34	
55	70		1.31	4.79	54.56	25.60
56	23				22.18	15.50
57	100		2.13		88.20	55.40
59					4.59	
60					2.0	
61						
62						
63						7.70

AMU	Welch 1407[a]	Fomblin Y-25[b]	DC-704[c]	DC-705[c]	Octoil-S[c]	Convalex-10[c]
64						8.30
65						7.10
66		2.23				
67	40				4.52	3.60
68	14				4.23	3.00
69	91	100.00		3.59		11.30
70	31	3.39			56.83	18.50
71	83			2.90	46.92	9.50
72					3.14	
73			1.65		2.98	
76						3.40
77	2		2.20			25.60
78	0.7			4.62		4.80
81	22	3.10		1.88	3.80	3.60
82	10				3.91	2.40
83	30			2.90	19.78	11.30
84	12				13.78	4.80
85		2.23		2.05	2.25	
87					1.82	
91		2.03	3.39	3.76		3.00
92						3.60
93	2				1.18	
94	7					3.60
95	3			2.05	2.87	2.40
96	10				2.11	
97		16.37		2.39	10.50	3.00
98					19.87	5.40
100		14.63				
101		14.63			2.00	
108						12.50
112					45.46	
113					25.01	61.30
119		13.88	3.87	3.24		
131		2.36				2.40
135		5.19	20.59	10.26		

[a] Data taken on UTI-100B quadrupole with $V_{EE} = -60$ V, only major peaks shown. Reprinted with permission from Uthe Technology Inc., 325 N. Mathilda Avenue, Sunnyvale, CA 94086;
[b] Sector data. Adapted with permission from *Vacuum*, **22**, p. 315, L. Holland, L. Laurenson, and P. N. Baker. Copyright 1972, Pergamon Press;
[c] Sector data. Reprinted with permission from *J. Vac. Sci. Technol.*, **6**, p. 871, G. M. Wood, Jr., and R. J. Roenig. Copyright 1969, The American Vacuum Society.
Note. Only peaks up to 135 AMU are shown for the data taken from source *c*; the largest mass peak (100%) occurs at a higher mass number. The data are *not* renormalized for the range tabulated here.

Appendix E.2   Cracking Patterns of Gases

AMU	Hydrogen[a] H_2	Helium[b] He	Neon[b] Ne	Carbon Monoxide[a] CO	Nitrogen[a] N_2	Oxygen[a] O_2	Argon[a] Ar	Carbon Dioxide[a] CO_2
1	2.7							
2	100	0.12						
3	0.31							
4		100						
6				0.0008				0.0005
7					0.0006			
8				0.0001		0.0013		0.0005
12				3.5				6.3
13								0.063
14				1.4	9			
15					0.026			
16				1.4		14		16
17						0.0052		
18						0.028	0.071	0.0088
19							0.016	
20			100				5.0	
21			0.33					
22			9.9					0.52
22.5								0.0047
23								0.0012
28				100	100			15
29				1.2	0.71			0.15
30				0.2	0.0014			0.029
32						100		
33						0.074		
34						0.38		
36						0.0023	0.36	
38							0.068	
40							100	
44								100
45								1.2
46								0.38
47								0.0034
48								0.0005

[a] Data taken on UTI-100C-02 quadrupole residual gas analyzer. Typical parameters, $V_{EE} = 70$ V, $V_{IE} = 15$ V, $V_{FO} = -20$ V $I_E = 2.5$ mA, resolution potentiometer = 5.00. Reprinted with permission from Uthe Technology Inc., 325 N. Mathilda Avenue, Sunnyvale, CA 94086.

[b] Sector data. Reprinted with permission from E. I. du Pont de Nemours & Co., Wilmington, DE 19898

Appendix E.3   Cracking Patterns of Common Vapors

AMU	Water Vapor[a] H_2O	Methane[b] CH_4	Acetylene[b] C_2H_2	Ethylene[b] C_2H_4	Ethane[b] C_2H_6	Cyclo-propane[b] C_3H_6
1	0.1	3.8	3.8	6.4	3.2	1.4
2		0.64	1.2	1.1	0.93	32.
3		0.009	0.002	0.022	0.15	0.10
6		0.0003	0.0006	0.0002		
7		0.0013		0.0018		
12		2.1	4.5	2.3	0.47	0.85
13		7.4	7.6	4.0	1.1	1.6
14		15.	0.86	8.1	3.4	5.6
14.5					0.24	
15		83.			5.7	8.1
16	3.07	100.			0.53	2.0
17	27.01	1.3				0.07
18	100.					
19	0.19					2.7
19.5						1.3
20						2.3
20.5						0.68
24			7.1	3.2	0.52	0.35
25			23.	12.	3.5	2.1
26			100.	61.	24.	17.
27			2.5	59.	33.	46.
28				100.	100.	18.
29				2.8	21.	11.
30					24.	0.29
31					0.54	
36						1.4
37						11.
38						15.
39						69.
40						30.
41						100.
42						90.
43						18.

[a] Sector data. Reprinted with permission from E. I. du Pont de Nemours & Co., Wilmington, DE 19898;

[b] Quadrupole data, same conditions as given in Appendix E1a. Reprinted with permission from Uthe Technology Inc, 325 N. Mathilda Avenue, Sunnyvale CA 94086.

Appendix E.4   Cracking Patterns of Common Solvents

AMU	Methyl Alcohol[a]	Ethyl Alcohol[a]	Acetone[a]	Isopropyl Alcohol[a]	Trichloro-ethylene[a]	Gentron-142B[b]
2						3.4
12						2.9
13						3.2
14						9.2
15						16.0
18	1.9	5.5				
19		2.3		6.6		3.3
20						1.7
25						6.5
26		8.3	5.8			17.0
27		23.9	8.0	15.7		4.4
28	6.4	6.9				
29	67.4	23.4	4.3	10.1		
30	0.8	6.0				
31	100.	100.		5.6		17.0
32	66.7					1.0
35					39.9	5.2
36						1.8
37			2.1		12.8	1.6
38		2.3				0.9
39			3.8	5.7		
41			2.1	6.6		
42		2.9	7.0	4.0		
43		7.6	100.	16.6		0.5
45		34.4		100.		53.0
46		16.5				3.5
47					25.8	1.1
48						0.5
49						1.4
50						2.9
51						1.9
58			27.1			
59				3.4		
60					64.9	0.6
62					20.9	0.5
63						3.4
64						8.6
65						100.0
66						2.5
87						1.8
95					100.0	
97					63.9	
130					89.8	
132					84.8	
134					26.8	

[a] Sector data. Reprinted with permission from VG-Micromass Ltd., 3 Tudor Road, Altringham, Cheshire, TN34-1YQ, England.
[b] Quadrupole data, same conditions as given in Appendix E1a. Reprinted with permission from Uthe Technology Inc., 325 N. Mathilda Avenue, Sunnyvale, CA 94086.

AMU	Arsine AsH_3	Silane SiH_4	Phosphine PH_3	Disilane Si_2H_6	Diphosphine P_2H_4	Diborane B_2H_6
1						21.1
2						134.7
3						0.35
10						9.72
11						39.4
12						26.4
13						34.9
14.		0.4				2.23
14.5		0.5				
15		0.4				
15.5		0.1	0.23			
16			0.62			
16.5			0.13			
17			0.48			
20						0.22
21						1.85
22						11.8
23						48.5
24						94.0
25						57.7
26						100.0
27						95.2
28		28.				16.7
29		32.				
30		100.0				
31		80.	26.7			
32		7.3	100.0			
33		1.5	25.4			
34		0.2	76.7			
56				33.		
57				48.		
58				82.		
59				37.		
60				100.0		
61				40.		
62				42.	100.0	
63				5.7	58.8	
64				4.0	70.6	
65					26.5	
66					1.5	
75	38.5					
76	100.0					
77	28.8					
78	92.3					

[a] Quadrupole data, same conditions as given in Appendix E1a. Reprinted with permission from Uthe Technology Inc., 325 N. Mathilda Avenue, Sunnyvale, CA 94086.

APPENDIX F

Pump Fluid Properties

Appendix F.1   Compatibility of Elastomers and Pump Fluids

Elastomer	Mineral Oil	Ester	Halo-carbon	Fluoro-carbon	Poly-siloxane
Butyl	No	<100°C	No data	<90°C	No
Buna-N	<100°C	No	<100°C	<90°C	No
Buna-S	No	No	No	No	No
Neoprene	<120°C	No	<120°C	<90°C	No
EPR	Yes	<70°C[a]	Yes	<70°C[a]	No
Silicone	Yes	<175°C	No	<150°C	No
Viton	Yes	<145°C	Yes	<200°C	Yes
Teflon	Yes	<175°C	Yes	<200°	Yes
Kalrez	Yes	<175°C	Yes	Yes	Yes

[a] No data available for $T > 70$°C.

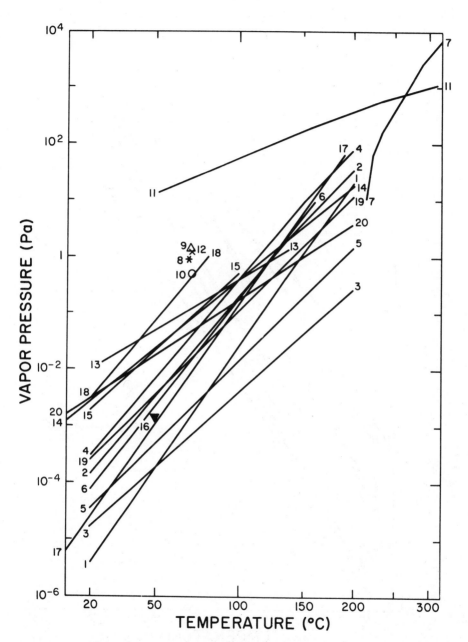

(1) Fomblin Y-H vac 18/8 [1], (2) Fomblin Y-L vac 25/5 [1], (3) Krytox 1525 [2]; (4)
Fomblin Y-L vac 06/6 [1]; (5) Krytox 1514 [2]; (6) Krytox 1506 [2]; (7) Fyrquel-220
[3]; (8) Halovac-100 [4]; (9) Halovac-125 [4]; (10) Halovac-190 [4]; (11) Versilube F-50
[5]; (12) Synlube [6]; (13) Inland-77 [7]; (14) Inland-19 [7]; (15) Balzers P-3 [8]; (16)
Welch Duo-Seal 1407 [9]; (17) Convoil-20 [10]; (18) Balzers T-11 [8]; (19) Invoil-20 [7];
(20) Dow Corning FS-1265 [11].

Appendix F.3   Vapor Pressure of Diffusion Pump Fluids

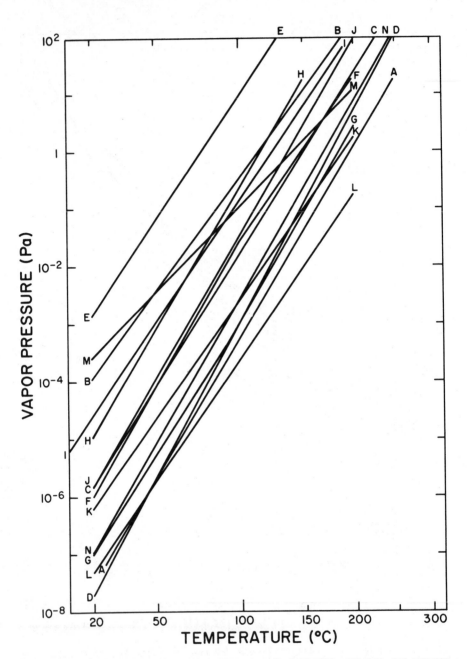

(A) Santovac 5 [12]; (B) DC-702 [11]; (C) DC-704 [11], Neo Vac-SY [13]; (D) DC-705 [11]; (E) butyl phthalate; (F) Fomblin Y-H vac 18/8 [1]; (G) Fomblin Y-H vac 25/9 [1]; (H) Octoil [10]; (I) Convoil-20 [10]; (J) Octoil-S [10]; (K) Krytox 1618 [2]; (L) Krytox 1625 [2]; (M) Invoil-20 [7]; (N) Edwards L9 [14]. The fluids grouped under (C) are not the same, but they have the nearly the same vapor pressure.

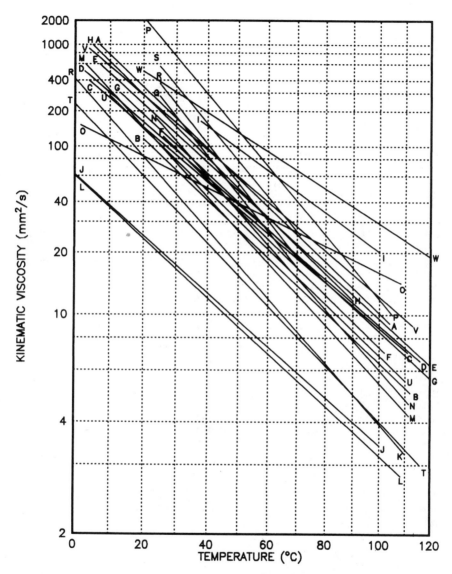

(A) Balzers P-3 [4], Kinney KV-100 [15]; (B) Convoil-20 [10]; (C) Inland-19 [7]; (D) Invoil-20 [7]; (E) Kinney OCR [15], Krytox 1525 and 1625 [2]; (F) Kinney Super X [15]; (G) Welch Duo-Seal 1407 [12], Krytox 1618 [2], Fomblin Y-H vac 18/8 [1]; (H) Inland-77 [7]; (I) Synlube [6]; (J) Balzers T-11 [8]; (K) Octoil [10], Invoil [7]; (L) Octoil-S [10]; (M) Fyrquel-220 [3]; (N) Kinlube 300 [15]; (O) Versilube F-50 [5]; (P) Santovac 5 [12]; (Q) Halovac 100 [4]; (R) Halovac 125 [4]; (S) Halovac 190 [4]; (T) Fomblin Y-L vac 06/6 [1], Krytox 1506 [2]; (U) Fomblin Y-L vac 14/6 [1], Krytox 1514 [2]; (V) Fomblin Y-L vac 25/5, Y-H vac 25/9 [1]; (W) Dow Corning FS-1265 [11].   The fluids grouped under a common heading are not the same, but happen to have similar viscosities.

Appendix F.5 Kinematic Viscosity Conversion Factors

Kinematic Viscosity (mm²/s)	Saybolt Universal Seconds (SUS)			Redwood Seconds at			Engler Degrees at all Temps.
	100°F (37.8°C)	130°F (54.4°C)	210°F (98.8°C)	70°F (21.1°C)	140°F (60°C)	210°F (93.3°C)	
5.0	42.3	42.4	42.6	37.9	38.5	38.9	1.40
6.0	45.5	45.6	45.8	40.5	41.0	41.5	1.48
7.0	48.7	48.8	49.0	43.2	43.7	44.2	1.56
8.0	52.0	52.1	52.4	46.0	46.4	46.9	1.65
9.0	55.4	55.5	55.8	48.9	49.1	49.7	1.75
10.0	58.8	58.9	59.2	51.7	52.0	52.6	1.84
12.0	65.9	66.0	66.4	57.9	58.1	58.8	2.02
14.0	73.4	73.5	73.9	64.4	64.6	65.3	2.22
16.0	81.1	81.3	81.7	71.0	71.4	72.2	2.43
18.0	89.2	89.4	89.8	77.9	78.5	79.4	2.64
20.0	97.5	97.7	98.2	85.0	85.8	86.9	2.87
22.0	106.0	106.2	106.7	92.4	93.3	94.5	3.10
24.0	114.6	114.8	115.4	99.9	100.9	102.2	3.34
26.0	123.3	123.5	124.2	107.5	108.6	110.0	3.58
28.0	132.1	132.4	133.0	115.3	116.5	118.0	3.82
30.0	140.9	141.2	141.9	123.1	124.4	126.0	4.07
32.0	149.7	150.0	150.8	131.0	132.3	134.1	4.32
34.0	158.7	159.0	159.8	138.9	140.2	142.2	4.57
36.0	167.7	168.0	168.9	146.9	148.2	150.3	4.83
38.0	176.7	177.0	177.9	155.0	156.2	158.3	5.08
40.0	185.7	186.0	187.0	163.0	164.3	166.7	5.34
42.0	194.7	195.1	196.1	171.0	172.3	175.0	5.59
44.0	203.8	204.2	205.2	179.1	180.4	183.3	5.85
46.0	213.0	213.4	214.5	187.1	188.5	191.7	6.11
48.0	222.2	222.6	223.8	195.2	196.6	200.0	6.37
50.0	231.4	231.8	233.0	203.3	204.7	208.3	6.63
60.0	277.4	277.9	279.3	243.5	245.3	250.0	7.90
70.0	323.4	324.0	325.7	283.9	286.0	291.7	9.21
80.0	369.6	370.3	372.2	323.9	326.6	333.4	10.53
90.0	415.8	416.6	418.7	364.4	367.4	375.0	11.84
100.0[a]	462.0	462.9	465.2	404.9	408.2	416.7	13.16

[a] At higher values use the same ratio as above for 100 mm²/s.

REFERENCES

1. Montedison, USA, Inc., 1114 Ave. of the Americas, New York, NY 10036
2. Du Pont and Co., Chemicals and Pigments Department, Wilmington, DE 19898.
3. Stauffer Chemical Company, Specialty Chemical Division, Westport, CN 06880
4. Fluoro-Chem Corporation, 82 Burlews Court, Hackensack, NJ 07601
5. General Electric Company, Silicone Products Department, Waterford, NY 12188
6. Synthatron Corp., 50 Intervale Rd., Parsippany, NJ 07054.
7. IVACO, Inc., 35 Howard Ave., Churchville, NY 14428
8. Balzers High Vacuum, Furstentum, Liechtenstein.
9. Sargent-Welch Scientific Co., Vacuum Products Division, 7300 N. Linder Ave. Skokie, IL 60077
10. CVC Products, Inc., 525 Lee Rd. Rochester, NY 14603
11. Dow Corning Company, Inc., 2030 Dow Center, Midland, MI 48640
12. Monsanto Company, 800 N. Lindbergh Blvd. St. Louis, MO 63166
13. Varian Associates, Lexington Vacuum Division, 121 Hartwell Ave., Lexington, MA
14. Edwards High Vacuum, Ltd. Manor Royal, Crawley, West Sussex, England.
15. Kinney Vacuum Co., 3529 Washington St., Boston, MA 02130.

Index

C

O

P